Oocyte Physiology and Development in Domestic Animals

Oocyte Physiology and Development in Domestic Animals

Edited by
REBECCA L. KRISHER

WILEY-BLACKWELL
A John Wiley & Sons, Inc., Publication

Wiley-Blackwell is an imprint of John Wiley & Sons, formed by the merger of Wiley's global Scientific, Technical and Medical business with Blackwell Publishing.

Editorial offices: 2121 State Avenue, Ames, Iowa 50014-8300, USA
The Atrium, Southern Gate, Chichester, West Sussex, PO19 8SQ, UK
9600 Garsington Road, Oxford, OX4 2DQ, UK

For details of our global editorial offices, for customer services and for information about how to apply for permission to reuse the copyright material in this book please see our website at www.wiley.com/wiley-blackwell.

Library of Congress Cataloging-in-Publication Data is available upon request

A catalogue record for this book is available from the British Library.

Wiley also publishes its books in a variety of electronic formats. Some content that appears in print may not be available in electronic books.

Cover design by Nicole Teut

Set in 10.5/12 pt Times by Aptara® Inc., New Delhi, India
Printed and bound in Malaysia by Vivar Printing Sdn Bhd

1 2013

This book is dedicated to my parents, Frederick and Mary Ellen Krisher,
for their unwavering support, belief and willingness to allow me
to pursue discovery of all kinds.

Contents

Color plate is located between pages 144 and 145.

Contributors

Mourad Assidi	Centre de Recherche en Biologie de la Reproduction, Université Laval, Québec City, Québec, Canada
Claudia Baumann	Department of Physiology and Pharmacology, College of Veterinary Medicine, University of Georgia, Athens, GA, USA
Rabindranath De La Fuente	Department of Physiology and Pharmacology, College of Veterinary Medicine, University of Georgia, Athens, GA, USA
Robert B. Gilchrist	Research Centre for Reproductive Health, Robinson Institute, School of Paediatrics and Reproductive Health, The University of Adelaide, Adelaide, Australia
Jason R. Herrick	National Foundation for Fertility Research, Lone Tree, CO, USA
Md M. Hossain	Institute of Animal Science, University of Bonn, Bonn, Germany and Department of Animal Breeding and Genetics, Bangladesh Agricultural University, Mymensingh, Bangladesh
Joshua Johnson	Yale School of Medicine, Department of OB/GYN & Reproductive Sciences, New Haven, CT, USA
Rebecca L. Krisher	National Foundation for Fertility Research, Lone Tree, CO, USA
Pat Lonergan	School of Agriculture and Food Science, University College Dublin, Belfield, Dublin, Ireland
Zoltan Machaty	Department of Animal Sciences, Purdue University, West Lafayette, IN, USA
David G. Mottershead	Research Centre for Reproductive Health, Robinson Institute, School of Paediatrics and Reproductive Health, The University of Adelaide, Adelaide, Australia
Melissa Pepling	Department of Biology, Syracuse University, Syracuse, NY, USA
Karl Schellander	Institute of Animal Science, University of Bonn, Bonn, Germany

Masayuki Shimada	Laboratory of Reproductive Endocrinology, Graduate School of Biosphere Science, Hiroshima University, Higashi-Hiroshima, Hiroshima, Japan
Marc-André Sirard	Centre de Recherche en Biologie de la Reproduction, Université Laval, Québec City, Québec, Canada
George W. Smith	Laboratory of Mammalian Reproductive Biology and Genomics, Departments of Animal Science and Physiology, Michigan State University, East Lansing, MI USA
Dawit Tesfaye	Institute of Animal Science, University of Bonn, Bonn, Germany
Jeremy G. Thompson	Research Centre for Reproductive Health, Robinson Institute, School of Paediatrics and Reproductive Health, The University of Adelaide, Adelaide, Australia
Swamy K. Tripurani	Laboratory of Animal Biotechnology and Genomics, Division of Animal and Nutritional Sciences, West Virginia University, Morgantown, WV, USA
Maria M. Viveiros	Department of Physiology and Pharmacology, College of Veterinary Medicine, University of Georgia, Athens, GA, USA
Jianbo Yao	Laboratory of Animal Biotechnology and Genomics, Division of Animal and Nutritional Sciences, West Virginia University, Morgantown, WV, USA

Preface

The oocyte is a fascinating cell. It is the foundation of embryonic growth, from a newly fertilized single cell zygote to a fully formed and functioning multicellular organism capable of independent survival. As such, it is incredibly complex. The growth and maturation of the oocyte is itself a highly coordinated process, which can ultimately have long-term consequences for both female fertility and the health of resulting offspring. Our challenge has been and continues to be to understand oocyte physiology, and to elucidate the molecular mechanisms that control oocyte developmental competence.

We have come a long way in this endeavor, as the information presented within the chapters of this book clearly demonstrates. We have made great strides in understanding the biochemical mechanisms at play within the oocyte, from early oocyte growth following follicular activation to communication between the oocyte and the follicle cells, control of nuclear maturation, regulation of gene expression, metabolic requirements, activation, and coordination of subsequent embryonic development. With this knowledge come advances in biomarker discovery to identify high-quality oocytes. We are also learning to appreciate on a molecular and systems biology level the considerable stress placed on this delicate balancing act when we attempt to complete these processes in vitro. This knowledge has substantial implication for assisted reproductive technologies in livestock species, both for agricultural production and biomedical application. In addition, much of what we learn in domestic and laboratory models has direct application to improvements in human medicine.

One of the aims of this book is to present a comprehensive overview that summarizes our current scientific knowledge of oocyte physiological and biochemical mechanisms. Last, but certainly not least, it is my hope that this information will stimulate conversation, collaboration, and critical thinking among current and future scientists, consequently encouraging development and advancement in this exciting field of research.

Acknowledgments

I am first and foremost very grateful to the authors who have contributed to this book. They have provided an excellent array of historical information as well as cutting-edge thought and theory in multiple focus areas of oocyte physiology and across many species. They have been gracious and patient with me as a first-time volume editor. I feel honored to be the coordinator of such an in-depth, significant examination of the mammalian oocyte. The final product is, I believe, a great addition to the scientific knowledge in the field, and I am indebted to the contributors for sharing their expertise.

I am also grateful to Justin Jeffryes, my editor at Wiley-Blackwell, without whom this book would not have come to fruition. He has fielded a mountain of inquiries and carefully guided me throughout the editorial process. He has been incredibly patient, understanding, and supportive, and most importantly encouraging. Justin has always believed in the timeliness and importance of this book. As I see the book coming together in its final form, I am convinced he was right.

1 Oocyte Development before and during Folliculogenesis

Melissa Pepling

1.1 Introduction

This chapter will focus on female germ cell development from the time the cells arrive at and populate the gonad through primordial follicle formation and initial activation as well as cyclical activation. The same basic events occur in most species but with variation in the timing as shown in Table 1.1. The migration of the primordial germ cells (PGCs) to the gonad will not be discussed here. Sex determination will also not be covered. A great deal of work on female germ cell development has been performed in rodents; thus the state of knowledge in rodents will be discussed first followed by information from domestic species.

1.2 Germ Cell Cyst and Ovigerous Cord Formation

In the mouse, PGCs arrive at the genital ridge starting at 10.5 days post coitum (dpc) and divide by mitosis until 13.5 dpc (Figure 1.1a) (Monk & McLaren, 1981). During this time germ cells are classified as oogonia and develop as clusters of interconnected cells called germ cell cysts (Figures 1.1b and 1.3a) (Pepling & Spradling, 1998). Germ cell cysts have been well studied in male and female invertebrates (de Cuevas et al., 1997). In the Drosophila female, cysts are formed from germline stem cells that divide to produce a daughter stem cell and a cyst forming cell called a cystoblast. The cystoblast undergoes four synchronous mitotic cell cycles. However, after each division, cytokinesis is incomplete so that the cells remain connected by intercellular bridges. Only one cell of the cyst will become an oocyte, the remaining cells serve as nurse cells, supplying nutrients to the oocyte through the intercellular bridges.

Mouse female germ cells share several characteristics of the Drosophila germline cysts, including synchronous divisions, incomplete cytokinesis, intercellular bridge connections, and transport of molecules and organelles across bridges (Pepling & Spradling, 1998). Unlike Drosophila, the number of cells per cyst seems to be variable, and synchrony is lost in some dividing cyst cells. It is also unclear if some of the germ cells of the cyst serve as nurse cells in the mouse. In Drosophila, the future oocyte increases in size as it receives cytoplasm from the nurse cells (de Cuevas et al., 1997) but this does not appear to happen in the mouse (Pepling & Spradling, 2001). As the oogonia divide and form germ cell cysts, the cell clusters become enclosed in ovigerous cords consisting of epithelial pregranulosa cells surrounded by a basal lamina (Mazaud et al., 2005). There are three

Oocyte Physiology and Development in Domestic Animals, First Edition. Edited by Rebecca L. Krisher.

Table 1.1 Timing of female germ cell development in humans, mice, and several domestic species (days of gestation).

	Cow	Sheep	Pig	Goat	Horse	Mouse	Human
Arrival at gonad	35 (Erickson, 1966)	23 (Juengel et al., 2002)	18 (Black & Erickson, 1968)	25 (Lee et al., 1998)	22 (Curran et al., 1997)	10.5 (Monk & McLaren, 1981)	28 (Witschi, 1948)
Germ cell cysts /ovigerous cords	57–90 (Garverick et al., 2010; Russe, 1983)	38–75 (Juengel et al., 2002; Sawyer et al., 2002)	20–50 (Black & Erickson, 1968)	35–90 (Pailhoux et al., 2002)	*	10.5–13.5 (Mazaud et al., 2005; Pepling & Spradling, 1998)	50–140 (Hartshorne et al., 2009)
Meiotic entry	75–82 (Erickson, 1966; Russe, 1983)	55 (Sawyer et al., 2002)	47 (Bielanska-Osuchowska, 2006)	55 (Pailhoux et al., 2002; Pannetier et al., 2006)	60 (Deanesly, 1977)	13.5 (McLaren, 2000)	70 (Hartshorne et al., 2009)
Follicle formation	90 (Yang & Fortune, 2008)	66–75 (Juengel et al., 2002; Russe, 1983)	56–68 (Bielanska-Osuchowska, 2006; Oxender et al., 1979)	90 (Pailhoux et al., 2002; Pannetier et al., 2006)	102 (Deanesly, 1977)	17.5 (Pepling et al., 2010)	140 (Gillman, 1948; Gondos et al., 1971; Witschi, 1963)
Follicle development (first wave)	140 (Yang & Fortune, 2008)	100 (Sawyer et al., 2002)	75–90 (Ding et al., 2010; Oxender et al., 1979)	*	*	17.5 (Pepling et al., 2010)	150 (Hartshorne et al., 2009)
Gestation	280	145	112	150	340	19.5	280

*indicates unknown.

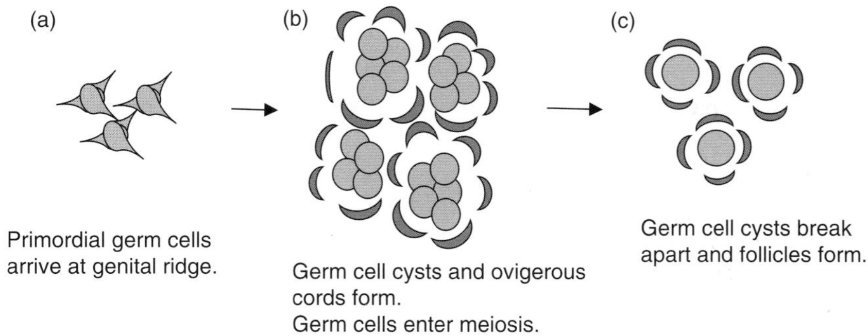

Figure 1.1 The sequence of events in mammalian oocyte development leading to primordial follicle formation. (a) The PGCs form outside the future gonad and then migrate to their final position during embryonic development. (b) The germ cells then divide to form germ cell cysts and somatic cells surround the cysts to form ovigerous cords. While in cysts, the germ cells enter meiosis and arrest at the end of prophase I in the diplotene stage. When the germ cells enter meiosis they are referred to as oocytes. (c) The germ cell cysts break apart, some oocytes die and the remaining oocytes become surrounded by somatic cells forming primordial follicles.

possible sources of pregranulosa cells in the mouse: the rete ovarii that connects to the mesonephrous, mesenchymal cells of the gonad, or ovarian surface epithelium (Liu et al., 2010). The pregranulosa cells may come from one or all three of these sources and may vary depending on the species (Sawyer et al., 2002).

The molecular control of ovigerous cord and germ cell cyst formation is not well understood. As the PGCs arrive at the gonad, several signaling pathways control their numbers such as Kit signaling, Fibroblast growth factor (FGF) signaling, and the interleukin pathway (see Table 1.2) (Farini et al., 2005; Merkwitz et al., 2011; Takeuchi et al., 2005). In addition, both Oct4 and Nanos3 block germ

Table 1.2 Genes involved germ cell survival.

Gene	Protein/ Function	Female Mutant Phenotype	References
bcl-x	Anti-apoptotic B-cell leukemia/lymphoma 2 (Bcl2) family member.	Germ cell loss by 15.5 dpc.	(Rucker et al., 2000)
β-catenin	Wnt signaling pathway.	Germ cell loss starting at 16.5 dpc, sex reversal.	(Liu et al., 2009)
fgf2r-IIIb	FGF signaling pathway.	Reduced number of germ cells at 11.5 dpc.	(Takeuchi et al., 2005)
follistatin	Activin antagonist, TGFβ family member.	Germ cell loss starting at 16.5 dpc, sex reversal.	(Yao et al., 2004)
kit	Kit oncogene, receptor tyrosine kinase.	Reduced number of germ cells	(Merkwitz et al., 2011)
kitl	Kit ligand, Stem cell factor.	Reduced number of germ cells	(Merkwitz et al., 2011)
nanos3	Nanos family of RNA binding proteins	Reduced number of germ cells at 10.5 dpc.	(Suzuki et al., 2008)
oct4	Pou domain transcription factor	Reduced number of germ cells at 10.5 dpc.	(Kehler et al., 2004)
rspo1	R-spondin homolog 1.	Germ cells do not enter meiosis, sex reversal.	(Chassot et al., 2008)
wnt4	Wnt, secreted glycoprotein family, wnt signaling pathway.	Germ cell loss starting at 16.5 dpc, sex reversal.	(Tomizuka et al., 2008)

cells from undergoing apoptosis (Kehler et al., 2004; Suzuki et al., 2008) whereas TGFβ1 and activin prevent proliferation of PGCs in culture (Richards et al., 1999). Several genes have also been implicated in the control of germ cell survival later, after arrival at the ovary (Table 1.2). Two B-cell leukemia/lymphoma 2 (Bcl2) family members, Bcl-x and Bax, have been implicated in the regulation of germ cell survival (Rucker et al., 2000). *bcl-x* hypomorphs lose their germ cells by 15.5 dpc but when mice lack both *bcl-x* and *bax*, germ cell numbers are restored. Other cell death regulators such as Bcl2 and Caspase 2 have been implicated in oocyte survival in the adult ovary (Bergeron et al., 1998; Ratts et al., 1995). There are also several genes that affect germ cell survival in the ovary slightly later, with loss starting at 16.5 dpc in mouse mutants of *β-catenin*, *follistatin*, *r-spondin homolog 1(rspo1)*, and *wnt4* (Chassot et al., 2008; Liu et al., 2009; Tomizuka et al., 2008; Yao et al., 2004). In addition, testes like characteristics are observed in mutants lacking these genes.

In cattle, germ cells begin to arrive at the gonad at approximately 35 days of gestation (Erickson, 1966). From arrival at the gonad to follicle formation, germ cell numbers increase from 16,000 to 2,700,000. Like mice, the bovine germ cells exhibit some of the key characteristics of germ cell cysts such as synchronous divisions and intercellular bridge connections (Russe, 1983). The germ cell clusters are observed starting at approximately 57 to 60 days of gestation (Garverick et al., 2010; Russe, 1983). The developing oogonia are surrounded by epithelial cells to form ovigerous cords starting at approximately 60 days of development (Garverick et al., 2010).

In a similar manner, in sheep, PGCs arrive at the gonad at about 23 days of development (Juengel et al., 2002). Again, ovine germ cells develop in nests and appear similar in morphology to mouse germ cell cysts. The germ cells continue to divide and reach their maximum number of 805,000 at day 75 (Smith et al., 1993). Ovigerous cords form from 38 to 75 days (Sawyer et al., 2002). Somatic cells contact germ cells by desmasomes and the germ cell somatic cell complexes progressively fuse to form the ovigerous cords. The somatic cells surround the germ cells and secrete a basal lamina similar to other species. There has been some question as to the source of the somatic cells that form the ovigerous cords in sheep. Cells from the mesonephrous stream in to the developing ovary, and it was thought that these cells were the cells that surrounded the germ cells (Sawyer et al., 2002). However, ovarian surface epithelial cells are also thought to be a source of somatic cord cells in the sheep, and it is likely that both cell populations contribute.

Oogonia in pigs, goats, and horses have not been as well studied as other species but there are a few reports describing this stage of development. Porcine germ cells are observed in the region of the genital ridge as early as embryonic day 18 (Black & Erickson, 1968). Mitotic divisions begin at approximately 20 days, and germ cells increase in number from 5,000 cells to 1,100,000 by 50 days. As in other species, oogonia develop in clusters and by electron microscopy have been shown to be connected by intercellular bridges at 47 days (Bielanska-Osuchowska, 2006). In goats, oogonia clusters are observed in ovigerous cords between about 35 to 90 days (Pailhoux et al., 2002). Germ cells have also been observed to develop in clusters in the horse (Deanesly, 1977).

1.3 Meiotic Entry and Progression

In the mouse, oogonia stop dividing and begin to enter meiosis at 13.5 dpc and are then considered oocytes. The oocytes remain in germ cell cysts as they enter meiosis. Oocytes progress through the stages of prophase I of meiosis (leptotene, zygotene, and pachytene) and arrest in the diplotene stage. Oocytes begin to enter the diplotene stage at 17.5 dpc, and most have reached diplotene by PND5 (Borum, 1961). The oocytes remain arrested in the diplotene stage, and meiosis is not resumed until right before ovulation in response to a surge in luteinizing hormone (LH).

Table 1.3 Genes involved in meiotic entry and progression.

Gene	Protein/Function	Female Mutant Phenotype	References
atm	Ataxia-telangiectasia mutated homolog, involved in recombination and mismatch repair.	Sterile, germ cells arrest at pachytene and eventually die.	(Barlow et al., 1998)
cyp26b1	Cytochrome P450, family 26, subfamily B, degrades retinoic acid.	Perinatal lethal. Female germ cells prematurely express Strat8.	(Bowles et al., 2006; MacLean et al., 2001)
dmc1	Disrupted meiotic cDNA 1 (recA homolog), involved in recombination and mismatch repair.	Sterile, germ cells arrest at pachytene and eventually die.	(Pittman et al., 1998; Yoshida et al., 1998)
msh4	MutS homolog 4, involved in recombination and mismatch repair.	Sterile, germ cells arrest at pachytene and eventually die.	(Kneitz et al., 2000)
msh5	MutS homolog 5, involved in recombination and mismatch repair.	Sterile, germ cells arrest at pachytene and eventually die.	(de Vries et al., 1999)
stra8	Stimulated by Retinoic acid, gene 8.	Sterile, germ cells do not enter meiosis.	(Baltus et al., 2006; Menke et al., 2003)
sycp1	Synaptonemal complex protein 1.	Sterile, lack oocytes.	(de Vries et al., 2005)
sycp3	Synaptonemal complex protein 3.	Reduced fertility, defective meiotic chromosome segregation.	(Yuan et al., 2002)

The mechanisms controlling meiotic entry have started to be uncovered (Table 1.3). Entry into meiosis is thought to be regulated by retinoic acid, which induces female cells to begin meiosis (Bowles et al., 2006; Koubova et al., 2006). When the Retinoic Acid Receptor is blocked using an antagonist in ovary organ culture, female germ cells do not enter meiosis. Males express Cytochrome P450, family 26, subfamily B (Cyp26b1), which degrades retinoic acid, thereby preventing male germ cells from entering meiosis. In the ovary, meiotic entry occurs in a wave from anterior to posterior (Bullejos & Koopman, 2004; Menke et al., 2003). Retinoic acid upregulates a cytoplasmic protein called Stimulated by Retinoic Acid, gene 8 (Stra8) in females (Baltus et al., 2006). Stra8 is expressed first in the anterior moving posterior reflecting the wave of meiotic entry (Menke et al., 2003). Stra8 plays a role in premeiotic DNA replication and in chromosome cohesion and synapsis (Baltus et al., 2006).

Defects in female germ cell development are observed in mutants of several genes involved in DNA mismatch repair and recombination including *ataxia-telangiectasia mutated homolog (atm)*, *disrupted meiotic cDNA 1 (dmc1)*, *mutS homolog 4 (msh4)*, and *msh5* (Table 1.3) (Barlow et al., 1998; de Vries et al., 1999; Kneitz et al., 2000; Pittman et al., 1998; Yoshida et al., 1998). Males and females are sterile, and female germ cells arrest in the pachytene stage of meiotic prophase at 16.5 dpc. Eventually, the germ cells are lost in the mutants. During meiotic prophase, homologous chromosomes are held together by the synaptonemal complex. Two structural components of the synaptonemal complex, Synaptonemal complex protein (Scyp) 1 and Scyp3, are required for normal fertility. *scyp1* mutants are sterile, and females lack oocytes (de Vries et al., 2005). In rats, inhibition of Scyp1 caused premature arrival at the diplotene stage and premature primordial follicle assembly, suggesting a link between cell cycle stage and primordial follicle development (Paredes et al., 2005). *scyp3* mutants have reduced fertility, and although the oocytes appear to develop normally, chromosome segregation does not occur properly (Yuan et al., 2002).

In bovine ovaries, germ cells begin to enter meiosis starting at 75 to 82 days of gestation (Erickson, 1966; Russe, 1983). Entry into meiosis seems to be a gradual and prolonged process with some cells still found in mitosis even at birth. In sheep, meiotic germ cells are first observed at 55 days though mitotic germ cells are still observed up to 90 days (Juengel et al., 2002; Sawyer et al., 2002). The germ cells farthest from the surface epithelium are the first to enter meiosis. In porcine ovaries, germ cells begin to enter meiosis at 47 days of gestation (Bielanska-Osuchowska, 2006), in the goat at 55 days (Pailhoux et al., 2002; Pannetier et al., 2006), and in the horse at 60 days of gestation (Deanesly, 1977).

1.4 Follicle Formation

The oocytes in germ cell cysts eventually separate, a process termed cyst breakdown, and become enclosed in primordial follicles consisting of one oocyte and several somatic granulosa cells (Figures 1.1c and 1.3b) (Pepling & Spradling, 2001). During this process, some cells in each cyst die by programmed cell death, with only a third of the total surviving. In one model, one cell of a cyst dies and breaks the large cyst into smaller cysts. This is repeated until a few individual oocytes remain. Thus, programmed cell death would be required for oocytes to break apart. Some cyst cells may support oocytes and eventually die, analogous to nurse cells in Drosophila. Programmed cell death during female germ cell development is common in many species including domestic species (Buszczak & Cooley, 2000). In mouse, cyst breakdown and oocyte loss occur concurrently, suggesting they are part of a regulated process. Mechanisms governing oocyte death remain uncharacterized. Work examining mutants lacking the programmed cell death regulator, Bax (a pro-death protein) suggests that oocyte cell death is required for cyst breakdown (Greenfeld et al., 2007). *bax* mutant ovaries have more oocytes than wild-type ovaries still in cysts, supporting the idea that programmed cell death is required for cyst breakdown.

In the mouse, oocyte loss and cyst breakdown begin in the medullary region of the ovary as early as 17.5 dpc (De Felici et al., 1999; Ghafari et al., 2007; McClellan et al., 2003; Pepling et al., 2010). In addition, follicles begin to form in the innermost region of the ovary at 17.5 dpc. The somatic pregranulosa cells surrounding the germ cells to form the ovigerous cords now begin to surround individual oocytes and become granulosa cells. In addition, before follicle formation, pregranulosa cells extend cytoplasmic protrusions between oocytes and may be involved in separating oocytes (Pepling & Spradling, 2001). There are regional differences in oocyte development, and oocytes located in the inner cortex and medullary regions of the ovary enter meiosis and start to grow first (Nandedkar et al., 2007; Peters, 1969). This regional pattern is set up between 13.5 and 16.5 dpc in the mouse concurrent with meiotic entry (Byskov et al., 1997). Oocytes in the resultant primordial follicles are thought to represent the entire pool available to a female during her reproductive life (Kezele et al., 2002).

Several mouse mutants have been generated that have ovaries with multiple oocyte follicles (MOFs) consisting of abnormal follicles with more than one oocyte (see Table 1.4). The oocytes in these follicles are believed to be remnants of germ cell cysts that did not completely break apart, resulting in more than one oocyte being enclosed in a follicle (Jefferson et al., 2006). This suggests that the genes disrupted in these mutants play a role in promoting cyst breakdown and primordial follicle formation. MOFs are observed in mutants of two members of the Transforming Growth Factor β (TGFβ) family, *bone morphogenetic protein 15* (*bmp15*) and *growth and differentiation factor 9* (*gdf9*) (Yan et al., 2001). Treatment of ovaries with another TGFβ family member, Activin A, promotes follicle formation (Bristol-Gould et al., 2006). In contrast, overexpression of

Table 1.4 Genes involved in primordial follicle formation.

	Protein/Function	Female Mutant Phenotype	References
ahr	Aryl Hydrocarbon Receptor, basic helix-loop-helix transcription factor.	Reduced fertility, accelerated primordial follicle formation.	(Benedict et al., 2000; Robles et al., 2000)
akt	Serine/threonine kinase, also known as Protein Kinase B (PKB).	Multiple oocyte follicles.	(Brown et al., 2010)
bax	Proapoptotic Bcl2 family member.	Increased germ cell numbers, reduced follicle formation.	(Greenfeld et al., 2007)
bmp15	Bone Morphogenetic Protein 15, TGFβ family member.	Multiple oocyte follicles.	(Yan et al., 2001)
dax	Dosage-sensitive sex reversal, adrenal hypoplasia critical region on chromosome X, gene 1, orphan steroid hormone receptor.	Multiple oocyte follicles.	(Yu et al., 1998)
figla	Factor in the germ line alpha, folliculogenesis specific basic helix-loop-helix.	Sterile, defective follicle formation, perinatal oocyte loss.	(Soyal et al., 2000)
follistatin	Activin antagonist, TGFβ family member.	Reduced fertility, reduced follicle formation.	(Kimura et al., 2011)
foxl2	Forkhead box L2, winged helix transcription factor.	Sterile, defective follicle formation, oocyte loss.	(Schmidt et al., 2004; Uda et al., 2004)
gdf9	Growth differentiation factor 9, TGFβ family member.	Multiple oocyte follicles.	(Yan et al., 2001)
lunatic fringe	Regulator of Notch signaling.	Multiple oocyte follicles.	(Hahn et al., 2005)
ngf	Nerve growth factor, neurotrophin signaling.	Reduced follicle formation.	(Dissen et al., 2001)
nobox	Newborn ovary homeobox-encoding gene.	Sterile, delayed follicle formation, oocyte loss.	(Rajkovic et al., 2004)
ntrk1	NGF receptor, neurotrophin signaling.	Reduced follicle formation.	(Kerr et al., 2009)
ntrk2	Receptor for NT-4 and BDNF, neurotrophin signaling.	Reduced follicle formation, reduced number of germ cells.	(Kerr et al., 2009; Spears et al., 2003)
p27	Cyclin-dependent kinase (Cdk) inhibitor 1, downstream of PI3K signaling.	Progressive loss of fertility, accelerated primordial follicle formation.	(Rajareddy et al., 2007)

the Activin antagonist, Inhibin B, causes an increase in MOFs (McMullen et al., 2001). In addition, *follistatin* mutants are subfertile and have a delay in cyst breakdown and follicle formation (Kimura et al., 2011). Mutation of a regulator of Notch signaling, *lunatic fringe*, also results in the appearance of MOFs, suggesting that the Notch signaling pathway may be important in cyst breakdown and primordial follicle assembly (Hahn et al., 2005). Supporting this idea, inhibition of Notch signaling in culture caused a reduction in primordial follicle formation (Trombly et al., 2008). Thus, TGFβ and Notch signaling pathways play a role in cyst breakdown and primordial follicle formation.

There is also evidence for Neurotrophin signaling in the regulation of cyst breakdown and primordial follicle formation (Table 1.4). Mutation of the Neurotrophin, Nerve Growth Factor (NGF) resulted in females with fewer oocytes enclosed in primordial follicles and more oocytes still in germ cell cysts at 1 week (Dissen et al., 2001). Blocking two other neurotrophins, Neurotrophin 4 (NT4) and Brain-derived Neurotrophic Factor (BDNF), with antibodies in neonatal ovary organ culture caused a reduction in oocyte survival (Spears et al., 2003). Mutation of the receptor for

NT4 and BDNF, *neurotrophic tyrosine kinase receptor type 2* (*ntrk2*), also resulted in a reduction of oocytes. Recent studies of ovaries from homozygous mutants of *ntrk1*, encoding the receptor for NGF, as well as *ntrk2,* report a reduction in the number of oocytes enclosed in follicles at 1 week supporting the role of Neurotrophins in germ cell cyst breakdown and primordial follicle formation (Kerr et al., 2009).

Mutations in at least three different genes encoding transcription factors cause female infertility resulting from altered primordial follicle assembly and subsequent oocyte loss (Table 1.4). *forkhead box l2* (*foxl2*) encodes a winged helix transcription factor, and granulosa cells do not properly surround oocytes to form primordial follicles in mutant females (Schmidt et al., 2004; Uda et al., 2004). Many germ cells were still in germ cell cysts at 1 week though some follicles did form, and many dying oocytes were observed by 8 weeks. *newborn ovary homeobox-encoding gene* (*nobox*) mutants have a similar phenotype with many more germ cells still in cysts at 1 week after birth (Rajkovic et al., 2004). However, oocyte loss was even more pronounced than in *foxl2* mutants with most oocytes lost by 2 weeks after birth. The third transcription factor with a similar mutant phenotype is Factor in the germ line alpha (Figla), also known as Folliculogenesis specific basic helix-loop-helix (Soyal et al., 2000). The *figla* phenotype was the most severe with no primordial follicles formed and loss of most oocytes by 1 week after birth.

Mutants have also been identified that form follicles at a faster rate than normal. Aryl hydrocarbon receptor (Ahr) is a basic helix-loop-helix protein. *ahr* mutants have accelerated follicle formation, though by 8 days after birth the number of follicle was the same as wild-type (Benedict et al., 2000; Robles et al., 2000). Another mutant with accelerated follicle formation is *p27,* which is also known as *cyclin-dependent kinase (cdk) inhibitor 1* (Rajareddy et al., 2007). The p27 protein also plays a role in follicle activation, as described in the next section.

According to several groups, bovine follicles begin to form at approximately 90 days of gestation (Dominguez et al., 1988; Russe, 1983; Yang & Fortune, 2008), although there is some controversy about the exact timing of follicle formation, with researchers observing primordial follicles as early as 74 days (Nilsson & Skinner, 2009; Tanaka et al., 2001) and one not observing follicles until 130 days (Erickson, 1966). As in rodents, follicle formation begins with the innermost region first (Russe, 1983). There is also a large number of apoptotic cells present as follicles are forming (Erickson, 1966; Garverick et al., 2010), which has been observed in most mammalian species that have been examined (Baker, 1972). Oocytes not surrounded by granulosa cells are thought to degenerate (Adams et al., 2008; Smitz & Cortvrindt, 2002).

In the sheep there is also some variation about the timing of follicle formation, with one study reporting the first follicles at 66 days and another at 75 days of gestation (Juengel et al., 2002; Russe, 1983). Similar to bovine primordial follicle formation, the first follicles that form are located at the interface of the ovarian cortex and medulla, and follicles form progressively toward the outer cortex (McNatty et al., 2000; Sawyer et al., 2002). A large number of germ cells, over 75%, undergo apoptosis as the follicles are forming (Sawyer et al., 2002; Smith et al., 1993). It is thought that one reason for oocyte death is that the loss of germ cells allows more pregranulosa cells to associate with an individual surviving oocyte. The pregranulosa cells also extend cytoplasmic protrusions between the oocytes, which may help to separate oocytes in cysts (Sawyer et al., 2002).

There is less information available regarding follicle formation in other domestic animals. In the pig, follicles begin to form at approximately 56 days of gestation (Bielanska-Osuchowska, 2006). As with cows and sheep, the first follicles form in the deepest part of the ovary. Primordial follicles are first observed in the goat at 90 days of gestation (Pailhoux et al., 2002; Pannetier et al., 2006) and in the horse at 102 days (Deanesly, 1977).

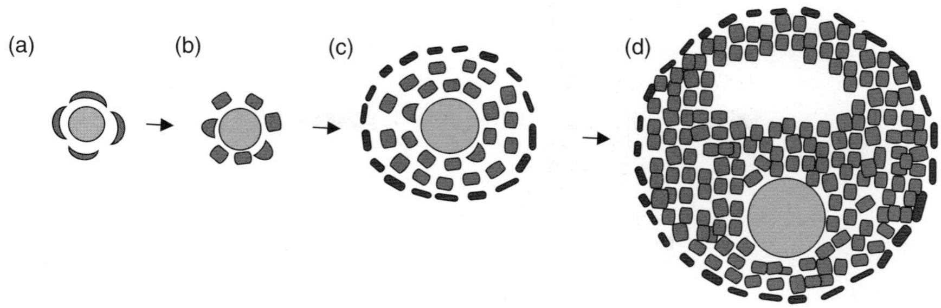

Figure 1.2 Primordial follicle activation and development. Groups of primordial follicles (a) consisting of one oocyte (light grey) and several flattened granulosa cells (medium grey) are activated to grow and progress to the primary follicle stage (b) indicated by a change in shape of the granulosa cells from flat to cuboidal. The oocyte also increases in diameter. Follicle development proceeds to the secondary or preantral stage (c) as granulosa cells multiply and surround the oocyte with multiple layers. Theca cells (dark grey) surround the preantral follicles. The follicles reach the antral stage (d) when a fluid filled space or antrum forms.

1.5 Follicle Development

Primordial follicles, each consisting of an oocyte and several granulosa cells that exhibit a flattened shape, remain dormant for varying amounts of time until activated to grow (Figure 1.2). A change in the morphology of the granulosa cells from flattened to cuboidal is indicative of follicle activation, and at this point the oocyte and associated granulosa cells are referred to as a primary follicle (Figures 1.2b and 1.3c). The primary follicle is enclosed in a basal lamina (Aerts & Bols, 2010). The oocyte remains arrested in the diplotene stage of prophase I of meiosis as the follicle grows. In addition, during follicle growth, the oocyte itself also grows, and in the mouse increases in size 300-fold in a 2- or 3-week period (Lintern-Moore & Moore, 1979). Furthermore, RNA content increases by 300-fold and protein synthesis by 38-fold (Wassarman & Albertini, 1994). As the granulosa cells of primary follicles divide to produce multiple cell layers, secondary or preantral follicles are formed (Figures 1.2c and 1.3d). Theca cells, which form from fibroblast-like cells in the ovarian stroma, become associated with the follicles at this stage (Hirshfield, 1991). The theca and granulosa cells support the oocytes and also produce hormones (Erickson et al., 1985). The preantral follicles eventually gain a fluid-filled space, and are then classified as antral follicles (Figure 1.2d). Many follicles do not survive past this stage. The surviving follicles are termed preovulatory follicles. Meiotic arrest is released just prior to ovulation in response to gonadotropins (the LH surge) (Jamnongjit & Hammes, 2005). The oocyte proceeds through meiosis and is then arrested a second time, in metaphase II, until fertilization.

In the mouse, some follicles begin to develop almost immediately after forming and are referred to as the first wave of developing follicles (Hirshfield & DeSanti, 1995). These first follicles that form are located in the core of the ovary. The follicles reach the antral stage by 3 weeks after birth and then become atretic and die because there is no gonadotropic surge to rescue them (Mazaud et al., 2002; McGee et al., 1998; Rajah et al., 1992). It is unclear why this first wave of follicular development occurs.

Follicle activation and development can be divided into two phases: initial recruitment and cyclic recruitment (McGee & Hsueh, 2000). Initial recruitment is a continuous process referring to the activation of groups of primordial follicles. It is thought that there are inhibitory factors that suppress the activation of follicles, and a few inhibitory proteins have been identified so far (Adhikari & Liu, 2009). Little is known about the mechanisms that control the selection of follicles that are activated.

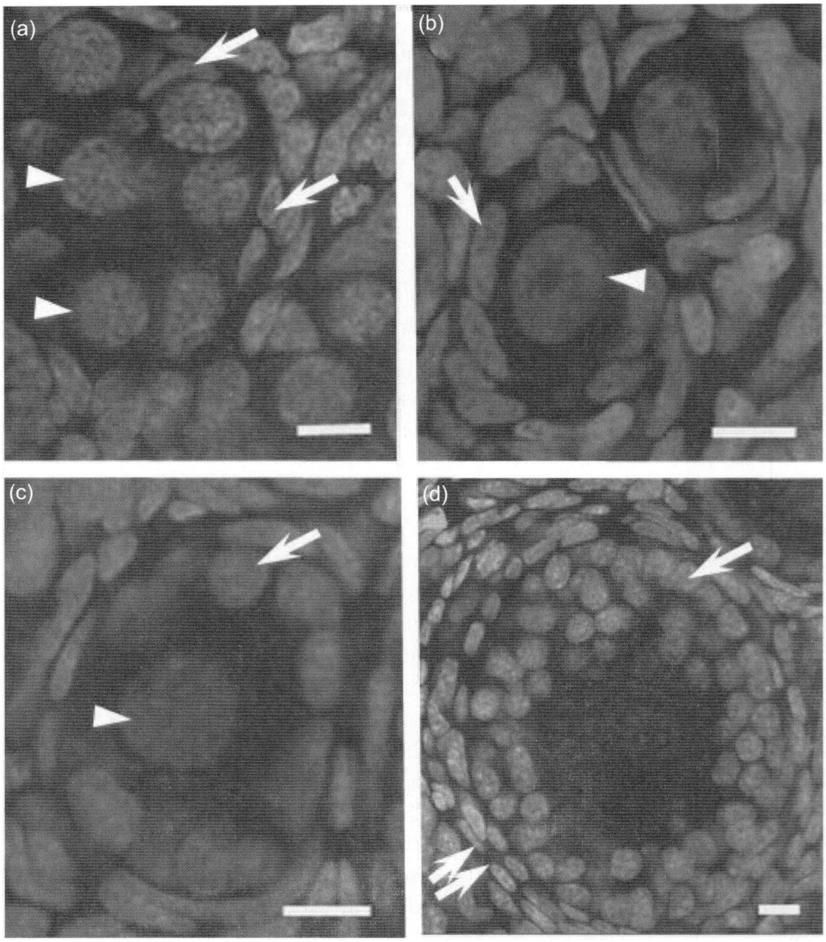

Figure 1.3 Single confocal sections of propidium iodide-stained ovarian follicles illustrating different stages of oocyte and follicle development. (a) Germ cell cysts within ovigerous cords. Two of the oocytes in the cyst are indicated by the arrowhead and one of the somatic cells is indicated by the arrow. (b) Primordial follicles each consisting of an individual oocyte (arrowhead) surrounded by several flattened granulosa cells (arrow). (c) A primary follicle with the oocyte indicated by an arrowhead surrounded by a single layer of cuboidal granulosa cells (arrow). (d) A secondary follicle with several layers of granulosa cells (arrow) and a layer of theca cells (double arrows). Scale bars = 10 μm.

One idea is that the follicles are recruited in the same order as they entered meiosis (Edwards et al., 1977). This idea is supported by the observation that oocytes in the medullary or inner cortex region of the ovary enter meiosis first and it is in this region where the first developing follicles are observed. During preantral growth, the granulosa cells multiply, the oocyte grows, the zona pellucida forms, the theca layer is made, and a vascular supply develops (McGee & Hsueh, 2000). Communication between the oocyte and the surrounding granulosa cells is very important during this phase of follicle growth (Tsafriri, 1997). This communication is in part through gap junctions connecting the oocyte to the granulosa cells.

The oocyte gains the ability to resume meiosis about the time the antrum begins to form in the follicle and is designated as meiotic competence (Mehlmann, 2005). At this point the oocyte has the required level of maturation-promoting factor (MPF) necessary to resume meiosis. MPF is a complex consisting of CDK1 and cyclin B. However, even though the oocyte is capable of resuming meiosis, meiotic arrest is maintained until the LH surge. Oocytes from antral follicles will spontaneously resume meiosis if removed from the follicle, and thus a signal from the granulosa cells is important. This signal results in high cAMP levels in the oocyte, which are important for maintenance of meiotic arrest.

Cyclic recruitment refers to the selection of only a few follicles to reach the preovulatory stage. Before puberty, follicles that grow eventually undergo atresia. After puberty, at the antral stage, only a few follicles continue to grow while the rest die. Both the oocyte and the granulosa cells of the atretic follicles undergo apoptosis (Pesce & De Felici, 1994). In response to FSH, one oocyte (or several if the species is polyovulatory) grows faster and becomes the dominant follicle (Zeleznik & Benyo, 1994). The dominant follicle then makes estrogen and inhibin, which suppresses FSH and inhibits the remaining follicles from growing. Dominant follicle selection has been most well studied in the cow and horse; it is discussed below.

Several genes have been implicated in the regulation of primordial follicle activation (see Table 1.5). One common mutant phenotype is the premature activation of all primordial follicles, leading to the eventual loss of all oocytes. The mice are initially fertile but eventually become sterile as the oocytes are lost. This suggests that this group of genes is involved in blocking primordial follicle activation. Several of the mouse mutations with this phenotype are associated with the phosphatidylinositol 3 kinase (PI3K) signaling pathway. One PI3K signaling mediator is 3-phosphoisnositide dependent protein kinase 1 (PDPK1), which phosphorylates AKT serine/threonine kinases (Reddy et al., 2009). *pdpk1* mutant females have a gradual loss of fertility due to the premature activation of follicles. However, *akt1* mutants are less severe with reduced fertility and premature activation of only a subset of follicles (Brown et al., 2010). Another mutation where the primordial follicle pool becomes prematurely activated is *phosphatase and tensin homolog deleted on chromosome 10* (*pten*), a negative regulator of PI3K signaling (Reddy et al., 2008). Finally, mutants in three genes that are downstream of PI3K signaling, *foxo3a*, *p27,* and *ribosomal protein s6* (*rsp6*) also prematurely activate all follicles and become sterile (Castrillon et al., 2003; Rajareddy et al., 2007; Reddy et al., 2009).

Premature loss of follicles is observed in mutants of at least two genes that have not been associated with the PI3K signaling pathway (Table 1.5). First, the transcription factor, FoxL2 (which plays a role in primordial follicle formation), is also important in regulating follicle activation (Schmidt et al., 2004; Uda et al., 2004). All follicles are activated by 2 weeks after birth in *foxl2* mutants. Second, although *anti-mullerian hormone* (*amh*) mutant females are reported to be fertile, there is a premature reduction in the pool of primordial follicles (Durlinger et al., 1999).

Mutation of several transcription factors causes infertility with arrest at the primordial follicle stage and eventual oocyte depletion (Table 1.5). This phenotype suggests that these molecules are required for the activation of follicles and progression to the primary follicle stage. The gene encoding LIM homeobox protein 8 (Lhm8) when mutant causes arrest of follicles at the primordial follicle stage (Choi et al., 2008a; Pangas et al., 2006). Mutations in *nobox* also have follicles arrested at the primordial stage (Rajkovic et al., 2004). Two basic helix-loop-helix encoding genes, *spermatogenesis and oogenesis-specific basic helix-loop-helix 1* (*sohlh1*) and *sohlh2,* also belong to this class of mutations that are arrested at the primordial follicle stage (Choi et al., 2008b; Pangas et al., 2006).

Another class of mutants has follicles that are activated but arrest at the primary follicle stage (Table 1.5). For example, *gdf9* mutants cannot progress farther than the primary follicle stage and are therefore sterile (Dong et al., 1996). Similarly, some mutants in the receptor tyrosine kinase, *kit*

Table 1.5 Genes involved in follicle activation and early follicle development.

	Protein/Function	Female Mutant Phenotype	References
akt	Serine/threonine kinase, also known as Protein Kinase B (PKB), PI3K signaling pathway.	Reduced fertility, premature reduction of primordial follicle pool.	(Brown et al., 2010)
amh	Anti-Mullerian hormone, TGFβ family member.	Fertile, premature reduction of primordial follicle pool.	(Durlinger et al., 1999)
foxl2	Forkhead box L2, winged helix transcription factor.	Progressive loss of fertility, premature activation of all follicles.	(Schmidt et al., 2004; Uda et al., 2004)
Foxo3a	Forkhead box O3, winged helix transcription factor, downstream of PI3K signaling.	Progressive loss of fertility, premature activation of all follicles.	(Castrillon et al., 2003)
gdf9	Growth differentiation factor 9, TGFβ family member.	Sterile, arrest at primary follicle stage, oocyte loss.	(Dong et al., 1996)
kit	Kit oncogene, receptor tyrosine kinase.	Some mutants arrest at primary follicle stage.	(Yoshida et al., 1997)
kitl	Kit ligand, Stem cell factor.	Some mutants arrest at primary follicle stage.	(Bedell et al., 1995)
lhx8	LIM homeobox protein 8.	Sterile, arrest at primordial follicle stage, oocyte loss.	(Choi et al., 2008a; Pangas et al., 2006)
nobox	Newborn ovary homeobox-encoding gene.	Sterile, arrest at primordial follicle stage, oocyte loss.	(Rajkovic et al., 2004)
p27	Cyclin-dependent kinase (Cdk) inhibitor 1, downstream of PI3K signaling.	Progressive loss of fertility, premature activation of all follicles.	(Rajareddy et al., 2007)
pdpk1	3-Phosphoinositide-dependent protein kinase-1, serine/threonine kinase, PI3K signaling pathway.	Progressive loss of fertility, premature activation of all follicles.	(Reddy et al., 2009)
pten	Phosphatase and tensin homolog deleted on chromosome 10, negative regulator of PI3K signaling pathway.	Progressive loss of fertility, premature activation of all follicles.	(Reddy et al., 2008)
rps6	Ribosomal protein S6, downstream of PI3K signaling.	Progressive loss of fertility, premature activation of all follicles.	(Reddy et al., 2009)
sohlh1	Spermatogenesis and oogenesis-specific basic helix-loop-helix 1.	Sterile, arrest at primordial follicle stage, oocyte loss.	(Pangas et al., 2006)
sohlh2	Spermatogenesis and oogenesis-specific basic helix-loop-helix 2.	Sterile, arrest at primordial follicle stage, oocyte loss.	(Choi et al., 2008b)

as well as in the kit ligand, *stem cell factor* (SCF), arrest at the primary follicle stage (Bedell et al., 1995; Dong et al., 1996; Yoshida et al., 1997). Interestingly, kit signaling can activate the PI3K signaling pathway, and this may be how kit regulates follicle development.

In most domestic species, primordial follicles do not develop into primary follicles immediately after forming, and there is a significant delay until the first appearance of primary follicles. In the cow, although primordial follicles form at 90 days of gestation, the first primary follicles are not observed until day 140 (Yang & Fortune, 2008). It is believed that these primordial follicles are not capable of activating directly after they form. It has also been observed that oocytes do not arrest in the diplotene stage until approximately 141 days, suggesting that the oocyte must reach diplotene arrest before the primordial follicle can be activated to develop into a primary follicle.

In sheep, follicles start to form between 66 and 75 days gestation, and there seems to be some variability in the literature as to when they begin to develop (Juengel et al., 2002; Russe, 1983). The first developing follicles are observed at day 100 (Sawyer et al., 2002). Similarly, porcine follicles have been reported to form at 56, or 68 days of gestation, and the first developing follicles have been observed at 75 or 90 days (Bielanska-Osuchowska, 2006; Ding et al., 2010; Oxender et al., 1979).

The selection of the dominant follicle has been well studied in both bovine and equine species (Beg & Ginther, 2006; Fortune et al., 2004; Ginther et al., 2001). A wave of antral follicles is recruited to grow by a small increase in circulating FSH levels. After several days one of the follicles becomes larger than the other follicles and will likely become the dominant follicle while the other follicles become subordinate follicles and are eventually lost. The dominant follicle then synthesizes estradiol, and FSH levels decrease. It is believed that insulin-like growth factor (IGF) signaling may important for dominant follicle selection. Levels of free IGF are higher in the follicular fluid of the dominant follicle whereas IGF binding proteins that sequester IGF are low. In addition, in mice, *igf1* mutants arrest by the early antral stage, suggesting that IGF is required for further development of the follicles (Baker et al., 1996).

The LH surge triggers the resumption of meiosis as well as other changes, including a change in mitochondrial localization. Prior to the LH surge, mitochondria are located peripherally in the oocyte, but during the final stages of nuclear maturation, they become more clustered (Ferreira et al., 2009). After ovulation, the mitochondria are dispersed throughout the cytoplasm. These changes are thought to reflect the changing energy requirements of the oocyte.

1.6 Steroid Hormone Signaling in Oocyte Development

Steroid hormone signaling is thought to be important for controlling the ability of bovine follicles to activate. Follicle activation can be blocked by exogenous estrogen treatment of cultured fetal bovine ovary explants (Yang & Fortune, 2008). In addition, fetal estrogen levels drop at about 141 days of gestation coinciding with follicle activation. Steroid hormone signaling has also been implicated not only in primordial follicle activation but also in primordial follicle formation in the cow. Nilsson and Skinner found that fetal ovarian estrogen and progesterone levels drop when primordial follicles begin to form (Nilsson & Skinner, 2009). They also showed that progesterone treatment of bovine ovaries in organ culture significantly blocked the assembly of follicles.

Fetal sheep ovaries produce both estrogen and progesterone (Lun et al., 1998). Steroidogenic cells have been identified in the ovine ovaries and high estrogen has been suggested to be important for ovigerous cord formation, while the drop in estrogen levels correlates with meiotic entry (Juengel et al., 2002).

Early rodent follicle development was thought to be independent of hormones, but recent work from several labs including ours suggests that steroids hormones are important in regulating germ cell cyst breakdown and primordial follicle formation (Chen et al., 2009; Chen et al., 2007; Kezele & Skinner, 2003; Lei et al., 2010). Adult female mice treated as neonates with estrogen or estrogen like compounds (Iguchi et al., 1990; Iguchi et al., 1986; Jefferson et al., 2002; Suzuki et al., 2002) have more MOFs suggesting that estrogen plays a role in controlling cyst breakdown and primordial follicle assembly (Gougeon, 1981; Iguchi & Takasugi, 1986; Iguchi et al., 1986). Our model is that normally, exposure of fetal oocytes to maternal estrogen keeps oocytes in cysts and at birth estrogen levels drop resulting in cyst breakdown. When oocytes are exposed to estrogens, cyst breakdown is inhibited. Supporting this, our studies showed that during neonatal ovary development, mice treated with the phytoestrogen, genistein, had more oocytes in cysts compared to control mice (Jefferson et al., 2006). Work from our lab has shown that estrogen causes a delay in individualization of oocytes, supporting the idea that MOFs are cysts that did not break down (Chen et al., 2007). Several synthetic compounds with estrogenic activity including bisphenol A, diethylstilbestrol, and ethylene estradiol also block cyst breakdown (Karavan & Pepling, 2012). Neonatal treatment with progesterone also results in more MOFs (Iguchi et al., 1988), and progesterone and estrogen affect neonatal oocyte development in rats (Kezele & Skinner, 2003). Neonatal progesterone treatment reduced primordial follicle assembly, while both progesterone and estrogen reduced primordial follicle activation. Interestingly, mutations in the orphan steroid hormone receptor, *dosage sensitive sex reversal, adrenal hypoplasia critical region on chromosome X, gene 1* (*dax*) also have MOFs suggesting a role for the Dax protein in follicle formation (Yu et al., 1998).

In some species, estrogen seems to have a positive effect on follicle formation. In the hamster, estrogen promotes follicle assembly (Wang et al., 2008; Wang & Roy, 2007). In the baboon, if estrogen production is blocked, cyst breakdown and follicle formation are disrupted (Zachos et al., 2002). It is not clear why in some species estrogen promotes follicle formation and in other inhibits follicle formation. One possibility is that relatively high concentrations of estrogen inhibit follicle assembly while low concentrations promote assembly (Nilsson & Skinner, 2009). Alternatively there could be species differences in estrogen signaling.

1.7 Summary

Most mammals progress through similar stages of germ cell development, although there is wide variation in the timing of each step. Some domestic animals are well studied, but information is lacking on others. A more complete understanding of oogenesis in domestic species will lead to the development of techniques to improve reproductive capacity and herd quality. Comparative studies will also lead to a better understanding of human oogenesis and aid in treating infertility.

References

Adams, G. P., Jaiswal, R., Singh, J., & Malhi, P. (2008). Progress in understanding ovarian follicular dynamics in cattle. *Theriogenology, 69*(1), 72–80.

Adhikari, D., & Liu, K. (2009). Molecular mechanisms underlying the activation of mammalian primordial follicles. *Endocrine Review, 30*(5), 438–464.

Aerts, J. M., & Bols, P. E. (2010). Ovarian follicular dynamics: A review with emphasis on the bovine species. Part I: Folliculogenesis and pre-antral follicle development. *Reproduction in Domestic Animals, 45*(1), 171–179.

Baker, J., Hardy, M. P., Zhou, J., Bondy, C., Lupu, F., Bellve, A. R., & Efstratiadis, A. (1996). Effects of an Igf1 gene null mutation on mouse reproduction. *Molecular Endocrinology, 10*(7), 903–918.

Baker, T. G. (1972). *Reproductive Biology*. Amsterdam: Excerpta Medica.

Baltus, A. E., Menke, D. B., Hu, Y. C., Goodheart, M. L., Carpenter, A. E. et al. (2006). In germ cells of mouse embryonic ovaries, the decision to enter meiosis precedes premeiotic DNA replication. *Nature Genetics, 38*(12), 1430–1434.

Barlow, C., Liyanage, M., Moens, P. B., Tarsounas, M., Nagashima, K. et al. (1998). Atm deficiency results in severe meiotic disruption as early as leptonema of prophase I. *Development, 125*(20), 4007–4017.

Bedell, M. A., Brannan, C. I., Evans, E. P., Copeland, N. G., Jenkins, N. A., & Donovan, P. J. (1995). DNA rearrangements located over 100 kb 5′ of the Steel (Sl)-coding region in Steel-panda and Steel-contrasted mice deregulate Sl expression and cause female sterility by disrupting ovarian follicle development. *Genes & Development, 9*(4), 455–470.

Beg, M. A., & Ginther, O. J. (2006). Follicle selection in cattle and horses: Role of intrafollicular factors. *Reproduction, 132*(3), 365–377.

Benedict, J. C., Lin, T. M., Loeffler, I. K., Peterson, R. E., & Flaws, J. A. (2000). Physiological role of the aryl hydrocarbon receptor in mouse ovary development. *Toxicological Sciences, 56*(2), 382–388.

Bergeron, L., Perez, G. I., Macdonald, G., Shi, L., Sun, Y. et al. (1998). Defects in regulation of apoptosis in caspase-2-deficient mice. *Genes & Development, 12*(9), 1304–1314.

Bielanska-Osuchowska, Z. (2006). Oogenesis in pig ovaries during the prenatal period: Ultrastructure and morphometry. *Reproductive Biology, 6*(2), 161–193.

Black, J. L., & Erickson, B. H. (1968). Oogenesis and ovarian development in the prenatal pig. *Anatomical Record, 161*(1), 45–55.

Borum, K. (1961). Oogenesis in the mouse. A study of the origin of the mature ova. *Experimental Cell Research, 45*, 39–47.

Bowles, J., Knight, D., Smith, C., Wilhelm, D., Richman, J. et al. (2006). Retinoid signaling determines germ cell fate in mice. *Science, 312*, 596–600.

Bristol-Gould, S. K., Kreeger, P. K., Selkirk, C. G., Kilen, S. M., Cook, R. W. et al. (2006). Postnatal regulation of germ cells by activin: The establishment of the initial follicle pool. *Developmental Biology, 298*(1), 132–148.

Brown, C., LaRocca, J., Pietruska, J., Ota, M., Anderson, L. et al. (2010). Subfertility caused by altered follicular development and oocyte growth in female mice lacking PKB alpha/Akt1. *Biology of Reproduction, 82*(2), 246–256.

Bullejos, M., & Koopman, P. (2004). Germ cells enter meiosis in a rostro-caudal wave during development of the mouse ovary. *Molecular Reproduction & Development, 68*(4), 422–428.

Buszczak, M., & Cooley, L. (2000). Eggs to die for: Cell death during Drosophila oogenesis. *Cell Death & Differentiation, 7*(11), 1071–1074.

Byskov, A. G., Guoliang, X., & Andersen, C. Y. (1997). The cortex-medulla oocyte growth pattern is organized during fetal life: An in-vitro study of the mouse ovary. *Molecular Human Reproduction, 3*(9), 795–800.

Castrillon, D. H., Miao, L., Kollipara, R., Horner, J. W., & DePinho, R. A. (2003). Suppression of ovarian follicle activation in mice by the transcription factor Foxo3a. *Science, 301*(5630), 215–218.

Chassot, A. A., Ranc, F., Gregoire, E. P., Roepers-Gajadien, H. L., Taketo, M. M. et al. (2008). Activation of beta-catenin signaling by Rspo1 controls differentiation of the mammalian ovary. *Human Molecular Genetics, 17*, 1264–1277.

Chen, Y., Breen, K., & Pepling, M. E. (2009). Estrogen can signal through multiple pathways to regulate oocyte cyst breakdown and primordial follicle assembly in the neonatal mouse ovary. *Journal of Endocrinology, 202*, 407–417.

Chen, Y., Jefferson, W. N., Newbold, R. R., Padilla-Banks, E., & Pepling, M. E. (2007). Estradiol, progesterone, and genistein inhibit oocyte nest breakdown and primordial follicle assembly in the neonatal mouse ovary in vitro and in vivo. *Endocrinology, 148*, 3580–3590.

Choi, Y., Ballow, D. J., Xin, Y., & Rajkovic, A. (2008a). Lim homeobox gene, lhx8, is essential for mouse oocyte differentiation and survival. *Biology of Reproduction, 79*, 442–449.

Choi, Y., Yuan, D., & Rajkovic, A. (2008b). Germ cell-specific transcriptional regulator sohlh2 is essential for early mouse folliculogenesis and oocyte-specific gene expression. *Biology of Reproduction, 79*, 1176–1182.

Curran, S., Urven, L., & Ginther, O. J. (1997). Distribution of putative primordial germ cells in equine embryos. *Equine Veterinary Journal Supplement,* 72–76.

de Cuevas, M., Lilly, M. A., & Spradling, A. C. (1997). Germline cyst formation in Drosophila. *Annual Review of Genetics, 31*, 405–428.

De Felici, M., Di Carlo, A., Pesce, M., Iona, S., Farrace, M. G., & Piacentini, M. (1999). Bcl-2 and Bax regulation of apoptosis in germ cells during prenatal oogenesis in the mouse embryo. *Cell Death Differential, 6*, 908–915.

de Vries, F. A., de Boer, E., van den Bosch, M., Baarends, W. M., Ooms, M. et al. (2005). Mouse Sycp1 functions in synaptonemal complex assembly, meiotic recombination, and XY body formation. *Genes Development, 19*, 1376–1389.

de Vries, S. S., Baart, E. B., Dekker, M., Siezen, A., de Rooij, D. G. et al. (1999). Mouse MutS-like protein Msh5 is required for proper chromosome synapsis in male and female meiosis. *Genes Development, 13*, 523–531.

Deanesly, R. (1977). Germ cell proliferations in the fetal horse ovary. *Cell Tissue Research, 185*, 361–371.

Ding, W., Wang, W., Zhou, B., Zhang, W., Huang, P., Shi, F., & Taya, K. (2010). Formation of primordial follicles and immunolocalization of PTEN, PKB and FOXO3A proteins in the ovaries of fetal and neonatal pigs. *Journal of Reproductive Development, 56*, 162–168.

Dissen, G. A., Romero, C., Hirshfield, A. N., & Ojeda, S. R. (2001). Nerve growth factor is required for early follicular development in the mammalian ovary. *Endocrinology, 142*, 2078–2086.

Dominguez, M. M., Liptrap, R. M., & Basrur, P. K. (1988). Steroidogenesis in fetal bovine gonads. *Canadian Journal of Veterinary Research, 52*, 401–406.

Dong, J., Albertini, D. F., Nishimori, K., Kumar, T. R., Lu, N., & Matzuk, M. M. (1996). Growth differentiation factor-9 is required during early ovarian folliculogenesis. *Nature, 383*, 531–535.

Durlinger, A. L., Kramer, P., Karels, B., de Jong, F. H., Uilenbroek, J. T. et al. (1999). Control of primordial follicle recruitment by anti-Mullerian hormone in the mouse ovary. *Endocrinology, 140*, 5789–5796.

Edwards, R. G., Fowler, R. E., Gore-Langton, R. E., Gosden, R. G., Jones, E. C. et al. (1977). Normal and abnormal follicular growth in mouse, rat and human ovaries. *Journal of Reproduction and Fertility, 51*, 237–263.

Erickson, B. H. (1966). Development and senescence of the postnatal bovine ovary. *Journal of Animal Science, 25*, 800–805.

Erickson, G. F., Magoffin, D. A., Dyer, C. A., & Hofeditz, C. (1985). The ovarian androgen producing cells: A review of structure/function relationships. *Endocrinology Review, 6*, 371–399.

Farini, D., Scaldaferri, M. L., Iona, S., La Sala, G., & De Felici, M. (2005). Growth factors sustain primordial germ cell survival, proliferation and entering into meiosis in the absence of somatic cells. *Developmental Biology, 285*, 49–56.

Ferreira, E. M., Vireque, A. A., Adona, P. R., Meirelles, F. V., Ferriani, R. A., & Navarro, P. A., (2009). Cytoplasmic maturation of bovine oocytes: Structural and biochemical modifications and acquisition of developmental competence. *Theriogenology, 71*, 836–848.

Fortune, J. E., Rivera, G. M., & Yang, M. Y. (2004). Follicular development: The role of the follicular microenvironment in selection of the dominant follicle. *Animal Reproductive Science, 82–83*, 109–126.

Garverick, H. A., Juengel, J. L., Smith, P., Heath, D. A., Burkhart, M. N. et al. (2010). Development of the ovary and ontongeny of mRNA and protein for P450 aromatase (arom) and estrogen receptors (ER) alpha and beta during early fetal life in cattle. *Animal Reproductive Science. 117*, 24–33.

Ghafari, F., Gutierrez, C. G., & Hartshorne, G. M. (2007). Apoptosis in mouse fetal and neonatal oocytes during meiotic prophase one. *BMC Developmental Biology, 7*, 87.

Gillman, J. (1948). The development of the gonads in man, with consideration of the role of fetal endocrines and the histogenesis of ovarian tumors. *Contributions to Embryology Carnegie Institution, 32*, 81–131.

Ginther, O. J., Beg, M. A., Bergfelt, D. R., Donadeu, F. X., & Kot, K. (2001). Follicle selection in monovular species. *Biology of Reproduction, 65*, 638–647.

Gondos, B., Bhiraleus, P., & Hobel, C. J. (1971). Ultrastructural observations on germ cells in human fetal ovaries. *American Journal of Obstetrics and Gynecology, 110*, 644–652.

Gougeon, A. (1981). Frequent occurrence of multiovular follicles and multinuclear oocytes in the adult human ovary. *Fertility and Sterility, 35*, 417–422.

Greenfeld, C. R., Pepling, M. E., Babus, J. K., Furth, P. A., & Flaws, J. A. (2007). BAX regulates follicular endowment in mice. *Reproduction, 133*, 865–876.

Hahn, K. L., Johnson, J., Beres, B. J., Howard, S., & Wilson-Rawls, J. (2005). Lunatic fringe null female mice are infertile due to defects in meiotic maturation. *Development, 132*, 817–828.

Hartshorne, G. M., Lyrakou, S., Hamoda, H., Oloto, E., & Ghafari, F. (2009). Oogenesis and cell death in human prenatal ovaries: what are the criteria for oocyte selection? *Molecular Human Reproduction, 15*, 805–819.

Hirshfield, A. N. (1991). Theca cells may be present at the outset of follicular growth. *Biology of Reproduction, 44*, 1157–1162.

Hirshfield, A. N., & DeSanti, A. M. (1995). Patterns of ovarian cell proliferation in rats during the embryonic period and the first three weeks postpartum. *Biology of Reproduction, 53*, 1208–1221.

Iguchi, T., Fukazawa, Y., Uesugi, Y., & Taksugi, N. (1990). Polyovular follicles in mouse ovaries exposed neonatally to diethylstibestrol in vivo and in vitro. *Biology of Reproduction, 43*, 478–484.

Iguchi, T., & Takasugi, N. (1986). Polyovular follicles in the ovary of immature mice exposed prenatally to diethylstilbestrol. *Anatomical Embryology (Berlin), 175*, 53–55.

Iguchi, T., Takasugi, N., Bern, H. A., & Mills, K. T. (1986). Frequent occurrence of polyovular follicles in ovaries of mice exposed neonatally to diethylstilbestrol. *Teratology, 34*, 29–35.

Iguchi, T., Todoroki, R., Takasugi, N., & Petrow, V. (1988). The effects of an Aromatase Inhibitor and a 5α-Reductase Inhibitor upon the Occurrence of Polyovular Follicles, Persistent Anovulation, and Permanent Vaginal Stratification in Mice Treated Neonatally with Testosterone. *Biology of Reproduction, 39*, 689–697.

Jamnongjit, M., & Hammes, S. R. (2005). Oocyte maturation: The coming of age of a germ cell. *Seminars in Reproductive Medicine, 23*, 234–241.

Jefferson, W., Newbold, R., Padilla-Banks, E., & Pepling, M. (2006). Neonatal genistein treatment alters ovarian differentiation in the mouse: inhibition of oocyte nest breakdown and increased oocyte survival. *Biology of Reproduction, 74*, 161–168.

Jefferson, W. N., Couse, J. F., Padilla-Banks, E., Korach, K. S., & Newbold, R. R. (2002). Neonatal exposure to genestein induces estrogen receptor (ER)α-expression and multioocyte follicles in the maturing mouse ovary: Evidence for ERβ-Mediated and nonestrogenic actions. *Biology of Reproduction, 67*, 1285–1296.

Juengel, J. L., Sawyer, H. R., Smith, P. R., Quirke, L. D., Heath, D. A. et al. (2002). Origins of follicular cells and ontogeny of steroidogenesis in ovine fetal ovaries. *Molecular and Cell Endocrinology, 191*, 1–10.

Karavan, J. R., & Pepling, M. E. (2012). Effects of estrogenic compounds on neonatal oocyte development. *Reproductive Toxicology*.

Kehler, J., Tolkunova, E., Koschorz, B., Pesce, M., Gentile, L. et al. (2004). Oct4 is required for primordial germ cell survival. *EMBO Report, 5*, 1078–1083.

Kerr, B., Garcia-Rudaz, C., Dorfman, M., Paredes, A., & Ojeda, S. R. (2009). NTRK1 and NTRK2 receptors facilitate follicle assembly and early follicular development in the mouse ovary. *Reproduction, 138*, 131–140.

Kezele, P., Nilsson, E., & Skinner, M. K. (2002). Cell-cell interactions in primordial follicle assembly and development. *Frontiers of Bioscience, 7*, d1990–d1996.

Kezele, P., & Skinner, M. K. (2003). Regulation of ovarian primordial follicle assembly and development by estrogen and progesterone: Endocrine model of follicle assembly. *Endocrinology, 144*, 3329–3337.

Kimura, F., Bonomi, L. M., & Schneyer, A. L. (2011). Follistatin regulates germ cell nest breakdown and primordial follicle formation. *Endocrinology, 152*, 697–706.

Kneitz, B., Cohen, P. E., Avdievich, E., Zhu, L., Kane, M. F. et al. (2000). MutS homolog 4 localization to meiotic chromosomes is required for chromosome pairing during meiosis in male and female mice. *Genes Development, 14*, 1085–1097.

Koubova, J., Menke, D. B., Zhou, Q., Capel, B., Griswold, M. D., & Page, D. C. (2006). Retinoic acid regulates sex-specific timing of meiotic initiation in mice. *Proceedings of the National Academy of Sciences USA, 103*, 2474–2479.

Lee, C. K., Scales, N., Newton, G., & Piedrahita, J. A. (1998). Isolation and initial characterization of primordial cell (PGC)-derived cells from goats, rabbits and rats. *Theriogenology, 49*, 388, abstract.

Lei, L., Jin, S., Mayo, K. E., & Woodruff, T. K. (2010). The interactions between the stimulatory effect of follicle-stimulating hormone and the inhibitory effect of estrogen on mouse primordial folliculogenesis. *Biology of Reproduction, 82*, 13–22.

Lintern-Moore, S., & Moore, G. P. (1979). The initiation of follicle and oocyte growth in the mouse ovary. *Biology of Reproduction, 20*, 773–778.

Liu, C. F., Bingham, N., Parker, K., & Yao, H. H. (2009). Sex-specific roles of beta-catenin in mouse gonadal development. *Human Molecular Genetics, 18*, 405–417.

Liu, C. F., Liu, C., & Yao, H. H. (2010). Building pathways for ovary organogenesis in the mouse embryo. *Current Topics in Developmental Biology, 90*, 263–290.

Lun, S., Smith, P., Lundy, T., O'Connell, A., Hudson, N., & McNatty, K. P. (1998). Steroid contents of and steroidogenesis in vitro by the developing gonad and mesonephros around sexual differentiation in fetal sheep. *Journal of Reproduction and Fertility, 114*, 131–119.

MacLean, G., Abu-Abed, S., Dolle, P., Tahayato, A., Chambon, P., & Petkovich, M. (2001). Cloning of a novel retinoic-acid metabolizing cytochrome P450, Cyp26B1, and comparative expression analysis with Cyp26A1 during early murine development. *Mechanics of Development, 107*, 195–201.

Mazaud, S., Guigon, C. J., Lozach, A., Coudouel, N., Forest, M. G. et al. (2002). Establishment of the reproductive function and transient fertility of female rats lacking primordial follicle stock after fetal gamma-irradiation. *Endocrinology, 143*, 4775–4787.

Mazaud, S., Guyot, R., Guigon, C. J., Coudouel, N., Le Magueresse-Battistoni, B., & Magre, S. (2005). Basal membrane remodeling during follicle histogenesis in the rat ovary: Contribution of proteinases of the MMP and PA families. *Developmental Biology, 277*, 403–416.

McClellan, K. A., Gosden, R., & Taketo, T. (2003). Continuous loss of oocytes throughout meiotic prophase in the normal mouse ovary. *Developmental Biology, 258*, 334–348.

McGee, E. A., Hsu, S. Y., Kaipia, A., & Hsueh, A. J. (1998). Cell death and survival during ovarian follicle development. *Molecular and Cellular Endocrinology, 140*, 15–18.

McGee, E. A., & Hsueh, A. J. (2000). Initial and cyclic recruitment of ovarian follicles. *Endocrinology Review, 21*, 200–214.

McLaren, A. (2000). Germ and somatic cell lineages in the developing gonad. *Molecular and Cellular Endocrinology, 163*, 3–9.

McMullen, M. L., Cho, B. N., Yates, C. J., & Mayo, K. E. (2001). Gonadal pathologies in transgenic mice expressing the rat inhibin alpha-subunit. *Endocrinology, 142*, 5005–5014.

McNatty, K. P., Fidler, A. E., Juengel, J. L., Quirke, L. D., Smith, P. R. et al. (2000). Growth and paracrine factors regulating follicular formation and cellular function. *Molecular and Cellular Endocrinology, 163*, 11–20.

Mehlmann, L. M. (2005). Stops and starts in mammalian oocytes: Recent advances in understanding the regulation of meiotic arrest and oocyte maturation. *Reproduction, 130*, 791–799.

Menke, D. B., Koubova, J., & Page, D. C. (2003). Sexual differentiation of germ cells in XX mouse gonads occurs in an anterior-to-posterior wave. *Developmental Biology, 262*, 303–312.

Merkwitz, C., Lochhead, P., Tsikolia, N., Koch, D., Sygnecka, K. et al. (2011). Expression of KIT in the ovary, and the role of somatic precursor cells. *Progress in Histochemistry and Cytochemistry, 46*, 131–184.

Monk, M., & McLaren, A. (1981). X-chromosome activity in foetal germ cells of the mouse. *Journal of Embryology and Experimental Morphology, 63*, 75–84.

Nandedkar, T., Dharma, S., Modi, D., & Dsouza, S. (2007). Differential gene expression in transition of primordial to preantral follicles in mouse ovary. *Society of Reproduction Fertility Supplement, 63*, 57–67.

Nilsson, E. E., & Skinner, M. K. (2009). Progesterone regulation of primordial follicle assembly in bovine fetal ovaries. *Molecular and Cellular Endocrinology* 313, 9–16.

Oxender, W. D., Colenbrander, B., van deWiel, D. F., & Wensing, C. J. (1979). Ovarian development in fetal and prepubertal pigs. *Biology of Reproduction, 21*, 715–721.

Pailhoux, E., Vigier, B., Vaiman, D., Servel, N., Chaffaux, S., Cribiu, E. P., & Cotinot, C. (2002). Ontogenesis of female-to-male sex-reversal in XX polled goats. *Developmental Dynamics, 224*, 39–50.

Pangas, S. A., Choi, Y., Ballow, D. J., Zhao, Y., Westphal, H. et al. (2006). Oogenesis requires germ cell-specific transcriptional regulators Sohlh1 and Lhx8. *Proceedings of the National Academy of Sciences USA, 103*, 8090–8095.

Pannetier, M., Fabre, S., Batista, F., Kocer, A., Renault, L. et al. (2006). FOXL2 activates P450 aromatase gene transcription: Towards a better characterization of the early steps of mammalian ovarian development. *Journal of Molecular Endocrinology, 36*, 399–413.

Paredes, A., Garcia-Rudaz, C., Kerr, B., Tapia, V., Dissen, G. A. et al. (2005). Loss of synaptonemal complex protein-1, a synaptonemal complex protein, contributes to the initiation of follicular assembly in the developing rat ovary. *Endocrinology, 146*, 5267–5277.

Pepling, M. E., & Spradling, A. C. (1998). Female mouse germ cells form synchronously dividing cysts. *Development, 125*, 3323–3328.

Pepling, M. E., & Spradling, A. C. (2001). The mouse ovary contains germ cell cysts that undergo programmed breakdown to form follicles. *Developmental Biology, 234*, 339–351.

Pepling, M. E., Sundman, E. A., Patterson, N. L., Gephardt, G. W., Medico, L., Jr., & Wilson, K. I. (2010). Differences in oocyte development and estradiol sensitivity among mouse strains. *Reproduction, 139*, 349–357.

Pesce, M., & De Felici, M. (1994). Apoptosis in mouse primordial germ cells: a study by transmission and scanning electron microscope. *Anatomy of Embryology (Berlin), 189*, 435–440.

Peters, H. (1969). The effect of radiation in early life of the morphology and reproductive function of the mouse ovary. In A. McLaren (ed.), *Advances in Reproductive Physiology* (pp. 149–185). Edinburgh: Logos Press Limited.

Pittman, D. L., Cobb, J., Schimenti, K. J., Wilson, L. A., Cooper, D. M. et al. (1998). Meiotic prophase arrest with failure of chromosome synapsis in mice deficient for Dmc1, a germline-specific RecA homolog. *Molecules and Cells, 1*, 697–705.

Rajah, R., Glaser, E. M., & Hirshfield, A. N. (1992). The changing architecture of the neonatal rat ovary during histogenesis. *Developmental Dynamics, 194*, 177–192.

Rajareddy, S., Reddy, P., Du, C., Liu, L., Jagarlamudi, K. et al. (2007). p27kip1 (cyclin-dependent kinase inhibitor 1B) controls ovarian development by suppressing follicle endowment and activation and promoting follicle atresia in mice. *Molecular Endocrinology, 21*, 2189–2202.

Rajkovic, A., Pangas, S. A., Ballow, D., Suzumori, N., & Matzuk, M. M. (2004). NOBOX deficiency disrupts early folliculogenesis and oocyte-specific gene expression. *Science, 305*, 1157–1159.

Ratts, V. S., Flaws, J. A., Klop, R., Sorenson, C. M., & Tilly, J. L. (1995). Ablation of *bcl-2* gene expression decreases the number of oocytes and primordial follicles established in the postnatal female mouse gonad. *Endocrinology, 136*, 3665–3668.

Reddy, P., Adhikari, D., Zheng, W., Liang, S., Hamalainen, T. et al. (2009). PDK1 signaling in oocytes controls reproductive aging and lifespan by manipulating the survival of primordial follicles. *Human Molecular Genetics, 18*, 2813–2824.

Reddy, P., Liu, L., Adhikari, D., Jagarlamudi, K., Rajareddy, S. et al. (2008). Oocyte-specific deletion of Pten causes premature activation of the primordial follicle pool. *Science, 319*, 611–613.

Richards, A. J., Enders, G. C., & Resnick, J. L. (1999). Activin and TGFbeta limit murine primordial germ cell proliferation. *Developmental Biology, 207*, 470–475.

Robles, R., Morita, Y., Mann, K. K., Perez, G. I., Yang, S. et al. (2000). The aryl hydrocarbon receptor, a basic helix-loop-helix transcription factor of the PAS gene family, is required for normal ovarian germ cell dynamics in the mouse. *Endocrinology, 141*, 450–453.

Rucker, III, E. B., Dierisseau, P., Wagner, K. U., Garrett, L., Wynshaw-Boris, A. et al. (2000). Bcl-x and Bax regulate mouse primordial germ cell survival and apoptosis during embryogenesis. *Molecular Endocrinology, 14*, 1038–1052.

Russe, I. (1983). Oogenesis in cattle and sheep. *Bibliography of Anatomy, 24*, 77–92.

Sawyer, H. R., Smith, P., Heath, D. A., Juengel, J. L., Wakefield, S. J., & McNatty, K. P. (2002). Formation of ovarian follicles during fetal development in sheep. *Biology of Reproduction, 66*, 1134–1150.

Schmidt, D., Ovitt, C. E., Anlag, K., Fehsenfeld, S., Gredsted, L. et al. (2004). The murine winged-helix transcription factor Foxl2 is required for granulosa cell differentiation and ovary maintenance. *Development, 131*, 933–942.

Smith, P., O, W. S., Hudson, N. L., Shaw, L., Heath, D. A. et al. (1993). Effects of the Booroola gene (FecB) on body weight, ovarian development and hormone concentrations during fetal life. *Journal of Reproduction and Fertility, 98*, 41–54.

Smitz, J. E., & Cortvrindt, R. G. (2002). The earliest stages of folliculogenesis in vitro. *Reproduction, 123*, 185–202.

Soyal, S. M., Amleh, A., & Dean, J. (2000). FIGalpha, a germ cell-specific transcription factor required for ovarian follicle formation. *Development, 127*, 4645–4654.

Spears, N., Molinek, M. D., Robinson, L. L., Fulton, N., Cameron, H. et al. (2003). The role of neurotrophin receptors in female germ-cell survival in mouse and human. *Development, 130*, 5481–5491.

Suzuki, A., Sugihara, A., Uchida, K., Sato, T., Ohta, Y. et al. (2002). Developmental effects of perinatal exposure to bisphenol-A and diiethylstilbestrol on reproductive organs in female mice. *Reproductive Toxicology, 16*, 107–116.

Suzuki, H., Tsuda, M., Kiso, M., & Saga, Y. (2008). Nanos3 maintains the germ cell lineage in the mouse by suppressing both Bax-dependent and -independent apoptotic pathways. *Developmental Biology, 318*, 133–142.

Takeuchi, Y., Molyneaux, K., Runyan, C., Schaible, K., & Wylie, C. (2005). The roles of FGF signaling in germ cell migration in the mouse. *Development, 132*, 5399–5409.

Tanaka, Y., Nakada, K., Moriyoshi, M., & Sawamukai, Y. (2001). Appearance and number of follicles and change in the concentration of serum FSH in female bovine fetuses. *Reproduction, 121*, 777–782.

Tomizuka, K., Horikoshi, K., Kitada, R., Sugawara, Y., Iba, Y. et al. (2008). R-spondin1 plays an essential role in ovarian development through positively regulating Wnt-4 signaling. *Human Molecular Genetics, 17*, 1278–1291.

Trombly, D. J., Woodruff, T. K., & Mayo, K. E. (2008). Suppression of notch signaling in the neonatal mouse ovary decreases primordial follicle formation. *Endocrinology*.

Tsafriri, A. (1997). Follicular development: impact on oocyte quality. In B. Fauser (ed.), *FSH action and intraovarian regulation* (pp. 83–105). New York: Parthenon Press.

Uda, M., Ottolenghi, C., Crisponi, L., Garcia, J. E., Deiana, M. et al. (2004). Foxl2 disruption causes mouse ovarian failure by pervasive blockage of follicle development. *Human Molecular Genetics, 13*, 1171–1181.

Wang, C., Prossnitz, E. R., & Roy, S. K. (2008). G protein-coupled receptor 30 expression is required for estrogen stimulation of primordial follicle formation in the hamster ovary. *Endocrinology, 149*, 4452–4461.

Wang, C., & Roy, S. K. (2007). Development of primordial follicles in the hamster: Role of estradiol-17beta. *Endocrinology, 148*, 1707–1716.

Wassarman, P. M., & Albertini, D. F. (1994). The mammalian ovum. In E. Knobil & J. D. Neill (eds.), *The physiology of reproduction, Vol. 1* (pp. 79–122). New York: Raven Press.

Witschi, E. (1948). Migration of the germ cells of human embryos from the yolk sac to the primitive gonadal folds. *Contributions to Embryology Carnegie Institution, 32*, 67–80.

Witschi, E. (1963). Embryology of the ovary. In H. a. S. Grady (ed.), *The ovary* (pp. 1–10). Baltimore, MD: Williams and Wilkens.

Yan, C., Wang, P., DeMayo, J., DeMayo, F. J., Elvin, J. A. et al. (2001). Synergistic roles of bone morphogenetic protein 15 and growth differentiation factor 9 in ovarian function. *Molecular Endocrinology, 15*, 854–866.

Yang, M. Y., & Fortune, J. E. (2008). The capacity of primordial follicles in fetal bovine ovaries to initiate growth in vitro develops during mid-gestation and is associated with meiotic arrest of oocytes. *Biology of Reproduction, 78*, 1153–1161.

Yao, H. H., Matzuk, M. M., Jorgez, C. J., Menke, D. B., Page, D. C. et al. (2004). Follistatin operates downstream of Wnt4 in mammalian ovary organogenesis. *Developmental Dynamics, 230*, 210-5.

Yoshida, H., Takakura, N., Kataoka, H., Kunisada, T., Okamura, H., & Nishikawa, S. I. (1997). Stepwise requirement of c-kit tyrosine kinase in mouse ovarian follicle development. *Developmental Biology, 184*, 122–137.

Yoshida, K., Kondoh, G., Matsuda, Y., Habu, T., Nishimune, Y., & Morita, T. (1998). The mouse RecA-like gene Dmc1 is required for homologous chromosome synapsis during meiosis. *Molecules and Cells, 1*, 707–718.

Yu, R. N., Ito, M., Saunders, T. L., Camper, S. A., & Jameson, J. L. (1998). Role of Ahch in gonadal development and gametogenesis. *Nature Genetics, 20*, 353–357.

Yuan, L., Liu, J. G., Hoja, M. R., Wilbertz, J., Nordqvist, K., & Hoog, C. (2002). Female germ cell aneuploidy and embryo death in mice lacking the meiosis-specific protein SCP3. *Science, 296*, 1115–1118.

Zachos, N. C., Billiar, R. B., Albrecht, E. D., & Pepe, G. J. (2002). Developmental regulation of baboon fetal ovarian maturation by estrogen. *Biology of Reproduction, 67*, 1148–1156.

Zeleznik, A. J., & Benyo, D. F. (1994). Control of follicular development, corpus luteum function, and the recognition of pregnancy in higher primates. In E. Knobil & J. D. Neill (eds.), *The physiology of reproduction* (pp. 751–782). New York: Raven Press.

2 The *In Vitro* Culture of Ovarian Follicles: A Brief History and Current Considerations

Bahar Uslu and Joshua Johnson

2.1 Introduction

2.1.1 Ovarian Follicles

The ovarian follicle is a defined developmental unit consisting of a single oocyte surrounded by layers of somatic granulosa and theca cells. During postnatal life, most oocytes exist in dormant primordial follicles within a single layer of a few "flat" squamous, nonproliferating granulosa cells. Upon growth activation, the granulosa cells switch to a cuboidal morphology and begin to proliferate while the oocyte initiates a growth phase. Follicle development then consists of continued proliferation and differentiation of the granulosa layers, acquisition of an additional cell layer outside the basement membrane (termed the theca), and continued oocyte growth. Eventually, a fluid-filled cavity forms within the granulosa layers (termed the antrum), giving the follicle an asymmetrical character. By this time, the granulosa cells in closest association with the oocyte are physically separate from those that make up the follicle wall; the granulosa cells in these locations are definitive lineages: the steroidogenic mural population (in the wall) and the cumulus population (surrounding the oocyte). Should the follicle survive to finish the journey of growth and maturation, the egg within its cumulus layers is expelled from the surface of the ovary during ovulation.

As elaborate as the development of follicles is *in vivo*, it is possible to recapitulate the process *in vitro* under defined culture conditions. It has long been appreciated that mammalian ovarian follicles can be isolated, intact, from mechanically disrupted ovaries. The immediate question was whether these structures could survive and continue to develop outside of the ovary, thereby bypassing physiological/cyclical limits on egg production.

The earliest investigators who attempted *in vitro* follicle culture would have immediately recognized that many more immature follicles could be isolated from ovaries than would normally be ovulated in a lifetime. For example, in a young adult mouse, approximately 3500 immature follicles are present. In young women, as many as five hundred thousand are present. Thus if an ideal

Oocyte Physiology and Development in Domestic Animals, First Edition. Edited by Rebecca L. Krisher.

standardized protocol for follicle maturation *in vitro* could be developed, each immature follicle could theoretically produce a normal mature egg.

2.1.2 Follicle Culture and Fertility Preservation

A revolution has begun in the clinical realm of fertility preservation in the last 10 years as it is now commonplace to cryopreserve human and animal ovarian tissue and oocytes with highly efficient survival and excellent performance post-thaw (see Johnson & Patrizio, 2011; Paffoni et al., 2011; Trokoudes, Pavlides, & Zhang, 2011 for reviews). When mature eggs are available, they can be cryopreserved with only a minimal decrease in their ability to give rise to offspring versus fresh eggs (Cobo et al., 2011; Trokoudes et al., 2011). This strategy is particularly useful in the case of girls and adolescents, and women who face potentially sterilizing therapeutic treatments but do not have a male partner. Further, egg cryopreservation is increasingly being used by women who seek to "freeze" their reproductive potential at a younger age. If eggs are cryopreserved, their likelihood of giving rise to a healthy offspring remains constant even while the woman herself ages (Cobo et al., 2012; Stoop et al., 2012). This strategy will not, however, help girls and women who cannot produce mature eggs, or those who repeatedly generate eggs of poor quality. The use of cryopreserved ovarian cortex, which can contain thousands of immature follicles per cubic centimeter, is one way that we can overcome this obstacle.

Ovarian cortex can be cryopreserved, and several groups have demonstrated high rates of post-thaw follicle survival (Amorim et al., 2012; McLaughlin et al., 2011; Nisolle et al., 2000; Oktem et al., 2011). In what has been a boon for cancer patients and other women positive for parameters of ovarian insufficiency, previously cryopreserved pieces of ovarian cortex have been thawed and sutured back to the surface of the patient's ovary. Those pieces of cortex can be capable of normal function, producing waves of follicle development that culminate in a restoration of menstrual cyclicity and ovulation. Multiple successful pregnancies and term deliveries have been achieved by this technique (Camboni et al., 2008; Dolmans et al., 2009; see Donnez, Silber et al. 2011 for a review). Having achieved pregnancies for a limited subset of women, the major remaining obstacle in the field is follicle survival efficiency. Indeed, the surgical delivery of pieces of ovarian cortex function so normally that the limited number of redelivered follicles dwindles due to what appears to be a normal rate of atresia (Dolmans et al., 2009; Donnez et al., 2006; Donnez, Squifflet et al., 2011). But again, what of the girls and women who cannot produce mature eggs, or those who repeatedly generate eggs of poor quality? The preservation and redelivery of ovarian cortex might allow the resumption of cyclicity for a short period of time, but the same issues with producing normal eggs in useful numbers would remain. If even a limited number of ovarian follicles are present in isolated ovarian cortex, improvements in *in vitro* follicle culture may be able to foster their development to mature, preovulatory stages (Johnson & Patrizio, 2011). Certainly, if we improve the rate that follicles survive and develop *in vitro*, we can vastly increase the number of normal, mature eggs available to use for conception.

In addition to providing cornerstone basic information about the development and death of follicles, follicle culture systems continue to be invaluable tools for the dissection of endocrine, paracrine, and intrafollicular signaling events in females. Here, we review the historical literature, starting with the earliest accounts of the handling and culture of ovarian follicles, and discuss how those experiments have informed our understanding of the ovary. Next, we provide an up-to-date accounting of current follicle culture protocols, with an emphasis on clinical relevance. Finally, we discuss experimental techniques that will complement and/or compete with ovarian follicle culture in the near and distant future.

2.2 A Brief Historical Review of *In Vitro* Follicle Culture

There is a long history of using *in vitro* culture of ovarian follicles to address basic questions about ovarian physiology and fertility, and also to develop clinical tools to support animal and human reproduction. The earliest reports of culturing ovarian tissue utilized different approaches and different model organisms. The culture of denuded mammalian oocytes (and cumulus-oocyte-complexes) will not be covered here; suffice it to say that studies of *in vitro* oocyte maturation is a subject worthy of its own many extensive review(s) (Reinblatt et al., 2011; Sirard, 2011; Smitz, Thompson, & Gilchrist, 2011). Further, beyond superficial mention we do not discuss the incredible amount of work done on isolated or co-cultured somatic cell types of the ovary. We will, however, briefly discuss attempts to culture either entire rodent or rabbit ovaries (or large pieces of those ovaries) as this work is relevant to so-called "two-step" protocols later.

Removal of intact mammalian ovaries tends to be a very simple procedure, requiring only a few incisions to liberate the robust organs. In the mid 1950s, D. L. Ingram of the University of Birmingham cultured both rat and mouse ovaries at time points between fetal life and sexual maturity and observed epithelial cell outgrowths onto his culture plates (Ingram, 1956). The small (8–10 mm) rodent ovary is capable of passive nutrient and gas exchange in simple plate or drop cultures. Ingram's histological assessment of both cultured ovaries and those fixed at the time of collection showed a remarkable maintenance of tissue architecture and an admirable retention of intact follicles after culture. Soon afterward, mouse, rat, and rabbit ovary culture were being applied to broader endocrinological questions, including the determination of the primarily gonadal source of steroid hormones (Planel et al., 1964) (see Forleo & Collins, 1964 for human studies) and the effects of those (and peptide) hormones on gonadal and other tissues (Schriefers, Kley, & Brodesser, 1969; Neal & Baker, 1974a, 1974b; see Baker & Neal, 1974 for human studies). These and other landmark studies were soon followed by those in which follicles were mechanically isolated and cultured.

Intact follicles of many different sizes can be mechanically isolated from ovaries by simply puncturing the organs with needles. With some care, large numbers of primordial follicles can be isolated from single ovaries. Enzymatic digestion protocols have also been used to isolate follicles; these have tended to only slightly compromise subsequent follicle growth and maturation (Carrell et al., 2005; Demeestere et al., 2002). Most often, studies have pooled follicles of one approximate developmental stage by size, and have then assessed their growth and development toward maturity. Initial attempts were performed using tissue culture vessels, either in the wells of plates or in droplets under oil. This strategy was used in the mid 1970s by T. G. Baker and colleagues in an important series of manuscripts. Factors that might direct antral follicles to survive, grow, and develop toward ovulation were initially evaluated (Baker & Neal, 1974b; Baker, Hunter, & Neal, 1975; Baker et al., 1975; Neal & Baker, 1975). The system was also successfully applied to the culture of large preantral follicles (Vanderhyden, Telfer, & Eppig, 1992).

Although antral and preantral follicle culture continued to be refined in several laboratories, a trend toward working with more and more immature follicles over time began to emerge. Successful culture of "early preantral" follicles (Cortvrindt & Smitz, 1998; Cortvrindt, Smitz, & Van Steirteghem, 1997; Smitz, Cortvrindt, & Van Steirteghem, 1996) was achieved, and follicle survival and growth after cryopreservation (Smitz & Cortvrindt, 1998) was reported. Cryopreservation added a new dimension to studies of *in vitro* follicle development, and hinted at an enormously powerful future clinical fertility preservation tool.

The recapitulation of follicle development and the system's use as a tool to characterize intrafollicular signaling were rewarding, but what of the production of mature, fertilizable oocytes? Spears et al. (1994) reported the generation of live pups after the two-dimensional culture of primary (small preantral) follicles to maturity, *in vitro* fertilization, and transfer of embryos to pseudopregnant females. More recent work has shown that offspring-competent oocytes can be generated from follicles that were cryopreserved and thawed (dela Pena et al., 2002; Kagawa et al., 2007; Liu et al., 2002). Thus, the straightforward culture of immature follicles is capable of supporting complete oogenesis as well. The caveat here, however, was that the efficiency of producing mature eggs was quite low as hundreds of follicles were cultured to produce just a few litters of pups. This issue was even more challenging in attempts to culture primordial follicles into mature, ovulatory follicles to produce mature eggs.

The pool of growth arrested primordial follicles (Adhikari & Liu, 2009) is the largest potential resource for the generation of mature eggs in any ovary. Directing their development through the entire process of follicle development, from primordial follicle to mature egg, was a remarkable challenge to overcome. In the first published report, Eppig and colleagues started by culturing the ovaries of newborn mice that ostensibly contained only primordial follicles. After 8 days of culture, oocytes were isolated and cultured further so that they could complete their maturation. In the end, the group was able to generate 192 two-cell embryos after fertilizing the resulting mature eggs. Of those 192 embryos, a single live pup was generated (Eppig & O'Brien, 1996). A short time later, an improved culture method was provided by the same group (O'Brien, Pendola, & Eppig, 2003). Here, beginning with thousands of primordial follicles, 1,160 two-cell embryos were generated that gave rise to 59 surviving pups. Although pup survival was slightly lower than in a control group using mature eggs that developed *in vivo*, the survivors themselves were reported to be normal. This achievement is in no way diminished by highlighting the poor efficiency of the process. Why are so few primordial oocytes capable of producing an offspring in current *in vitro* culture systems? Focusing on improving the efficiency of production of high-quality eggs capable of giving rise to healthy offspring is our highest priority.

2.3 State-of-the-Art *In Vitro* Follicle Culture

Follicle culture methods are becoming more precisely tailored to the organism and its ovarian and follicular structure(s) (Figure 2.1; see McLaughlin et al., 2011; Smitz et al., 2010 for reviews). In terms of media composition, there has been a trend toward serum-free conditions (Oktem & Oktay, 2007; Ola et al., 2008; Senbon & Miyano, 2002; Xiet at al., 2004; see Hartshorne, 1997 for a review). Several groups continue to use "two-dimensional" culture of follicles on tissue culture plates or inserts. Although effective, these systems tend to result in artifacts of follicle growth where the adherence of theca and granulosa cells to the dish deform the follicles and could result in early release of the oocyte. Three-dimensional culture systems that prevent substrate attachment were developed to address this problem

Initially, culturing follicles on or within collagen matrices (Combelles et al., 2005; Loret de Mola et al., 2004; Telfer, Torrance, Gosden, 1990; Torrance, Telfer, & Gosden, 1989) resulted in improved morphological features of follicle growth, and very recently this has been used to characterize the establishment of the theca layers (Itami et al., 2011). Alginate matrices (Kreeger et al., 2005; protocol accompanied by video demonstration in Shikanov et al., 2011) have been used to produce large antral follicles from small preantral follicles at a higher rate than two-dimensional culture and to generate offspring (Xu et al., 2006), again at an improved rate. This technique has also been applied to follicles of the rhesus monkey (Xu et al., 2010; Xu et al., 2011). Recently, follicle culture

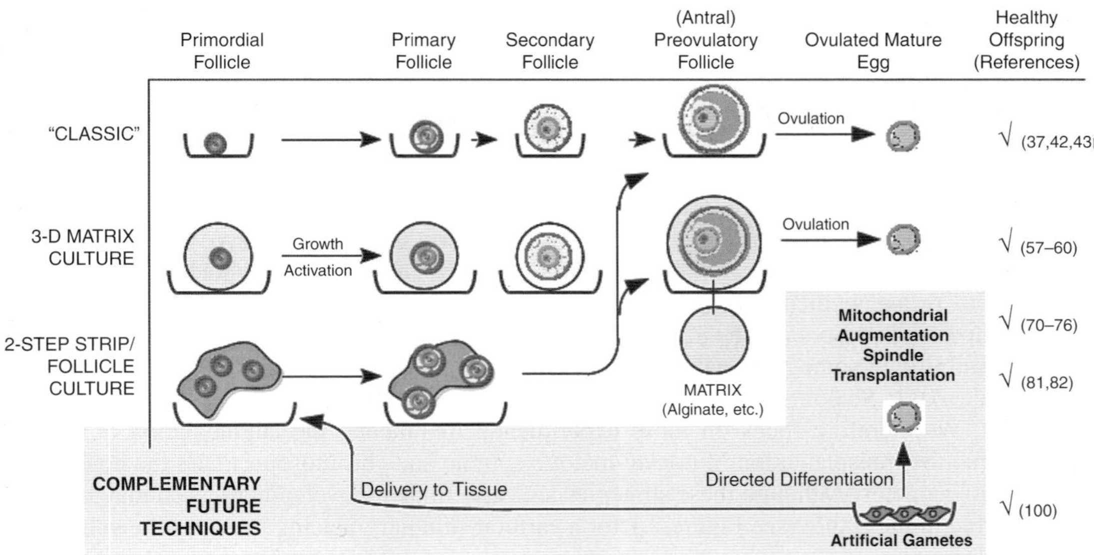

Figure 2.1 Present and future *in vitro* follicle culture techniques. In "CLASSIC" follicle culture, ovarian follicles are mechanically isolated and cultured singly in culture dishes, optionally coated with extracellular matrix components. 3-dimensional ("3-D MATRIX") matrix culture protocols have been developed where follicles are embedded in biocompatible materials; follicle growth and architecture is generally enhanced compared to "classic" culture. The 2-STEP strip-then-follicle culture protocol makes use of the culture of intact cortical strips until follicles grow such that they can be isolated singly. Individual follicles are then cultured further in attempts to produce mature eggs. Complementary future techniques are overlaid with existing strategies, highlighted within a gray box. Mature eggs produced either *in vivo* or *in vitro* could be manipulated using mitochondrial augmentation or spindle transformation in order to improve their quality. Separately, "artificial" eggs may be produced by delivering stem cells to ovarian tissue, or, by inducing them to differentiate *in vitro* into mature eggs. Modified from Johnson and Patrizio, 2011.

within a hyaluronan matrix (Desai et al., 2012) has been reported. Promising features of these biomaterials are the ability to change their concentration and thus permeability and mechanical rigidity, and to have additional extracellular matrix components mixed within that might enhance follicle development. Overall, such three-dimensional systems may allow the closer recapitulation of the intra-ovarian environment for follicles grown *in vitro*.

The second technique, the "two-step" culture first of ovarian cortex, followed by culture of individual follicles, has likewise showed impressive results. Here, ovarian cortical strips can be optionally cryopreserved, then cultured *in vitro*. When growing follicles are large enough, they can be isolated from the adjacent cortical tissue for further culture, often using engineered biomaterials as mentioned above. Telfer *et al.* (McLaughlin & Telfer, 2010; Telfer et al., 2008) have shown that such a two-step procedure can result in the generation of large antral follicles. Primordial follicles are notoriously difficult to mechanically isolate from the surrounding cortical tissue; their growth, first within the cortical strip, helps address this problem.

This technique has the benefit of allowing follicles to initially develop in their native ovarian tissue, more like conditions *in vivo*. And although the process is, again, extremely inefficient, with very few mature follicles having been produced from many thousands of primordial follicles, the promise of generating mature, healthy eggs has been realized. Having summarized the long, quite successful history of ovarian follicle culture and considered the state of the art, we now turn to experimental strategies under development.

2.3.1 Alternative and Complementary Strategies

Although follicle culture will continue to have a central role in experimental and clinical settings, it is important to consider complementary and alternative strategies to increase the number of available mature eggs of high quality. As embryo, and more recently, egg cryopreservation have become mainstream treatments in the clinic, we only briefly mention them in terms of their limitations and the need for new techniques. Embryo cryopreservation is limited to those women who have male partners or are amenable to the use of a donor source of sperm. Egg (and embryo) cryopreservation is limited to those women who can produce them for retrieval after hormonal/medical stimulation. Of course, there are women whose reserve of oocytes is so low as to make retrieval of useful numbers impossible. Also, both the need to begin cancer treatment immediately upon detection and the incompatibility of some malignancies with standard ovarian stimulation (Bedoschi et al., 2010; Fatemi et al., 2011; Shalom-Paz et al., 2010; Sonmezer et al., 2011) can render these techniques unavailable. We therefore focus on more experimental treatments: the augmentation of oocyte mitochondria, the transplantation of oocyte meiotic spindle and chromosomes, and the generation of "artificial" gametes. Although these strategies are early in their development, each may have an important role in the future (see Figure 2.1 for a cartoon summary placing these future techniques into context).

Oocyte mitochondrial augmentation, as its name implies, is intended to correct compromised mitochondrial function, including *declining* mitochondrial function with age. The decline in egg "quality" and function with aging as measured by meiotic progression and the development of normal offspring has long been associated with declining mitochondrial function (particularly, declining ATP production (Barnett & Bavister, 1996a, 1996b; see Van Blerkom, Sinclair, & Davis, 1998 for a review). Mitochondrial augmentation is performed using transplantation (e.g., withdrawal of cytoplasm containing mitochondria from a donor oocyte and delivery to a recipient oocyte via microinjection). This technique is perhaps especially useful to consider in the case of women who have a reduced ovarian reserve and/or have been shown to produce eggs of poor quality.

Initially, oocyte mitochondrial transplantation that led to the generation of heteroplasmic embryos and mice was performed as a proof of concept and to address questions of mitochondrial inheritance (Meirelles & Smith, 1997, 1998; Van Blerkom, Sinclair, & Davis, 1998) to successive generations. This model also allowed the testing of the hypothesis linking mitochondrial performance and oocyte quality (Harvey et al., 2007; Palermo, Takeuchi, & Rosenwaks, 2002; Van Blerkom, Sinclair, & Davis, 1998). Indeed, transplantation of oocyte cytoplasm containing intact mitochondria into aged or damaged oocytes resulted in improved developmental progression and reduced aneuploidy (Palermo, Takeuchi, &Rosenwaks, 2002), supporting human studies. Mitochondrial transplantation was thus applied to human oocytes, and in a 2001 publication, Jacques Cohen's group demonstrated that mitochondrial heteroplasmy was compatible with the development of normal offspring (Barritt, Willadsen et al., 2001). Assessment of the first cohort of women treated with the technique showed an improvement beyond the expected pregnancy rate without detectable defects in their children (Barritt, Brenner et al., 2001). Those authors cited prior work showing that mitochondrial heteroplasmy occurs in normal families (Bendall, & Sykes, 1995; Howell et al., 1992; Wilson et al., 1997), arguing that its introduction during assisted reproduction would be unlikely to result in abnormalities in offspring. Even so, ethical issues mainly having to do with the unique parentage status (maternal, paternal, and mitochondria donor) have been raised (Bredenoord, Pennings, & de Wert, 2008), and so it will remain experimental in nature at least until an autologous source of mitochondria can be developed. The next technique, spindle/chromosome transplantation, is essentially the converse of the delivery of mitochondria to oocytes.

In spindle/chromosome transplantation, the meiotic spindle bearing oocyte chromosomes is removed from one oocyte of questionable quality and transplanted to an oocyte that has had its own meiotic spindle removed (Tachibana, Sparman, & Mitalipov, 2010, 2012). The recipient oocyte could, as with mitochondria donors above, be selected from either a younger donor or one with proven fertility. In this way, the genetic material of the donor would be housed in an improved cytoplasm with improved chances of development. The work cited here was performed using mature rhesus monkey eggs, and it would seem that with minimal modification(s), the technique could be applied to those of the human as another way to generate eggs of improved quality.

The last future technique to consider is the production of "artificial" gametes, in our case, oocytes. Here, the central approach has been the generation of oocytes from an autologous stem or progenitor cell source. If oocytes capable of giving rise to offspring can be produced from autologous stem cells, there will no longer be a limit on the number of eggs that a woman (or other mammal) can produce. What source(s) of stem cells might be used to produce eggs? Several stem cell types are being evaluated, each is in its own stage of experimental development.

First, embryonic stem cells (ES cells) derived from the inner cell mass of blastocysts have been directed to develop into oocyte-like cells (Geijsen et al., 2004; Hubner et al., 2003; Kehler et al., 2005; Nicholas et al., 2009; Qing et al., 2007; Tedesco, Farini, & De Felici, 2011; Toyooka et al., 2003) and cells of the male germline (Easley et al., 2012; Li et al., 2012; Nayernia et al., 2006) *in vitro*. Second, a genetic engineering technique has been developed that can generate pluripotent stem cells from somatic lineages (often skin-derived fibroblasts); these are referred to as *i*nduced *P*luripotent *S*tem *Cells* (iPS cells) (Meissner, Wernig, & Jaenisch, 2007; Takahashi & Yamanaka, 2006). The generation of female (Imamura et al., 2010; Medrano et al., 2012; Panula et al., 2011) and male (Easley et al., 2012) germline cells from iPS cells have also been reported. Although morphological and gene expression data suggested that these gamete-like cells approximated normal development, the proof of their function would necessarily wait until the generation of offspring.

Very recently, a protocol has been demonstrated where ES or iPS can be directed to develop into mature eggs (Hayashi et al., 2012). First, ES or iPS cells bearing germline-specific fluorescent markers are cultured *in vitro* and monitored for differentiatation into primordial germ cell-like cells. Afterward, reconstitution of these germ cells with embryonic gonadal somatic cells results in an "artificial" ovary. Delivery of the reconstituted ovaries either under the kidney capule or ovarian bursa of recipient mice allows the entire range of follicle development to occur *in vivo*, including the development of mature, fertilization-competent eggs. Normal offspring were generated using these stem cell-derived eggs, showing that despite their origins, their functionality was intact. Given the replication of this protocol by other groups and further advances in directing the development of iPS cells into mature eggs, one can envision such a strategy being used for minimally invasive (and embryo-sparing) production of "artificial" oocytes at some point in the future.

Using a strategy that has attracted some controversy, multiple laboratories have isolated cells termed "oogonial stem cells" (OSC) from adult mouse (Johnson et al., 2004; Pacchiarotti et al., 2010; White et al., 2012; Zou et al., 2009; Zou et al., 2011) ovaries, and one group has reported the cells' isolation and characterization from human (White et al., 2012) ovaries. These cells have been shown to express an extensive list of genes associated with specific germ cell identity (Uslu, Wallace, & Johnson, unpublished observations; White et al., 2012; Zou et al., 2009; Zou et al., 2011) and, like ES and iPS cells, are capable of proliferating long-term in culture. Mouse and human OSC have also been shown to spontaneously develop into large spherical cells that are highly reminiscent of oocytes when cultured alone *in vitro* and that take on the central position of oocytes within follicle-like structures when co-cultured with somatic ovarian cells. Strikingly, fertilization of mouse oocyte-like cells generated from OSC resulted in the development of structures indistinguishable from hatching blastocysts (White et al., 2012). Further, transplantation of labeled transgenic mouse OSC (e.g., mouse OSC stably expressing green fluorescent protein) into wild-type recipient ovaries resulted in the generation of transgenic offspring (Zou et al., 2009). To date, two groups have produced what appear to be embryos after fertilization (Zou et al., 2009; White et al., 2012), and one group has published evidence of mature, offspring-competent eggs (Zou et al., 2009) – this latter example occurred after OSC transplantation into ovaries as performed using iPS cells (Hayashi et al., 2012). Although our group has confirmed the gene expression profile and long term culture capabilities of one OSC line, a recent report has been published that OSC can neither be identified nor isolated from adult ovaries (Zhang et al., 2012). This means that these data, along with the iPS data above, await replication by multiple laboratories to clarify the properties of stem cells capable of generating "artificial" eggs.

The clinical question is whether any stem cell type can be (i) delivered to ovaries in order to augment or (ii) extend ovarian function, or, (iii) produce mature, offspring-competent eggs whether *in vitro* or *in vivo*. Inducing these and other (see Virant-Klun, Stimpfel, & Skutella, 2011; Woods & Tilly, 2012 for reviews) stem cell types to efficiently give rise to "artificial" oocytes that proceed through meiosis normally and *efficiently* in a manner that would be relevant to the clinical treatment of women is an enormous remaining hurdle (Tedesco et al., 2011). Even so, continued rapid progress in this area may result in multiple stem cell–based approaches to support fertility and/or ovarian function in the clinic.

2.4 The Future of Ovarian Follicle Culture

The ultimate goal of follicle culture is the recapitulation of physiological folliculogenesis, including complete oogenesis. In terms of oogenesis, we are nearing this goal: the consistent production of fertilizable metaphase II eggs capable of giving rise to offspring. However, significant challenges remain, particularly as regards the low efficiency of the process. How can we overcome this issue and use follicle culture to inform the remaining major questions in follicle and tissue biology?

Today, the predominant outcome of the culture of pieces of ovarian cortex is *the loss*, or atresia, of the majority of follicles in order to produce even one healthy mature egg. Thus although one desired outcome is achieved, the efficiency of the system is inarguably poor, as in the ovary *in vivo*, most follicles are destined to die. The underlying goal of *in vitro* systems is therefore the unlocking of the reserve of follicles in the ovarian cortex by determining the factors that sentence their majority to death. At the same time, we will address the question of how follicle selection is achieved mechanistically.

Practically speaking, overcoming the limitation imposed by follicle selection would obliterate the current limits on egg availability that some women face in the clinic. What if a piece of cortex containing 100 primordial follicles could be cultured in an ideal manner that produced *100* mature, fertilizable eggs? The likelihood of conception increases as the number of available eggs increases. This is particularly relevant to patients who are not able to undergo ovarian stimulation by gonadotrophins to increase the number of mature eggs retrieved (e.g., those afflicted by estrogen-responsive cancers).

In addition, by fostering the survival of large numbers of immature follicles, we will be able to assess the relative "quality" of their enclosed oocytes: that is, determining whether each primordial oocyte is capable of giving rise to a mature egg that in turn is capable of giving rise to a healthy offspring. Conversely, we may find that a fraction of the immature pool of oocytes is not competent to develop normally under any circumstances. It may be that the poor efficiency seen after large-scale primordial follicile culture (O'Brien et al., 2003) cannot be overcome. Additionally, we can determine whether the intrinsic developmental potential of primordial oocytes worsens with age, or whether oocyte extrinsic, "environmental" factors are responsible for the decline in oocyte "quality" (Thomson, Fitzpatrick, & Johnson, 2010). We may then be able to determine whether improved culture systems or pharmacological additives can "rescue" the development of oocytes that would normally have limited "quality" due to aging or other factors. These questions will be addressed in part by the mentioned attempts to augment oocyte quality by manipulating mitochondrial number and function, but can only be definitively tested by increasing follicle survival. Our ability to address these key questions appears to be on the horizon.

Ovarian follicle culture is an indispensable technique that continues to unlock the secrets of the ovary, both in the clinic and in terms of its basic physiology. Follicle culture using model organisms and human tissue are examples of translational research of the highest order. Thanks to the solid foundation built by a long line of committed researchers, we will be able to address the crucial basic questions mentioned and overcome cornerstone hurdles in reproductive endocrinology, fertility, and fertility preservation.

Acknowledgments

We would like to thank Drs. Evelyn Telfer and Marie McLaughlin for scientific and editorial critiques of the manuscript while in preparation. JJ is supported by the Yale Department of Obstetrics, Gynecology, and Reproductive Sciences Research Funds and the Donaghue Yale Women's Health Research Program.

References

Adhikari, D., & Liu, K. (2009). Molecular mechanisms underlying the activation of mammalian primordial follicles. *Endocrinology Review, 30*, 438–464.

Amorim, C. A., Dolmans, M. M., David, A., Jaeger, J., Vanacker, J., Camboni, A. et al. (2012). Vitrification and xenografting of human ovarian tissue. *Fertility and Sterility*.

Baker, T. G., Hunter, R. H., & Neal, P. (1975). Studies on the maintenance of porcine graafian follicles in organ culture. *Experientia, 31*, 133–135.

Baker, T. G., & Neal, P. (1974a). Oogenesis in human fetal ovaries maintained in organ culture. *Journal of Anatomy, 117*, 591–604.

Baker, T. G., & Neal, P. (1974b). Organ culture of cortical fragments and Graafian follicles from human ovaries. *Journal of Anatomy, 117*, 361–371.

Barnett, D. K., & Bavister, B. D. (1996a). What is the relationship between the metabolism of preimplantation embryos and their developmental competence? *Molecular Reproduction and Development, 43*, 105–133.

Barnett, D. K., & Bavister, B. D. (1996b). Inhibitory effect of glucose and phosphate on the second cleavage division of hamster embryos: Is it linked to metabolism? *Human Reproduction, 11*, 177–183.

Barritt, J. A., Brenner, C. A., Malter, H. E., & Cohen, J. (2001). Mitochondria in human offspring derived from ooplasmic transplantation. *Human Reproduction, 16*, 513–516.

Barritt, J., Willadsen, S., Brenner, C., & Cohen, J. (2001). Cytoplasmic transfer in assisted reproduction. *Human Reproduction Update, 7*, 428–435.

Bedoschi, G. M., de Albuquerque, F. O., Ferriani, R. A., & Navarro, P. A. (2010). Ovarian stimulation during the luteal phase for fertility preservation of cancer patients: case reports and review of the literature. *Journal of Assistive Reproduction and Genetics, 27*, 491–494.

Bendall, K. E., & Sykes, B. C. (1995). Length heteroplasmy in the first hypervariable segment of the human mtDNA control region. *American Journal of Human Genetics, 57*, 248–256.

Bredenoord, A. L., Pennings, G., & de Wert, G. (2008). Ooplasmic and nuclear transfer to prevent mitochondrial DNA disorders: Conceptual and normative issues. *Human Reproduction Update, 14*, 669–678.

Camboni, A., Martinez-Madrid, B., Dolmans, M. M., Nottola, S., Van Langendonckt, A., & Donnez, J. (2008). Autotransplantation of frozen-thawed ovarian tissue in a young woman: Ultrastructure and viability of grafted tissue. *Fertility and Sterility, 90*, 1215–1218.

Carrell, D. T., Liu, L., Huang, I., & Peterson, C. M. (2005). Comparison of maturation, meiotic competence, and chromosome aneuploidy of oocytes derived from two protocols for in vitro culture of mouse secondary follicles. *Journal of Assistive Reproduction and Genetics, 22*, 347–354.

Cobo, A., de Los Santos, M. J., Castello, D., Gamiz, P., Campos, P., & Remohi, J. (2012). Outcomes of vitrified early cleavage-stage and blastocyst-stage embryos in a cryopreservation program: Evaluation of 3,150 warming cycles. *Fertility and Sterility*.

Cobo, A., Remohi, J., Chang, C. C., & Nagy, Z. P. (2011). Oocyte cryopreservation for donor egg banking. *Reproductive Biomedicine Online, 23*, 341–346.

Combelles, C. M., Fissore, R. A., Albertini, D. F., & Racowsky, C. (2005). In vitro maturation of human oocytes and cumulus cells using a co-culture three-dimensional collagen gel system. *Human Reproduction, 20*, 1349–1358.

Cortvrindt, R., & Smitz, J. (1998). Early preantral mouse follicle in vitro maturation: Oocyte growth, meiotic maturation and granulosa-cell proliferation. *Theriogenology, 49*, 845–859.

Cortvrindt, R., Smitz, J., & Van Steirteghem, A. C. (1997). Assessment of the need for follicle stimulating hormone in early preantral mouse follicle culture in vitro. *Human Reproduction, 12*, 759–768.

dela Pena, E. C., Takahashi, Y., Katagiri, S., Atabay, E. C., & Nagano, M. (2002). Birth of pups after transfer of mouse embryos derived from vitrified preantral follicles. *Reproduction, 123*, 593–600.

Demeestere, I., Delbaere, A., Gervy, C., Van Den Bergh, M., Devreker, F., & Englert, Y. (2002). Effect of preantral follicle isolation technique on in-vitro follicular growth, oocyte maturation and embryo development in mice. *Human Reproduction, 17*, 2152–2159.

Desai, N., Abdelhafez, F., Calabro, A., & Falcone, T. (2012). Three dimensional culture of fresh and vitrified mouse pre-antral follicles in a hyaluronan-based hydrogel: A preliminary investigation of a novel biomaterial for in vitro follicle maturation. *Reproductive Biology and Endocrinology, 10*, 29.

Dolmans, M. M., Donnez, J., Camboni, A., Demylle, D., Amorim, C., Van Langendonckt, A. et al. (2009). IVF outcome in patients with orthotopically transplanted ovarian tissue. *Human Reproduction, 24*, 2778–2787.

Donnez, J., Dolmans, M. M., Demylle, D., Jadoul, P., Pirard, C., Squifflet, J. et al. (2006). Restoration of ovarian function after orthotopic (intraovarian and periovarian) transplantation of cryopreserved ovarian tissue in a woman treated by bone marrow transplantation for sickle cell anaemia: Case report. *Human Reproduction, 21*, 183–188.

Donnez, J., Silber, S., Andersen, C. Y., Demeestere, I., Piver, P., Meirow, D. et al. (2011). Children born after autotransplantation of cryopreserved ovarian tissue. A review of 13 live births. *Annals of Medicine, 43*, 437–450.

Donnez, J., Squifflet, J., Pirard, C., Demylle, D., Delbaere, A., Armenio, L. et al. (2011). Live birth after allografting of ovarian cortex between genetically non-identical sisters. *Human Reproduction, 26*, 1384–1388.

Easley, C. A., Phillips, B. T., McGuire, M. M., Barringer, J. M., Valli, H., Hermann, B. P. et al. (2012). Direct differentiation of human pluripotent stem cells into haploid spermatogenic cells. *Cell Reproduction, 2*, 440–446.

Eppig, J. J., & O'Brien, M. J. (1996). Development in vitro of mouse oocytes from primordial follicles. *Biology of Reproduction, 54*, 197–207.

Fatemi, H. M., Kyrou, D., Al-Azemi, M., Stoop, D., De Sutter, P., Bourgain, C. et al. (2011). Ex-vivo oocyte retrieval for fertility preservation. *Fertility and Sterility, 95*, 15–17.

Forleo, R., & Collins, W. P. (1964). Some aspects of steroid biosynthesis in human ovarian tissue. *Acta Endocrinologica, 46*, 265–278.

Geijsen, N., Horoschak, M., Kim, K., Gribnau, J., Eggan, K., & Daley, G. Q. (2004). Derivation of embryonic germ cells and male gametes from embryonic stem cells. *Nature, 427*, 148–154.

Hartshorne, G. M. (1997). In vitro culture of ovarian follicles. *Review of Reproduction, 2*, 94–104.

Harvey, A. J., Gibson, T. C., Quebedeaux, T. M., & Brenner, C. A. (2007). Impact of assisted reproductive technologies: A mitochondrial perspective of cytoplasmic transplantation. *Current Topics in Developmental Biology, 77*, 229–249.

Hayashi, K., Ogushi, S., Kurimoto, K., Shimamoto, S., Ohta, H., & Saitou, M. (2012). Offspring from oocytes derived from in vitro primordial germ cell-like cells in mice. *Science*.

Howell, N., Halvorson, S., Kubacka, I., McCullough, D. A., Bindoff, L. A., & Turnbull, D. M. (1992). Mitochondrial gene segregation in mammals: Is the bottleneck always narrow? *Human Genetics, 90*, 117–120.

Hubner, K., Fuhrmann, G., Christenson, L. K., Kehler, J., Reinbold, R., De La Fuente, R. et al. (2003). Derivation of oocytes from mouse embryonic stem cells. *Science, 300*, 1251–1256.

Imamura, M., Aoi, T., Tokumasu, A., Mise, N., Abe, K., Yamanaka, S. et al. (2010). Induction of primordial germ cells from mouse induced pluripotent stem cells derived from adult hepatocytes. *Molecular Reproduction and Development, 77*, 802–811.

Ingram, D. L. (1956). Observations on the ovary cultured in vitro. *Journal of Endocrinology, 14*, 155–159.

Itami, S., Yasuda, K., Yoshida, Y., Matsui, C., Hashiura, S., Sakai, A. et al. (2011). Co-culturing of follicles with interstitial cells in collagen gel reproduce follicular development accompanied with theca cell layer formation. *Reproductive Biology and Endocrinology, 9*, 159.

Johnson, J., Canning, J., Kaneko, T., Pru, J. K., & Tilly, J. L. (2004). Germline stem cells and follicular renewal in the postnatal mammalian ovary. *Nature, 428*, 145–150.

Johnson, J., & Patrizio, P. (2011). Ovarian cryopreservation strategies and the fine control of ovarian follicle development in vitro. *Annals of the New York Academy of Sciences, 1221*, 40–46.

Kagawa, N., Kuwayama, M., Nakata, K., Vajta, G., Silber, S., Manabe, N. et al. (2007). Production of the first offspring from oocytes derived from fresh and cryopreserved pre-antral follicles of adult mice. *Reproductive Biomedicine Online, 14*, 693–699.

Kehler, J., Hubner, K., Garrett, S., & Scholer, H. R. (2005). Generating oocytes and sperm from embryonic stem cells. *Seminars in Reproductive Medicine, 23*, 222–233.

Kreeger, P. K., Fernandes, N. N., Woodruff, T. K., & Shea, L. D. (2005). Regulation of mouse follicle development by follicle-stimulating hormone in a three-dimensional in vitro culture system is dependent on follicle stage and dose. *Biology of Reproduction, 73*, 942–950.

Li, W., Shuai, L., Wan, H., Dong, M., Wang, M., Sang, L. et al. (2012). Androgenetic haploid embryonic stem cells produce live transgenic mice. *Nature*.

Liu, J., Rybouchkin, A., Van der Elst, J., Dhont, M. (2002). Fertilization of mouse oocytes from in vitro-matured preantral follicles using classical in vitro fertilization or intracytoplasmic sperm injection. *Biology of Reproduction, 67*, 575–579.

Loret de Mola, J. R., Barnhart, K., Kopf, G. S., Heyner, S., Garside, W., & Coutifaris, C. B. (2004). Comparison of two culture systems for the in-vitro growth and maturation of mouse preantral follicles. *Clinical Experiments in Obstetrics and Gynecology, 31*, 15–19.

McLaughlin, M., Patrizio, P., Kayisli, U., Luk, J., Thomson, T. C., Anderson, R. A. et al. (2011). mTOR kinase inhibition results in oocyte loss characterized by empty follicles in human ovarian cortical strips cultured in vitro. *Fertility and Sterility, 96*, 1154–1159.

McLaughlin, M., & Telfer, E. E. (2010). Oocyte development in bovine primordial follicles is promoted by activin and FSH within a two-step serum-free culture system. *Reproduction, 139*, 971–978.

Medrano, J. V., Ramathal, C., Nguyen, H. N., Simon, C., & Reijo Pera, R. A. (2012). Divergent RNA-binding proteins, DAZL and VASA, induce meiotic progression in human germ cells derived in vitro. *Stem Cells, 30*, 441–451.

Meirelles, F. V., & Smith, L. C. (1997). Mitochondrial genotype segregation in a mouse heteroplasmic lineage produced by embryonic karyoplast transplantation. *Genetics, 145*, 445–451.

Meirelles, F. V., & Smith, L. C. (1998). Mitochondrial genotype segregation during preimplantation development in mouse heteroplasmic embryos. *Genetics, 148*, 877–883.

Meissner, A., Wernig, M., & Jaenisch, R. (2007). Direct reprogramming of genetically unmodified fibroblasts into pluripotent stem cells. *Nature Biotechnology, 25*, 1177–1181.

Nayernia, K., Nolte, J., Michelmann, H. W., Lee, J. H., Rathsack, K., Drusenheimer, N. et al. (2006). In vitro-differentiated embryonic stem cells give rise to male gametes that can generate offspring mice. *Developing Cell, 11*, 125–132.

Neal, P., & Baker, T. G. (1974a). Response of mouse ovaries in vivo and in organ culture to pregnant mare's serum gonadotrophin and human chorionic gonadotrophin. 3. Effect of age. *Journal of Reproduction and Fertility, 39*, 411–414.

Neal, P., & Baker, T. G. (1974b). Response of mouse ovaries in vivo and in organ culture to pregnant mare's serum gonadotrophin and human chorionic gonadotrophin. *Journal of Reproduction and Fertility, 37*, 399–404.

Neal, P., & Baker, T. G. (1975). Response of mouse graafian follicles in organ culture to varying doses of follicle-stimulating hormone and luteinizing hormone. *Journal of Endocrinology, 65*, 27–32.

Neal, P., Baker, T. G., McNatty, K. P., & Scaramuzzi, R. J. (1975). Influence of prostaglandins and human chorionic gonadotrophin on progesterone concentration and oocyte maturation in mouse ovarian follicles maintained in organ culture. *Journal of Endocrinology, 65*, 19–25.

Nicholas, C. R., Haston, K. M., Grewall, A. K., Longacre, T. A., & Reijo Pera, R. A. (2009). Transplantation directs oocyte maturation from embryonic stem cells and provides a therapeutic strategy for female infertility. *Human Molecular Genetics, 18*, 4376–4389.

Nisolle, M., Casanas-Roux, F., Qu, J., Motta, P., & Donnez, J. (2000). Histologic and ultrastructural evaluation of fresh and frozen-thawed human ovarian xenografts in nude mice. *Fertility and Sterility, 74*, 122–129.

O'Brien, M. J., Pendola, J. K., & Eppig, J. J. (2003). A revised protocol for in vitro development of mouse oocytes from primordial follicles dramatically improves their developmental competence. *Biology of Reproduction, 68*, 1682–1686.

Oktem, O., Alper, E., Balaban, B., Palaoglu, E., Peker, K., Karakaya, C. et al. (2011). Vitrified human ovaries have fewer primordial follicles and produce less antimÃ¼llerian hormone than slow-frozen ovaries. *Fertility and Sterility, 95*, 2661–2664.

Oktem, O., & Oktay, K. (2007). The role of extracellular matrix and activin-A in in vitro growth and survival of murine preantral follicles. *Reproductive Science, 14*, 358–366.

Ola, S. I., Ai, J. S., Liu, J. H., Wang, Q., Wang, Z. B., Chen, D. Y. et al. (2008). Effects of gonadotrophins, growth hormone, and activin A on enzymatically isolated follicle growth, oocyte chromatin organization, and steroid secretion. *Molecular and Reproductive Development, 75*, 89–96.

Pacchiarotti, J., Maki, C., Ramos, T., Marh, J., Howerton, K., Wong, J. et al. (2010). Differentiation potential of germ line stem cells derived from the postnatal mouse ovary. *Differentiation, 79*, 159–170.

Paffoni, A., Guarneri, C., Ferrari, S., Restelli, L., Nicolosi, A. E., Scarduelli, C. et al. (2011). Effects of two vitrification protocols on the developmental potential of human mature oocytes. *Reproductive Biomedicine Online, 22*, 292–298.

Palermo, G. D., Takeuchi, T., & Rosenwaks, Z. (2002). Technical approaches to correction of oocyte aneuploidy. *Human Reproduction, 17*, 2165–2173.

Panula, S., Medrano, J. V., Kee, K., Bergstrom, R., Nguyen, H. N., Byers, B. et al. (2011). Human germ cell differentiation from fetal- and adult-derived induced pluripotent stem cells. *Human Molecular Genetics, 20*, 752–762.

Planel, H., David, J. F., Demasles, N., & Soleilhavoup, J. P. (1964). Action of the ovary, kidney and adrenal gland on vaginal epithelium cultivated in vitro. *C R Seances of Society of Biology Files, 158*, 1598–1602.

Qing, T., Shi, Y., Qin, H., Ye, X., Wei, W., Liu, H. et al. (2007). Induction of oocyte-like cells from mouse embryonic stem cells by co-culture with ovarian granulosa cells. *Differentiation, 75*, 902–911.

Reinblatt, S. L., Son, W. Y., Shalom-Paz, E., & Holzer, H. (2011). Controversies in IVM. *Journal of Assistive Reproductive Genetics, 28*, 525–530.

Schriefers, H., Kley, H. K., & Brodesser, M. (1969). Androgen metabolism and estrogen production in explant cultures of rat ovaries following incubation with androstenedione and testosterone. *Hoppe-Seyler's Z Physiology and Chemistry, 350*, 999–1007.

Senbon, S., & Miyano, T. (2002). Bovine oocytes in early antral follicles grow in serum-free media: effect of hypoxanthine on follicular morphology and oocyte growth. *Zygote, 10*, 301–309.

Shalom-Paz, E., Almog, B., Shehata, F., Huang, J., Holzer, H., Chian, R. C. et al. (2010). Fertility preservation for breast-cancer patients using IVM followed by oocyte or embryo vitrification. *Reproductive Biomedicine Online, 21*, 566–571.

Shikanov, A., Xu, M., Woodruff, T. K., & Shea, L. D. (2011). A method for ovarian follicle encapsulation and culture in a proteolytically degradable 3 dimensional system. *Journal of Visual Experiences.*

Sirard, M. A. (2011). Follicle environment and quality of in vitro matured oocytes. *Journal of Assistive Reproductive Genetics, 28*, 483–488.

Smitz, J., & Cortvrindt, R. (1998). Follicle culture after ovarian cryostorage. *Maturitas, 30*, 171–179.

Smitz, J., Cortvrindt, R., & Van Steirteghem, A. C. (1996). Normal oxygen atmosphere is essential for the solitary long-term culture of early preantral mouse follicles. *Molecular Reproduction and Development, 45*, 466–475.

Smitz, J., Dolmans, M. M., Donnez, J., Fortune, J. E., Hovatta, O., Jewgenow, K. et al. (2010). Current achievements and future research directions in ovarian tissue culture, in vitro follicle development and transplantation: implications for fertility preservation. *Human Reproduction Update, 16*, 395–414.

Smitz, J. E., Thompson, J. G., & Gilchrist, R. B. (2011). The promise of in vitro maturation in assisted reproduction and fertility preservation. *Seminars in Reproductive Medicine, 29*, 24–37.

Sonmezer, M., Turkcuoglu, I., Cozkun, U., & Oktay, K. (2011). Random-start controlled ovarian hyperstimulation for emergency fertility preservation in letrozole cycles. *Fertility and Sterility, 95*, 9–11.

Spears, N., Boland, N. I., Murray, A. A., & Gosden, R. G. (1994). Mouse oocytes derived from in vitro grown primary ovarian follicles are fertile. *Human Reproduction, 9*, 527–532.

Stoop, D., De Munck, N., Jansen, E., Platteau, P., Van den Abbeel, E., Verheyen, G. et al. (2012). Clinical validation of a closed vitrification system in an oocyte-donation programme. *Reproductive Biomedicine Online, 24*, 180–185.

Tachibana, M., Sparman, M., & Mitalipov, S. (2010). Chromosome transfer in mature oocytes. *Nature Protocols, 5*, 1138–1147.

Tachibana, M., Sparman, M., & Mitalipov, S. (2012). Chromosome transfer in mature oocytes. *Fertility and Sterility, 97*, e16.

Takahashi, K., & Yamanaka, S. (2006). Induction of pluripotent stem cells from mouse embryonic and adult fibroblast cultures by defined factors. *Cell, 126*, 663–676.

Tedesco, M., Farini, D., & De Felici, M. (2011). Impaired meiotic competence in putative primordial germ cells produced from mouse embryonic stem cells. *International Journal of Developmental Biology, 55*, 215–222.

Telfer, E. E., McLaughlin, M., Ding, C., & Thong, K. J. (2008). A two-step serum-free culture system supports development of human oocytes from primordial follicles in the presence of activin. *Human Reproduction, 23*, 1151–1158.

Telfer, E., Torrance, C., & Gosden, R. G. (1990). Morphological study of cultured preantral ovarian follicles of mice after transplantation under the kidney capsule. *Journal of Reproduction and Fertility, 89*, 565–571.

Thomson, T. C., Fitzpatrick, K. E., & Johnson, J. (2010). Intrinsic and extrinsic mechanisms of oocyte loss. *Molecular Human Reproduction, 16*, 916–927.

Torrance, C., Telfer, E., & Gosden, R. G. (1989). Quantitative study of the development of isolated mouse pre-antral follicles in collagen gel culture. *Journal of Reproduction and Fertility, 87*, 367–374.

Toyooka, Y., Tsunekawa, N., Akasu, R., & Noce, T. (2003). Embryonic stem cells can form germ cells in vitro. *Proceedings of the National Academy of Sciences USA, 100*, 11457–11462.

Trokoudes, K. M., Pavlides, C., & Zhang, X. (2011). Comparison outcome of fresh and vitrified donor oocytes in an egg-sharing donation program. *Fertility and Sterility, 95*, 1996–2000.

Van Blerkom, J., Sinclair, J., & Davis, P. (1998). Mitochondrial transfer between oocytes: potential applications of mitochondrial donation and the issue of heteroplasmy. *Human Reproduction, 13*, 2857–2868.

Vanderhyden, B. C., Telfer, E. E., & Eppig, J. J. (1992). Mouse oocytes promote proliferation of granulosa cells from preantral and antral follicles in vitro. *Biology of Reproduction, 46*, 1196–1204.

Virant-Klun, I., Stimpfel, M., & Skutella, T. (2011). Ovarian pluripotent/multipotent stem cells and in vitro oogenesis in mammals. *Histology and Histopathology, 26*, 1071–1082.

White, Y. A., Woods, D. C., Takai, Y., Ishihara, O., Seki, H., & Tilly, J. L. (2012). Oocyte formation by mitotically active germ cells purified from ovaries of reproductive-age women. *Nature Medicine, 18*, 413–421.

Wilson, M. R., Polanskey, D., Replogle, J., DiZinno, J. A., & Budowle, B. (1997). A family exhibiting heteroplasmy in the human mitochondrial DNA control region reveals both somatic mosaicism and pronounced segregation of mitotypes. *Human Genetics, 100*, 167–171.

Woods, D. C., & Tilly, J. L. (2012). The next (re)generation of ovarian biology and fertility in women: is current science tomorrow's practice? *Fertility and Sterility, 98*, 3–10.

Xie, H., Xia, G., Byskov, A. G., Andersen, C. Y., Bo, S., & Tao, Y. (2004). Roles of gonadotropins and meiosis-activating sterols in meiotic resumption of cultured follicle-enclosed mouse oocytes. *Molecular and Cell Endocrinology, 218*, 155–163.

Xu, J., Bernuci, M. P., Lawson, M. S., Yeoman, R. R., Fisher, T. E., Zelinski, M. B. et al. (2010). Survival, growth, and maturation of secondary follicles from prepubertal, young, and older adult rhesus monkeys during encapsulated three-dimensional culture: Effects of gonadotropins and insulin. *Reproduction, 140*, 685–697.

Xu, M., Kreeger, P. K., Shea, L. D., & Woodruff, T. K. (2006). Tissue-engineered follicles produce live, fertile offspring. *Tissue Engineering, 12*, 2739–2746.

Xu, J., Lawson, M. S., Yeoman, R. R., Pau, K. Y., Barrett, S. L., Zelinski, M. B. et al. (2011). Secondary follicle growth and oocyte maturation during encapsulated three-dimensional culture in rhesus monkeys: Effects of gonadotrophins, oxygen and fetuin. *Human Reproduction, 26*, 1061–1072.

Zhang, H., Zheng, W., Shen, Y., Adhikari, D., Ueno, H., & Liu, K. (2012). Experimental evidence showing that no mitotically active female germline progenitors exist in postnatal mouse ovaries. *Proceedings of the National Academy of Sciences USA, 109*, 12580–12585.

Zou, K., Hou, L., Sun, K., Xie, W., & Wu, J. (2011). Improved efficiency of female germline stem cell purification using fragilis-based magnetic bead sorting. *Stem Cells Development, 20*, 2197–2204.

Zou, K., Yuan, Z., Yang, Z., Luo, H., Sun, K., Zhou, L. et al. (2009). Production of offspring from a germline stem cell line derived from neonatal ovaries. *Nature Cell Biology, 11*, 631–636.

3 Regulation of Oocyte Meiotic Resumption by Somatic Cells

Masayuki Shimada

Mammalian oocytes arrested at the diplotene stage of the first meiotic prophase are closely surrounded by somatic cells (designated pre-granulosa cells in primordial follicles and cumulus cells in growing follicles). During early stages of follicle growth, granulosa cells and cumulus cells transport energy sources and other factors into oocytes via numerous gap junctions, promoting oocyte growth to full development size (Downs & Eppig, 1984; Downs et al., 1986; Simon et al., 1997). After the preovulatory LH surge, the oocytes resume meiosis, complete germinal vesicle breakdown (GVBD), and progress to metaphase II (MII) stage before ovulation. Because LH receptors (LHCGR) are not present on the surface of oocytes, and because the expression in cumulus cells is minimal compared with that in granulosa cells (Peng et al., 1991), the LH surge dramatically and directly changes the pattern of gene expression and protein synthesis in granulosa cells (Richards, 1994). Factors induced and secreted rapidly from granulosa cells then act on cumulus cells to induce differentiation-dependent changes in cell function. Concomitantly with cumulus cell differentiation, meiotic resumption and maturation are also induced in the oocyte, along with developmental competence (Cross & Brinster, 1970). Thus, cumulus cells play a critical role in oocyte maturation; however there is less information about the mechanisms by which cumulus cells differentiate and how these changes in cumulus cells impact oocyte maturation at the molecular level. The mechanisms by which cumulus cells regulate oocyte meiotic resumption are the focus of this chapter.

3.1 Meiotic Resumption Is Negatively Regulated in a cAMP-Dependent Manner

Cyclic AMP (cAMP) is synthesized in cumulus cells and granulosa cells stimulated by FSH during follicular development (Schultz et al., 1983a, 1983b; Racowsky, 1985a; Mattioli et al., 1994). FSH receptor is a member of GTP binding protein coupling receptors, and activates adenylate cyclase to increase cAMP level in the FSH-stimulated cells (Grisworld et al., 1995). We revealed that intact porcine cumulus oocyte complexes (COC) just after collection from small antral follicles have a low level of cAMP, whereas it was increased by culture with FSH (Shimada & Terada, 2002a). In the study, intact COC just after collection have a low level of cAMP (0.15 + /-0.02 pmol/ COC; Fig. 3.1a). A significant increase of cAMP in intact COC cultured for 4 hr was noted (Fig. 3.1a).

Oocyte Physiology and Development in Domestic Animals, First Edition. Edited by Rebecca L. Krisher.

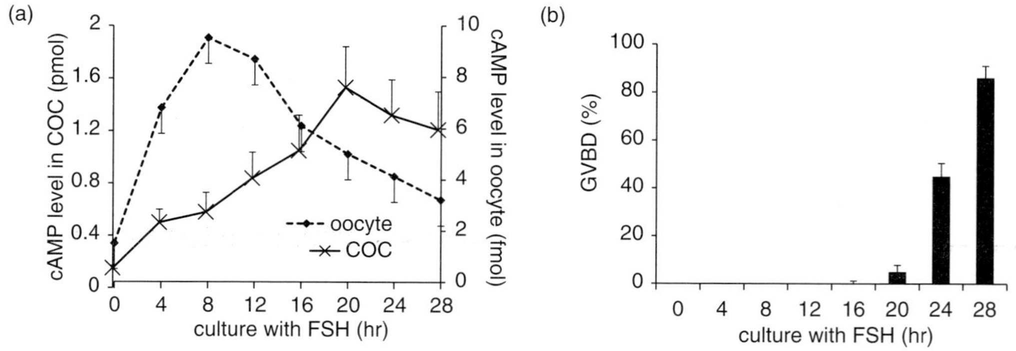

Figure 3.1 Time-dependent changes of cAMP levels either in oocytes separated from COC or in intact COC during oocyte meiotic resumption (exhibiting GVBD). (a) Porcine COC were recovered from their follicles and then cultured with FSH for up to 28 hr. After culture, some COC were separated into oocyte and cumulus cells, and the denuded oocytes and other COC were used for cAMP analysis by HPLC-UV detection according to Shimada and Terada (2002a). Values are mean +/– SEM of 3 replicates. (b) The kinetic changes of meiotic resumption (exhibiting GVBD) were observed. COCs were cultured with FSH for up to 28 hr.

Thereafter, the level of cAMP in intact COC increased significantly and reached a peak at 20 hr of COC culture (Fig. 3.1a). Although there was a tendency toward decline in cAMP levels with the incubation period, a high level of cAMP was maintained for up to 28 hr (Fig. 3.1a).

We also detected time-dependent changes in the cAMP level of oocytes separated from cultured porcine COC. The concentration of cAMP in the oocyte just after recovery from its follicle was 1.5 +/–0.5 fmol, and 4-hr culture of COC led to a significant increase in the cAMP level in the oocyte (Fig. 3.1a). The concentration of cAMP reached a peak at 8 hr of culture, but it dramatically decreased after the 12-hr point (Fig. 3.1a). After COC were cultured for 16 hr, the level of cAMP in the oocyte was significantly lower than the level of the peak at 8-hr culture. Under the above culture condition, GVBD was undetectable after 16 hr of COC culture. At 20 hr, GVBD was realized in 5% of COC (Fig. 3.1b). As cultured periods were further prolonged, the proportion of GVBD progressively increased; a period of up to 28 hr of culture was required for the majority of oocytes (86%) to accomplish GVBD (Fig. 3.1b).

Because the oocyte does not have any FSH receptor (Grisworld et al., 1995), and small molecules are transferred to the oocyte from cumulus cells via gap junctional communication (Grazul-Bilska et al., 1997), the one possibility is that cAMP is synthesized in FSH-stimulated cumulus cells, and then transferred to oocytes. However, in the oocyte but not in cumulus cells, the concentration of cAMP is returned to basal values around the initiation of GVBD, as shown in Fig. 3.1. When a fall of cAMP level in bovine oocytes was prevented by a phosphodiesterase inhibitor, 3-isobutyl-1-methylxanthine (IBMX), the majority of oocytes were arrested at the GV stage (Rose-Hellekant & Bavister, 1996; Luciano et al., 1999). Moreover, in the oocytes, the addition of adenylate cyclase activator, forskolin, to maturation medium led to an increase in the cAMP level and also suppressed meiotic resumption (Racowsky, 1985b; Rose-Hellekant & Bavister, 1996). In our study using porcine COC, the addition of IBMX to the maturation medium produced a significant increase in the cAMP content of oocytes and a significantly declined proportion of oocytes exhibiting GVBD; both results were observed as dose-dependent. Thus, an increased level of cAMP in oocytes arrests meiosis at the germinal vesicle stage (GV stage), and the drop of cAMP level from maximum is required for induction of meiotic resumption of oocytes (Shimada & Terada, 2002a).

3.2 The Regulation of cAMP Level in Mouse Oocytes

Recent mutant mouse models show that the regulation of cAMP level in oocytes is dependent on G protein coupling receptor family member, GPR3/GPR12, expressed in the oocyte itself (Mehlmann et al., 2004; Hinckley et al., 2005). In the *Gpr3* mutant mice, oocyte meiotic resumption was spontaneously exhibited before ovulation stimuli in small antral follicles (Mehlmann et al., 2004). Although the endogenous lignads for GPR3 remains unclear, it has been reported that sphingosine 1 phosphate (S1P) is potentially acting as a ligand for GPR3 (Hinckley et al., 2005). When S1P was added to the medium, cAMP level was increased in oocytes and the spontaneous meiotic resumption was suppressed in mice, suggesting that the mouse oocyte produces cAMP to prevent meiotic resumption before ovulation stimuli (Hinckley et al., 2005). However, in this theory, the mechanisms controlling how cAMP level is decreased in the oocyte after ovulation stimuli have remained unclear.

Although oocytes produce cAMP, the regulation of cAMP level in oocytes is dependent on cumulus cells, because when oocytes were removed from cumulus cell layers and the denuded oocytes were cultured, the level of cAMP quickly decreased (Mattioli et al., 1994). Most recently, Eppig and collaborators clearly showed that granulosa cells expressed and secreted natriuretic peptide precursor type C (NPPC) that acted on cumulus cells to produce cGMP (Zhang et al., 2010). The cGMP was transferred to oocytes via gap junctional communication. One of the functions of cGMP is to decrease PDE3 activity, which increases cAMP level in the cells. In oocytes, PDE3 is expressed and acted as a key factor to regulate oocyte meiotic resumption (Richard et al., 2001). Thus, in the mouse model, cGMP produced in cumulus cells is transferred to oocyte via gap junctions to maintain the level of cAMP that impacts meiotic arrest at the GV stage (Fig. 3.2).

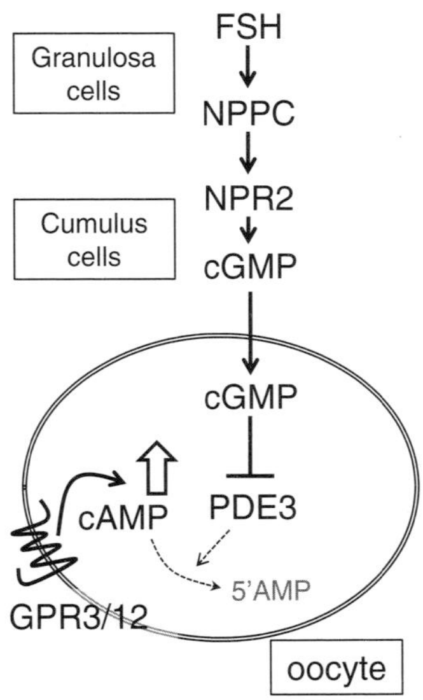

Figure 3.2 Model describing the expression and roles of NPPC (natriuretic peptide precursor type C) and its receptor, NPR2 in the maintenance of oocyte meiotic arrest at the GV stage in mice. NPPC is secreted from FSH-stimulated granulosa cells, and then acts on cumulus cells to stimulate cGMP production. The cGMP is transferred to the oocyte via gap junctional communication, which suppresses PDE3 activity that converts cAMP to 5′AMP. The cAMP is produced in the oocyte via a GPR3/12- dependent manner. The pathway increases the cAMP level in the oocyte during follicular development.

3.3 The Expression and Roles of PDEs in both Cumulus Cells and Oocytes in Domestic Animals

In domestic animals, although the molecular mechanisms in detail are not clearly understood, it is reported that some of them are similar but others are not common among the species, such as cAMP transportation from cumulus cells to oocytes. The marked difference in the pig compared with the mouse model is the localization and function of PDE3 that is expressed in both oocyte and cumulus cells in porcine COC (Sasseville et al., 2007). The role of PDEs in the oocyte is to arrest meiosis at the GV stage, described in the above sections. In porcine cumulus cells of COC, PDEs plays an important role in cell differentiation during oocyte maturation. In fact, the activity of PDE3 in cumulus cells is dependent on de novo transcribed mRNA that is induced in a cAMP-dependent manner (Sasseville et al., 2007). We investigated the dose-dependent effects of forskolin, an adenylate cyclase activator, to modify the cAMP level in cumulus cells on cell differentiation during *in vitro* maturation of porcine COC (Shimada et al., 2002). COC were cultured for 24 hr with 0 to 100 uM forskolin. The level of cAMP in cumulus cells was increased in a dose-dependent manner. Concomitantly with cAMP elevation, progesterone in the culture medium was increased by the addition of forskolin (Fig. 3.3a). The stimulatory effects reached their maximum at 5.0 uM, and the level of progesterone in the medium where COC had been cultured with 30 uM forskolin was significantly lower than the maximum level observed at 5.0 uM (Fig. 3.3a), indicating that cAMP

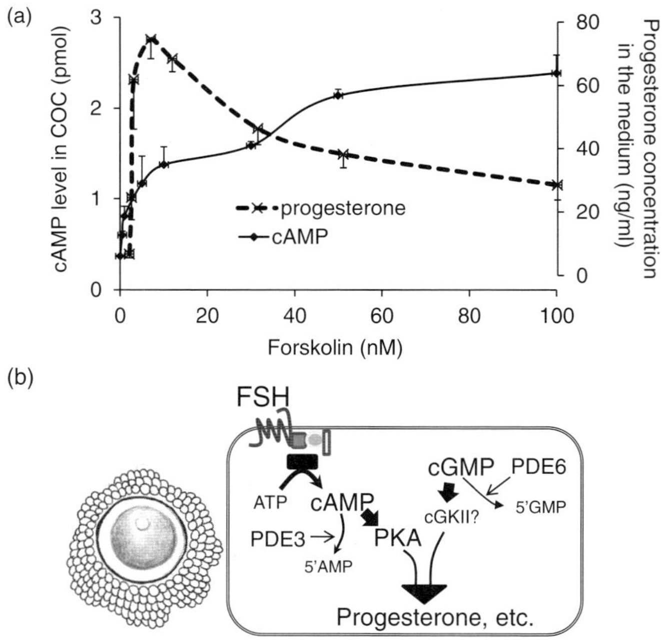

Figure 3.3 The relationship between cAMP level and progesterone production in cumulus cells of COC cultured with different doses of forskolin for 20 hr. (a) Porcine COC were cultured with 0 to 100 nM forskolin for 20 hr. The condition medium was recovered and then used for progesterone assay by EIA, and the cultured COC were used for cAMP analysis as described in the Figure 3.1 legend. (b) A schematic diagram with regard to the regulation of cAMP and cGMP levels in cumulus cells to induce differentiation. FSH produces cAMP to activate the PKA pathway. cGMP is also produced and accumulated in cumulus cells, which may activate the cGKII pathway. Both pathways are required to produce progesterone. The level of cAMP or cGMP is negatively regulated by specific enzyme PDE3 or PDE6 in cumulus cells. (For color detail, see color plate section.)

production is required for progesterone synthesis in cumulus cells whereas the high level of cAMP suppressed progesterone production. Moreover, porcine cumulus cells also express PDE type 6 that degenerated cGMP but not cAMP, and the inhibitor treatment decreases progesterone production by cumulus cells of COC (Sasseville et al., 2008). Thus, cAMP and cGMP are produced, and they are maintained at optimal levels in cumulus cells, where the functional changes of cumulus cells including progesterone production induces oocyte meiotic resumption (Fig. 3.3b).

The secreted progesterone stimulates its specific receptor (progesterone receptor, PGR) expressed on cumulus cell themselves (Shimada & Terada, 2002b). When 1, 5, 10, or 20 porcine COC were cultured in each well containing 300 ul of maturation medium for 20 hr, there was a significant positive correlation between the GVBD rate and the number of COC cultured in each well (Yamashita et al., 2003, Fig. 3.4a). The level of progesterone in the medium in which COC had been cultured was also significantly increased with an increase in the number of COC cultured in each well. A significantly lower proportion of GVBD in oocytes when 1 COC was cultured in each well for 20 hr was improved by the addition of progesterone (Fig. 3.4b). This proportion of GVBD was fully comparable to that of COC cultured in the absence of additional progesterone with 20 COC (Fig. 3.4b). Thus, progesterone secreted by COC plays a positive role in the induction of GVBD in porcine oocytes. Treatment with either PGR antagonist or progesterone synthesis inhibitor not only suppressed cumulus cell functions but also significantly increased the level of cAMP in oocytes, which results in the suppression of meiotic resumption (Shimada & Terada, 2002b; Yamashita et al., 2005, Fig. 3.4c).

Sasseville et al. (2009a) investigated which kind of PDE was expressed in either oocyte or cumulus cells in bovine. In bovine COC, PDE3 is localized only in the oocyte but not in cumulus cells; however cumulus cells express type 8 PDE. Type 3 inhibitors (milrinone or cilostamide) suppress meiotic resumption of bovine oocytes, whereas treatment with PDE8 inhibitor decreases GVBD of oocytes coincident with an increase of cAMP levels in cumulus cells. Thus, in bovine COC, the control of cAMP level by PDEs in both cumulus cells and oocytes is required for successful meiotic resumption using different PDEs than mouse or pig systems.

3.4 Closure of Gap Junctional Communication

How is the level of cAMP decreased in oocytes during meiotic maturation? In oocytes, phosphodiesterase type 3 is activated and cAMP is catabolized to 5'AMP before oocytes exhibit meiotic resumption, as described above. In amphibian oocytes, progesterone or insulin-like growth factor 1 (IGF1) activates the PI3-kinase-PKB pathway that phosphorylates PDE3 to increase enzyme activity (Chuang et al., 1993; Muslin et al., 1993; Conti et al., 2002). The PI3-kinase-PKB pathway is also activated in mouse oocytes before meiotic resumption (Han et al., 2006). This signaling pathway potentially upregulates PDE3 enzyme activity in oocytes; however PI3-kinase inhibitors do not block meiotic resumption of mouse oocytes (Hoshino et al., 2004). These results suggest that although PI3-kinase-PKB pathway is involved in cAMP degradation, this regulation is not the primary switch for oocyte maturation and may be working as an amplifying loop to decrease cAMP during oocyte meiotic resumption in mammals. Other factors are required for the initial reduction of cAMP and the induction of oocyte meiotic resumption.

The other possibility is that the amount of cAMP and/or cGMP transferred from cumulus cells to oocytes is decreased. It has been reported that the disruption of gap junctions in cumulus cells induces meiotic resumption of mouse, rat, and porcine oocytes, due to interruption of the conduction of meiosis inhibitory signals from the outer layers of cumulus cells to the oocytes (Larsen

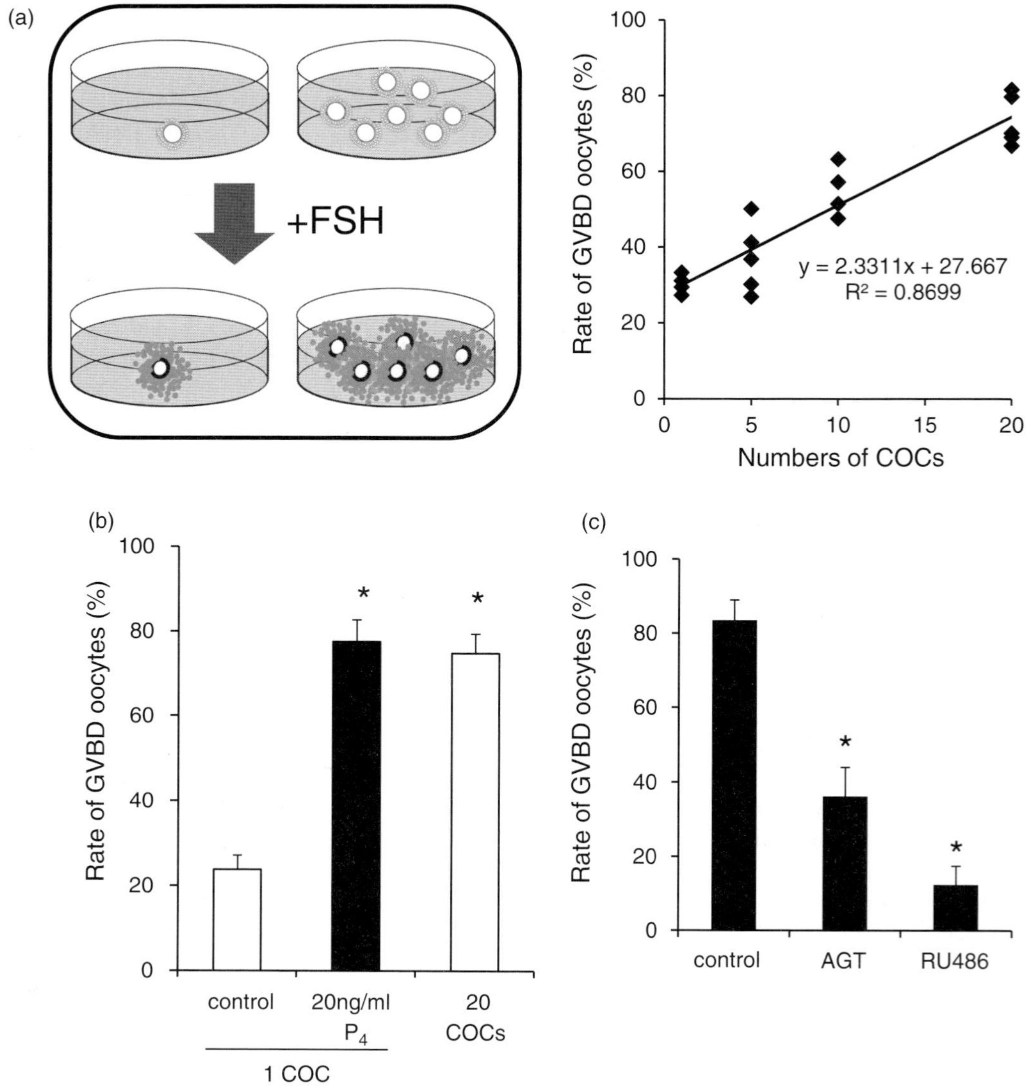

Figure 3.4 The role of progesterone secreted from cumulus cells in the induction of meiotic resumption of porcine oocytes *in vitro*. (a) The relationship between additional numbers of COC in each well and the proportion of oocytes undergoing GVBD when COC were cultured for 20 hr. A significant positive correlation was observed ($r = 0.869$, $p < 0.05$). (b) Effect of addition of 20 ng/ml progesterone to the medium on the proportion of oocytes undergoing GVBD when one COC was cultured in each well for 20 hr. 1 COC; one COC was transferred to each well and cultured alone for 20 hr with or without 20 ng/ml progesterone. 20COC; twenty COC were transferred to each well and cultured for 20 hr. *An asterisk indicates a significant difference from control-1COC ($p < 0.05$). Values are mean +/− SEM of 3 replicates. (c) The negative effects of aminogluthethimide (AGT), a P450scc inhibitor, or RU486, a PGR antagonist on the proportion of oocytes undergoing GVBD when 20 COC were cultured in each well for 20 hr. *An asterisk indicates a significant difference from control ($p < 0.05$). Values are mean +/− SEM of 3 replicates.

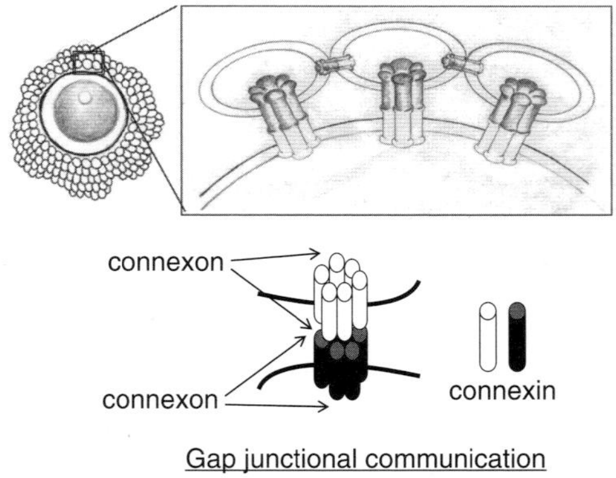

Figure 3.5 A schematic view of gap junctional communication, which is formed by two connexons located in the cytoplasmic membrane of two adjacent cells. The connexon is formed by hexametric structures consisting of 6 connexin molecules.

et al., 1986, 1987; Isobe et al., 1998; Sasseville et al., 2009b). Gap junctions, the specialized regions in opposite membranes between neighboring cells, are channels that pass low-molecular-weight substances and ions in order to enhance cellular interactions (Grazul-Bilska et al., 1997). These channels are formed by hexametric structures consisting of connexin molecules (connexon) in numerous tissues (Grazul-Bilska et al., 1997), as shown in Fig. 3.5. Porcine ovarian follicles have been reported to express five members of connexin genes: connexin-26, connexin-30.3, connexin-32, connexin-43, and connexin-60 (Itahana et al., 1996, 1998). The connexin-43 (Cx-43) protein is dominantly expressed in cumulus cells of porcine COC (Shimada et al., 2001a, 2004, Shimada & Terada, 2001). Cx-43 is also expressed in bovine cumulus cells, although Cx-37 selectively participates in gap junctional communications between cumulus cells and oocyte (Nuttinck et al., 2000).

Cx-43 has numerous phosphorylated sites, and these phosphorylations play a key role in regulatory mechanisms governing the assembly of connexons into gap junctions in the plasma membrane, and gating the formed gap junction (Musil et al., 1990; Hill et al., 1994; Lau et al., 1996). In particular, the phosphorylation of serine on Cx-43 is stimulated by mitogen-activated protein kinase (MAP kinase [ERK 1, 2]), closing gap junctional communication in rat liver cells (Hill et al., 1994; Warn-Cramer et al., 1996). In fact, the phosphorylation of Cx-43 is induced in an ERK1/2 dependent manner in mice and rat COC in culture (Sela-Abramovich et al., 2005), which results in the reduction of Cx-43 and then closure of gap junctional communication among the cumulus cells (Granot & Dekel, 1994). In porcine COC, at least three Cx-43 positive bands were detected by immunoblotting, and the upper bands disappeared following treatment with phosphatase, indicating that Cx-43 is phosphorylated in cumulus cells of porcine COC (Fig. 3.6a). Time-dependent changes were examined during culture of porcine COC with FSH for up to 28 hr. The staining intensities of faster (43 kDa) and moderate (45 kDa) migrating bands were significantly increased at 8 hr of culture compared to those of cumulus cells separated from COC immediately after collection. However, after culture of 16 hr up to 28 hr, significant reductions in intensity of the 43 kDa band of Cx-43 were noted (Fig. 3.6b). In contrast, the phosphorylated forms of 45 and 47 kDa Cx-43 were still detected at 16 hr up to 28 hr (Fig. 3.6b). Thus, within the first 4-hr culture, a high amount of Cx-43 was synthesized in all layers of the cumulus cells, whereas after 16 hr of culture, Cx-43 disappeared in cumulus cells, with increasing level of phosphorylated Cx-43.

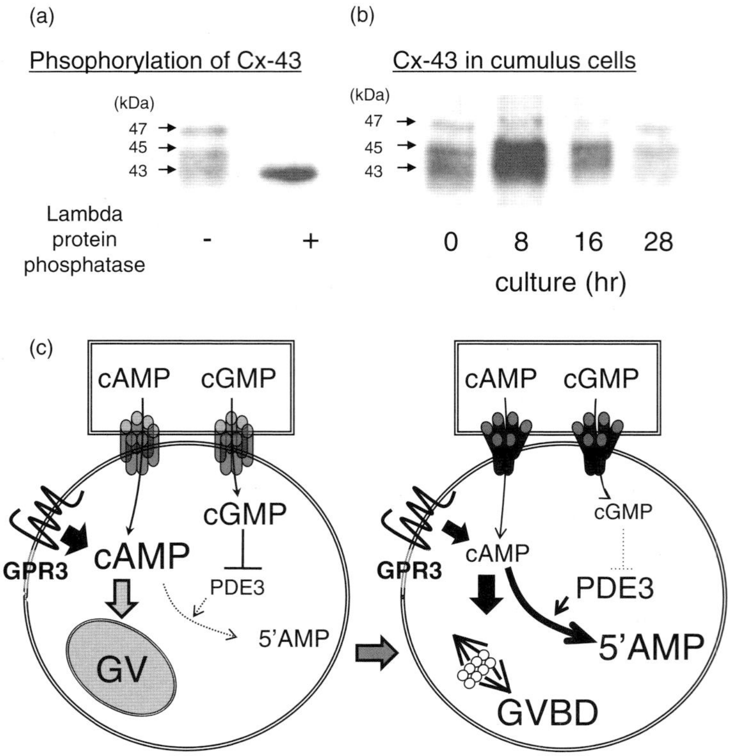

Figure 3.6 The close of gap junctional communication by the phosphorylation of connexin-43 (Cx-43) induces oocyte meiotic resumption. (a) Three different bands were detected by immunoblotting of connexin 43 in cumulus cells. To determine which band(s) was(were) the phosphorylated form(s), the cumulus cell lysate was treated with lambda protein phosphatase and $MnCl_2$ and then used for western blotting analysis. (b) Profiles of phosphorylation states of connexin-43 in cumulus cells of porcine COC cultured with FSH and LH. (c) A schematic diagram of the induction of meiotic resumption of porcine COC. The close of gap junctional communication reduces the amount of cAMP and cGMP transferred from cumulus cells to oocyte, which results in the induction of meiotic resumption of the oocyte.

At the time of reduction and phosphorylation of Cx-43, a majority of oocytes proceed from the GVII stage to the GVIII stage, with observed networks of filamentous bivalents in the germinal vesicle (Shimada et al., 2001a). It has been known that meiotic progression beyond the GVII stage is correlated with the reduction of cAMP level in oocytes (Funahashi et al., 1997; Shimada et al., 2003a). Funahashi et al. (1997) reported that when porcine COC were cultured with dbcAMP, meiosis was arrested at GVII. Additionally, we also observed an increased rate of oocyte arrest at the GVII stage when porcine COC were cultured with FSH and IBMX (Shimada et al., 2003a). It has already been mentioned in this chapter that the cAMP level decreases at 12 hr in oocytes of porcine COC cultured with FSH (Fig. 3.1). The gap junctional communication between outer layers of cumulus cells is closed around the same period (Isobe *et al.*, 1998, Isobe & Terada, 2001). Additionally, the sensitivity of Cx-43 to protease is increased by phosphorylation. This information suggests that the phosphorylation of Cx-43 in cumulus cells results in reducing Cx-43 protein and then closure of gap junctional communication before oocytes exhibit meiotic resumption.

In conclusion, transfer from cumulus cells to oocyte is shut down after ovulation stimuli via ERK1/2-induced phosphorylation of Cx-43 because gap junctional communication is closed by its phosphorylation. The reduction of cGMP activates PDE3 to degenerate cAMP produced in a GPR3-dependent manner and/or transferred from cumulus cells, and these pathways are essential for meiotic resumption (Fig. 3.6c).

3.5 How to Activate the ERK1/2 Pathway in Cumulus Cells of COC

In cumulus cells of porcine COC, ERK1/2 were activated when COC were cultured with FSH, forskolin, PGE2 or EGF, respectively (Shimada et al., 2001a; Yamashita et al., 2007, 2009, 2010). Induction was suppressed by a broad tyrosine kinase inhibitor, or specific tyrosine kinase inhibitor of the EGF receptor (EGFR) (Yamashita et al., 2007). Moreover, protein kinase A (PKA) or protein kinase C (PKC) inhibitor also decreases the phosphorylation level of ERK1/2 in cumulus cells of porcine COC (Yamashita et al., 2009), suggesting that the ERK1/2 upstream pathway in cumulus cells is a complex process.

In mice, the EGFR-RAS pathway is a key signaling pathway to induce the phosphorylation of ERK1/2 in cumulus cells (Fan et al., 2008; Fan & Richards, 2010). The EGF receptor (EGFR, ERBB1) is one member of the EGF receptor super family that is expressed in cumulus cells but not in the oocyte, and based on specific receptor tyrosine kinase inhibitor, is known to impact oocyte maturation in LH-stimulated preovulatory follicle cultures (Park et al., 2004; Hsieh et al., 2007). Specifically, the EGF-like factors Amphiregulin (*Areg*), Betacellulin (*Btc*), and Epiregulin (*Ereg*) are transiently expressed after LH stimulation in granulosa cells and act by binding to ERBB1 expressed on cumulus cells (Espey & Richards, 2002; Park et al., 2004; Shimada et al., 2006). Additionally, mutant mice null for *Areg* and homozygous for *Egfrwa2* (*Areg-/- Egfrwa2/wa2*) exhibit significantly reduced phosphorylation of ERBB1 in cumulus cells, impaired COC expansion, and oocyte meiotic arrest at the germinal vesicle (GV) stage (Hsieh et al., 2007). We also showed the expression of EGF-like factors in cumulus cells during the ovulation process. Because cumulus cells have low or undetectable levels of LH receptor (Peng et al., 1991), other stimulatory factors are required to induce expression of EGF-like factors in cumulus cells. In the mouse *Areg* gene promoter region, a putative cAMP responsible element (CRE) site is observed (Shao et al., 2004; Fan et al., 2010). The mutation of this region decreases promoter activity in the luciferase promoter assay using primary cultures of mouse granulosa cells, and CRE sequence binds to phosphoryltaed CREB (CRE binding protein) at 2 hr after LH stimulation (Fan et al., 2010). The expression of *Areg* is directly regulated by the cAMP-PKA-CREB cascade in granulosa cells and cumulus cells during the ovulation process (Fig. 3.7). It has been well-known that cumulus cells express G protein–coupled receptor subtypes EP2 (PTGER2) or EP4 (PTGER4), which, when activated, stimulate adenylate cyclase to produce cAMP (Fujino et al., 2005). EP2 and EP4 are receptors for prostaglandin E2 (PGE2) that is converted from arachidonic acid by the rate-limiting enzyme Prostaglandin Synthase 2 (PTGS2) (Sirois & Richards, 1992; Sirois et al., 1992, 1993). Using an *in vivo* approach we documented that *Areg* and *Ereg* expression levels were markedly reduced in COC and granulosa cells of *Ptgs2* null mice (Shimada et al., 2006). Thus, the initial induction of EGF-like factor expression is directly induced by LH via the cAMP-PKA-CREB pathway, and the expression is maintained in a PGE2 production-dependent manner (Fig. 3.7).

In cultured porcine COC, EGF-like factor expression was observed when COC were cultured with FSH (Yamashita et al., 2007; Kawashima et al., 2008). Induction was suppressed by a PKA inhibitor but not by a PKC inhibitor; however the activation of EGFR-ERK1/2 pathway was suppressed by both drugs (Yamashita et al., 2009). To understand why the PKC inhibitor suppressed the EGFR pathway in cumulus cells of porcine COC, we focused on the modification of EGF-like

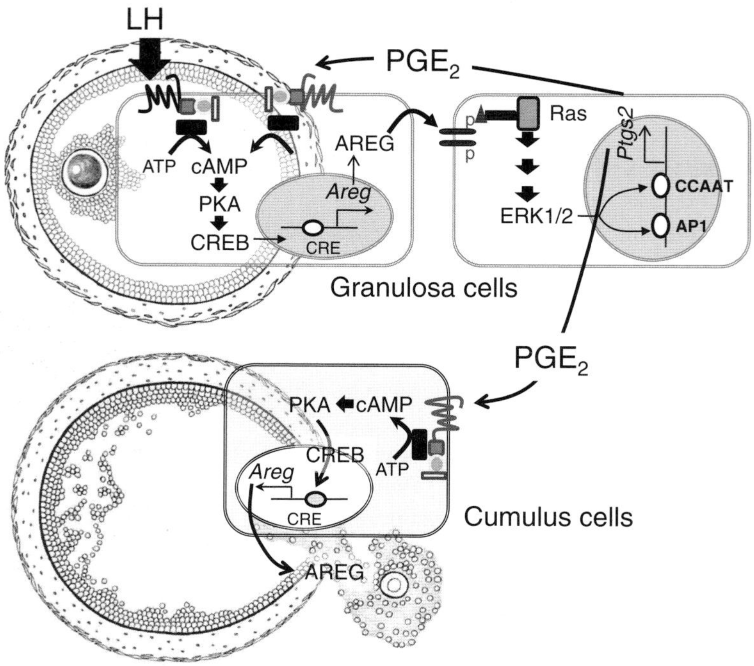

Figure 3.7 Schematic showing potential paracrine and autocrine pathways by which expression of amphiregulin (*Areg*) and *Ptgs2* is regulated in granulosa cells and cumulus cells of ovulating follicles. The proposed sequence is as follows: 1) LH binds to its cognate receptor localized to granulosa cells stimulating cAMP production to activate PKA-CREB pathway. Phosphorylated CREB binds to the cAMP responsible element (CRE) in the promoter region of the *Areg* gene. 2) As a consequence, *Areg* mRNAs are induced rapidly providing ligand that 3) bind EGF-receptors on granulosa cells (autocrine) leading to activation of ERK1/2 and induced expression of *Ptgs2*. In the promoter region of the *Ptgs2* gene, AP1 site and CCAAT site are essential transcription factor binding sites to upregulate gene expression. 4) The resulting increased production of prostaglandins (PGs; PGE) then provides ligands that bind EP2 on granulosa cells and cumulus cells (paracrine and autocrine) that (like the FSH and LH receptors) activate the cAMP-PKA-CREB pathway to increase the expression level of *Areg* mRNA. Thus, by the PGE/EP2 pathway, *Areg* mRNA can be induced in both cell types. (For color detail, see color plate section.)

factors. EGF-like factors are produced as inactive precursors spanning the cell membrane that consist of a signal sequence, transmembrane domain, and EGF domain (Peschon et al., 1998; Lee et al., 2003). The activation of these factors requires proteolytic shedding to release the mature form containing the EGF domain that activates the EGF receptor on target cells (Fig. 3.8a, Dong et al., 1999; Sahin et al., 2004). For AREG or EREG, disinteglin metaloprotease 17 (ADAM17) is the cleavage enzyme, and its enzyme activity is increased when the serine and tyrosine residues are phosphorylated by PKC and Src tyrosine kinase (Jackson et al., 2003; Sahin et al., 2004). Strikingly, levels of ADAM17 mRNA and protein are induced rapidly (within 5 hr) in cumulus cells of porcine COC cultured with FSH and LH (Fig. 3.8b). Importantly, ADAM17 enzyme activity was increased by FSH and LH in a similar temporal pattern and maintained during the 40-hr culture (Fig. 3.8c). When endogenous protease activity was reduced by the treatment with ADAM17 inhibitor TAPI-2, multiple downstream targets of EGFR activation were suppressed; including phosphorylation of ERK1/2, *Ptgs2, Has2*, and *Tnfaip6* mRNAs levels in cumulus cells, and meiotic maturation of oocytes (Yamashita et al., 2007). Moreover, the negative effects of TAPI-2 on these downstream targets of EGFR activation were each reversed at selected time intervals by the

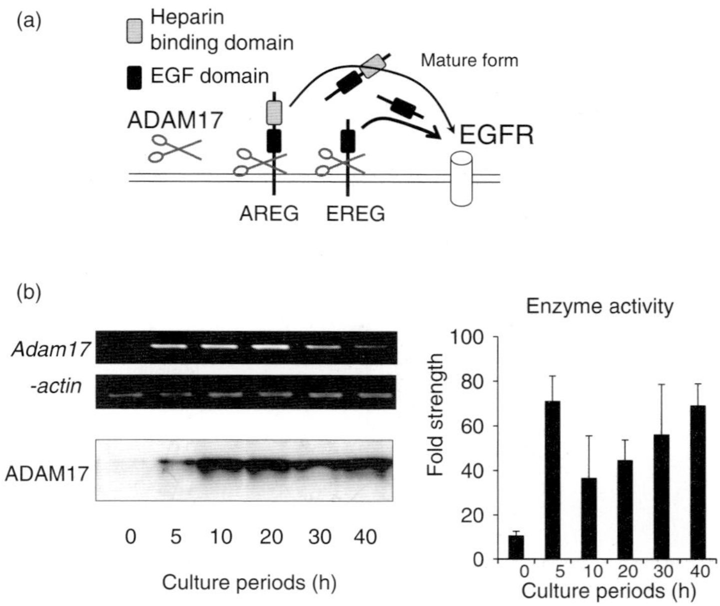

Figure 3.8 The expression and activation of ADAM17 in cumulus cells of porcine COC during in vitro maturation. (a) A schematic view of the mechanism by which EGF-like factors are matured by ADAM17 to activate EGF receptor (EGFR). (b) The expression of ADAM17 mRNA and protein levels in cumulus cells of porcine COC cultured with FSH and LH. (c) The time-dependent changes of ADAM17 enzyme activity in cumulus cells of porcine COC cultured with FSH and LH.

simultaneous addition of EGF. Thus, in porcine COC, ADAM17 is potentially activated by PKC and mediates EGF-like factor activation, with the released EGF domain working coordinately to achieve cumulus cell differentiation via an ERK1/2-dependent manner, which induces meiotic resumption of oocytes.

3.6 ERK1/2 in Cumulus Cells Is Required for Meiotic Resumption

When COC or intact follicles are cultured *in vitro*, the MEK inhibitor significantly suppresses meiotic resumption of oocytes, presumably by blocking the activity of ERK1/2 in cumulus and granulosa cells (Shimada et al., 2001b; Su et al., 2002). It has been known that the ERK1/2 pathway in oocytes is not essential for the resumption of meiosis (Su et al., 2002), suggesting that the ERK1/2 pathway in cumulus cells is required for meiotic resumption of oocytes. Moreover, the EGF-like factors can induce oocytes to resume meiosis and reach metaphase II, indicating that the activated EGF receptor pathway in cumulus cells is required for the resumption of meiosis (Park et al., 2004; Ashkenazi et al., 2005). Collectively these results indicate that the EGF-like factor/EGFR/ERK1/2 pathway in cumulus cells is critical for oocyte maturation (Fig. 3.9). This hypothesis has been supported recently by the total block of oocyte maturation in mutant mice lacking ERK1/2 in somatic cumulus/granulosa cells but not in the oocyte (Fan et al., 2009). On the other hand, the PI3-kinase-PKB/AKT pathway is also activated in cumulus cells of COC during the oocyte maturation process, and this pathway may mediate the delay in the resumption of meiosis in porcine and mouse oocytes (Shimada et al., 2003b; Noma et al., 2010). In our microarray analysis of rat COC, we found

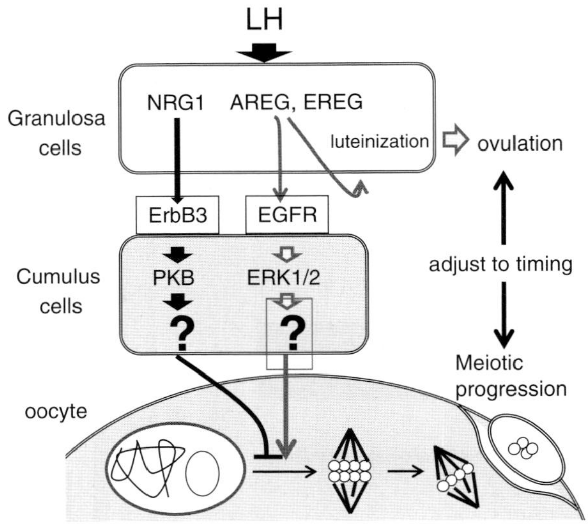

Figure 3.9 Schematic showing the role of AREG that acts on EGFR, and NRG1 that binds to ErbB3, to increase ERK1/2 or PKB pathways, respectively, in the regulation of meiotic progression in ovulating follicles of mice. Activated ERK1/2 or PKB pathways in cumulus cells indirectly regulate meiotic resumption of oocytes. However, the targets of both pathways remain unclear.

the expression of neuregulin (NRG1), another EGF-like factor that acted on ErbB2/3 hetero-dimmer expressed in cumulus cells. The activation of the hetero-dimmer was also observed in cumulus cells and granulosa cells during oocyte maturation (ovulation). When NRG1 was added to the maturation medium, the PI 3-kinase-PKB pathway was activated in a dose-dependent manner. Additionally, spontaneous meiotic resumption of oocytes was delayed by exogenous NRG1. The induction of meiosis initiated by AREG was also delayed by NRG1 when COCs were cultured in the presence of 4 mM hypoxanthine to suppress spontaneous meiotic resumption. Mature oocytes cultured with both NRG1 and AREG had significantly higher developmental competence as compared with that of oocytes cultured with AREG alone. The mechanisms by which the addition of NRG1 improves developmental competence remain unclear; however, one possibility is that the delay of meiotic resumption by NRG1 reduces the risk of oocyte-aging before fertilization.

The murine *Nrg1* gene has three promoter and transcriptional start sites (Meyer et al., 1997), and type I and type III are expressed in granulosa cells during ovulation (Noma et al., 2010). Expression of type III *Nrg1* was dramatically increased in granulosa cells by hCG whereas that of type I was not changed. The type III *Nrg1* promoter region has three putative C/EBP binding sites and a CRE site. Based on the transfection of specific promoter-reporter constructs in granulosa cells, the most distal C/EBP binding site appears to play a critical role in the increase of promoter activity after stimulation. Because we have shown recently that the ERK1/2-C/EBPb pathway is essential for inducing cell fate decisions in granulosa cells and cumulus cells of preovulatory follicles (Fan et al., 2009), it is clear that *Nrg1* expression is also dependent on this pathway, suggesting that the AREG-EREG-ERK1/2 pathway upregulates the negative regulator of meiotic resumption. Thus, the AREG-EGFR-ERK1/2 pathway accelerates meiotic resumption, whereas the activation of the NRG1-ErbB2/3-PKB pathway works as a brake to control the timing of meiotic resumption. Both pathways are required to regulate the timing of meiotic progression, which impacts oocyte developmental competence (Fig. 3.9).

3.7 Dynamic Changes of Kinase Activities within Oocytes

Oocyte maturation depends on the activation of maturation-promoting factor (MPF), which is composed of p34^{cdc2} kinase (CDK1) and Cyclin B (Masui & Markert, 1971; Lohka et al., 1988; Dunphy et al., 1988). Activated MPF induces meiotic resumption in amphibian oocytes (Swenson et al., 1986; Gautier et al., 1990), mouse oocytes (Choi et al., 1991), porcine oocytes (Naito & Toyoda, 1991 Naito et al., 1995), and bovine oocytes (Tatemoto & Horiuchi, 1995; Tatemoto & Terada, 1996). The association with Cyclin B is required for the activation of CDK1 as well as the dephosphorylation of its Thr 14 and Tyr 15 residues (De-Bondt et al., 1993; Kumagai & Dunphy, 1992; Nebreda et al., 1995). The phosphorylation status of these residues is regulated by the activity of two key enzymes, Wee1 kinase family and Cdc25 phosphatase (Kumagai & Dunphy, 1992; Morgan, 1995; Okamoto et al., 2002; Leise & Mueller, 2002, Fig. 3.10a).

The Wee1 kinase family consists of three members: Wee1A, Wee1B, and Myt1; Wee1B is selectively expressed in oocytes (Han et al., 2005). PKA directly phosphorylates Wee1B at least *in vitro* at Serine 15 in the N terminus, and the phosphorylation enhances the kinase enzyme activity of Wee1B that phosphorylates CDK1 in oocytes (Han et al., 2005). Thus, cAMP-PKA pathway phosphorylates Wee1B to suppress CDK1 activity, which results in meiotic arrest at the GV stage in mouse oocytes.

Three Cdc25 families (Cdc25A, Cdc25B, and Cdc25C) are expressed in mouse oocytes, whereas only Cdc25B is essential for the induction of meiosis in mice (Wu & Wolgemuth, 1995). Interestingly, Cdc25B is also phosphorylated by PKA, and the phosphorylated form is inactive Cdc25B, suggesting that the elevation of cAMP activates Wee1B, but decreases Cdc25B activity to markedly increase the phosphrylated inactive form of CDK1 in oocytes arrested at the GV stage (Han & Conti, 2006, Fig. 3.10b).

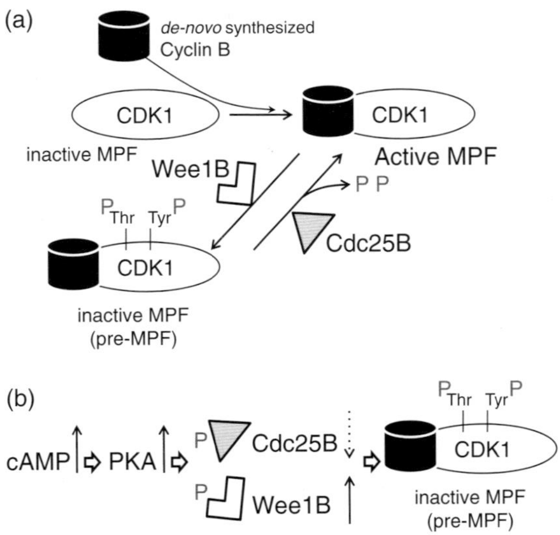

Figure 3.10 The mechanisms of maturation promoting factor (MPF) activation in Xenopus and mouse oocytes. (a) The association with Cyclin B is required for the activation of CDK1 as well as the dephosphorylation of its Thr 14 and Tyr 15 residues by Cdc25 phosphatase. Wee1 kinase family phosphorylates CDK1 to decrease CDK1 activity. (b) cAMP-PKA pathway phosphorylates both Wee1 kinase family to increase its enzyme activity and Cdc25 phosphatase to down-regulate the activity, which results in decreasing CDK1 activity.

The phosphorylated form of Wee1B is localized in the germinal vesicle, but the phosphorylated Cdc25B in the cytoplasm. The decrease of cAMP is a turning point to change the localization of these two key enzymes: dephospohrylated Wee1B moves to the cytoplasm, and dephosphorylated Cdc25B enters the nucleus from the cytoplasm. The nuclear localization of Wee1B is essential for suppression of meiotic progression, because when the nuclear localization signal was removed from the Wee1B sequence the injection of mutant Wee1B did not arrest meiosis at the GV stage (Oh et al., 2010). Similarly, the translocation of dephosphorylated Cdc25B to the nucleus is also an important event for activation of CDK1 and initiation of GVBD in mouse oocytes.

In most mammalian species except for rodents, protein synthesis is required for the induction of meiotic resumption (Fulka et al., 1986; Hunter & Moor, 1987; Tatemoto & Terada, 1995). In mouse oocytes, the concentration of Cyclin B accumulated within oocytes is about sevenfold higher than CDK1 (Han & Conti, 2006). However, this ratio is opposite in other species including amphibians; the expression level of Cyclin B in Xenopus oocytes is less than 5% compared with CDK1 (Han & Conti, 2006), suggesting that *de novo* synthesis of Cyclin B is an initial step of meiotic resumption of oocytes. In Xenopus oocytes, PKA suppressed the translation of Cyclin B before GVBD (Matten et al., 1994; Frank-Vaillant et al., 1999). Because CDK1 activity is dependent on the binding of Cyclin B1 and dephosphorylation at 14 Thr and 15 Tyr (Morgan, 1995), the cAMP-PKA pathway can strongly suppress the activity of CDK1 due to both phosphorylation of these sites and suppression of Cyclin B1 synthesis (Fig. 3.10b).

In pig oocytes, the activation of CDK1 is induced concomitantly with the reduction of cAMP level (Fig. 3.11). To examine the role of decreasing cAMP in oocytes before GVBD, the cAMP level and CDK1 activity were analyzed in the oocyte after a 28-hr culture in the presence of various concentrations of phosphodiesterase inhibitor, IBMX. The addition of IBMX to the maturation medium produced a significant increase in the cAMP content of oocytes and a significantly decreased proportion of oocytes exhibiting GVBD and CDK1 activity; both results were observed in a dose-dependent manner. Thus, a fall in porcine oocyte cAMP level is required for activation of CDK1, which then induces meiotic resumption similar to mouse oocytes. The protein level of Cyclin B is very low in oocytes arrested at the GV stage, whereas the level was increased after the onset of culture (Shimaoka et al., 2009). However, the injection of dbcAMP into oocytes dramatically suppressed the

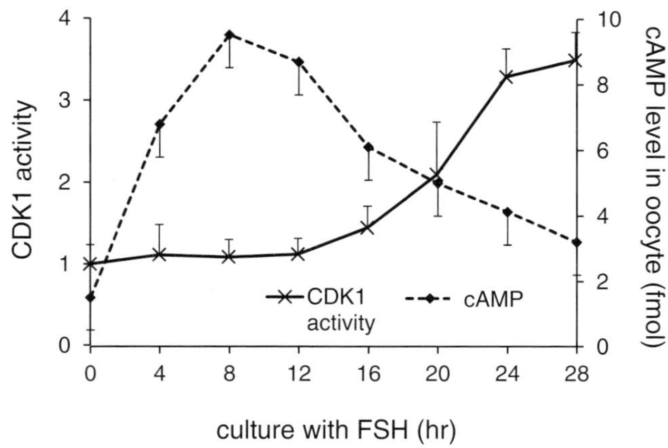

Figure 3.11 The relationship between the level of cAMP and CDK1 activity in porcine oocytes from COC cultured with FSH and LH. Porcine COCs were cultured with FSH and LH for up to 28 hr. Oocytes collected every 4 hr of culture were used for analysis of cAMP level in the oocyte, and CDK1 enzyme activity.

synthesis of Cyclin B, suggesting that the cAMP-PKA pathway prevents the translation of Cyclin B in porcine oocytes as well. Additionally, Wee1B and Cdc25C are also expressed in porcine oocytes, and play important roles in the regulation of meiotic resumption (Shimaoka et al., 2009). When cumulus cells were immediately removed from COC and then the denuded oocytes were cultured, activation of CDK1 was detected at a much earlier time point as compared with that in oocytes of COC (Shimada et al., 2001b). Therefore, cumulus cells regulate oocyte meiotic resumption via a cAMP-dependent mechanism, and CDK1 is maintained in an inactive form by three-way PKA dependent activation of Wee1B, inactivation of Cdc25C, and suppression of cyclin B synthesis.

3.8 Conclusion

In preovulatory follicles, oocytes are surrounded by numerous layers of cumulus cells as the cumulus cell-oocyte complex. During the follicular development stage, granulosa cells secrete NPPC by FSH stimulation. The NPPC acts on its receptor, NPR2, expressed on cumulus cells to produce cGMP, which is transferred to the oocyte via gap junctional communication. cGMP in the oocyte suppresses PDE3 activity to prevent meiotic resumption. After stimulation of ovulation by the LH surge, the oocyte resumes meiosis and progresses to metaphase II. However, the expression of LH receptors is not detected in the oocyte and is minimal in cumulus cells compared with granulosa cells. Cumulus cells express members of the EGF receptor family (ErbB family) that respond to specific ligands (EGF-like factor family members) that are secreted by granulosa cells during the ovulation process. By these intermediary steps, the cumulus cells mediate LH signaling from granulosa cells to induce oocyte meiotic resumption. One of the key signaling pathways in cumulus cells is the EGFR-ERK1/2 pathway that regulates gap-junctional communication among cumulus cells and between the cumulus cell and oocyte. Closure of gap-junctional communication by ERK1/2 decreases the level of cGMP and/or cAMP transferred into the oocyte, which activates phosphodiesterase type III (PDE3) in the oocyte. PDE3 breaks down cAMP to decrease PKA activity in the oocyte. Because the PKA pathway activates Wee1B, and down-regulates both Cdc25C activity and Cyclin B synthesis to decrease CDK1 activity, the decrease of PKA activity by PDE3 induces the activation of CDK1 to activate MPF and resume meiosis from the GV stage. Thus, the control of cAMP levels in oocytes by cumulus cells is essential for oocyte meiotic resumption.

References

Ashkenazi, Cao, Motola, Popliker, Conti, & Tsafriri. (2005). Epidermal growth factor family members: Endogenous mediators of the ovulatory response. *Endocrinology, 146*(1), 77–84.

Choi, Aoki, Mori, Yamashita, Nagahama, & Kohmoto. (1991). Activation of p34cdc2 protein kinase activity in meiotic and mitotic cell cycles in mouse oocytes and embryos. *Development, 113*(3), 789–795.

Chuang, Myers Jr., Backer, Shoelson, White, Birnbaum, & Kahn. (1993). Insulin-stimulated oocyte maturation requires insulin receptor substrate 1 and interaction with the SH2 domains of phosphatidylinositol 3-kinase. *Molecular and Cellular Biology, 13*(11), 6653–6660

Conti, Andersen, Richard, Mehats, Chun, Horner, Jin, & Tsafriri. (2002). Role of cyclic nucleotide signaling in oocyte maturation. *Molecular and Cellular Endocrinology, 187*(1–2), 153–159.

Cross & Brinster. (1970). In vitro development of mouse oocytes. *Biology of Reproduction, 3*(3), 298–307.

De Bondt, Rosenblatt, Jancarik, Jones, Morgan, & Kim. (1993). Crystal structure of cyclin-dependent kinase 2. *Nature, 363*(6430), 595–602.

Dong, Opresko, Dempsey, Lauffenburger, Coffey, & Wiley. (1999). Metalloprotease-mediated ligand release regulates autocrine signaling through the epidermal growth factor receptor. *Proceeding of the National Academy of Science USA, 96*(11), 6235–6240.

Downs, Coleman, & Eppig. (1986). Maintenance of murine oocyte meiotic arrest: uptake and metabolism of hypoxanthine and adenosine by cumulus cell-enclosed and denuded oocytes. *Developmental Biology, 117*(1), 174–183.

Downs & Eppig. (1984). Cyclic adenosine monophosphate and ovarian follicular fluid act synergistically to inhibit mouse oocyte maturation. *Endocrinology, 114*(2), 418–427.

Dunphy, Brizuela, Beach, & Newport. (1988). The Xenopus *cdc2* protein is a component of MPF, a cytoplasmic regulator of mitosis. *Cell, 54*(3), 423–431.

Espey & Richards. (2002). Temporal and spatial patterns of ovarian gene transcription following an ovulatory dose of gonadotropin in the rat. *Biology of Reproduction, 67*(16), 1662–1670.

Fan, Liu, Shimada, Sterneck, Johnson, Hedrick, & Richards. (2009). MAPK3/1 (ERK1/2) in ovarian granulosa cells are essential for female fertility. *Science, 324*(5929), 938–941.

Fan, O'Connor, Shitanaka, Shimada, Liu, & Richards. (2010). Beta-catenin (CTNNB1) promotes preovulatory follicular development but represses LH-mediated ovulation and luteinization. *Molecular Endocrinology, 24*(8), 1529–1542.

Fan & Richards. (2010). Minirevew: Physiological and pathological actions of RAS in the ovary. *Molecular Endocrinology, 24*(2), 286–298.

Fan, Shimada, Liu, Cahill, Noma, Wu, Gossen, & Richards. (2008). Selective expression of KrasG12D in granulosa cells of the mouse ovary causes defects in follicle development and ovulation. *Development, 135*(12), 2127–2137.

Frank-Vaillant, Jessus, Ozon, Maller, & Haccard. (1999). Two distinct mechanisms control the accumulation of cyclin B1 and Mos in Xenopus oocytes in response to progesterone. *Molecular Biology of the Cell, 10*(10), 3279–3288.

Fujino, Salvi, & Regan. (2005). Differential regulation of phosphorylation of the cAMP response element-binding protein after activation of EP2 and EP4 prostanoid receptors by prostaglandin E2. *Molecular Pharmacology, 68*(1), 251–259.

Fulka Jr, Motlík, Fulka, & Jílek. (1986). Effect of cycloheximide on nuclear maturation of pig and mouse oocytes. *Journal of Reproduction and Fertility, 77*(1), 281–285.

Funahashi, Cantley, & Day. (1997). Synchronization of meiosis in porcine oocytes by exposure to dibutyryl cyclic adenosine monophosphate improves developmental competence following in vitro fertilization. *Biology of Reproduction, 57*(1), 49–53.

Gautier, Minshull, Lohka, Grotzer, Hunt, & Maller. (1990). Cyclin is a component of maturation-promoting factor from Xenopus. *Cell, 60*(3), 487–494.

Granot & Dekel. (1994). Phosphorylation and expression of connexin-43 ovarian gap junction protein are regulated by luteinizing hormone. *Journal of Biological Chemistry, 269*(48), 30502–30509.

Grazul-Bilska, Reynolds, & Redmer. (1997). Gap junctions in the ovaries. *Biology of Reproduction, 57*(5), 947–957.

Griswold, Heckert, & Linder. (1995). The molecular biology of the FSH receptor. *Journal of Steroid Biochemistry and Molecular Biology, 53*(1–6), 215–218.

Han, Chen, Paronetto, & Conti. (2005) Wee1B is an oocyte-specific kinase involved in the control of meiotic arrest in the mouse. *Current Biology, 15*(18), 1670–1676.

Han & Conti. (2006) New pathways from PKA to the Cdc2/cyclin B complex in oocytes: Wee1B as a potential PKA substrate. *Cell Cycle, 5*(3), 227–231.

Han, Vaccari, Nedachi, Andersen, Kovacina, Roth, & Conti. (2006). Protein kinase B/Akt phosphorylation of PDE3A and its role in mammalian oocyte maturation. *EMBO Journal, 25*(24), 5716–5725

Hill, Oh, Schmidt, Clark, & Murray. (1994). Lysophosphatidic acid inhibits gap-junctional communication and stimulates phosphorylation of connexin-43 in WB cells: Possible involvement of the mitogen-activated protein kinase cascade. *Biochemical Journal, 303*(2), 475–479.

Hinckley, Vaccari, Horner, Chen, & Conti. (2005). The G-protein-coupled receptors GPR3 and GPR12 are involved in cAMP signaling and maintenance of meiotic arrest in rodent oocytes. *Developmental Biology, 287*(2), 249–261.

Hoshino, Yokoo, Yoshida, Sasada, Matsumoto, & Sato. (2004). Phosphatidylinositol 3-kinase and Akt participate in the FSH-induced meiotic maturation of mouse oocytes. *Molecular Reproduction and Development, 69*(1), 77–86.

Hsieh, Lee, Panigone, Horne, Chen et al. (2007). Luteinizing hormone-dependent activation of the epidermal growth factor network is essential for ovulation. *Molecular and Cellular Biology, 27*(5), 1914–1924.

Hunter & Moor. (1987) Stage-dependent effects of inhibiting ribonucleic acids and protein synthesis on meiotic maturation of bovine oocytes in vitro. *Journal of Dairy Science, 70*(8), 1646–1651.

Isobe, Maeda, & Terada. (1998). Involvement of meiotic resumption in the disruption of gap junctions between cumulus cells attached to pig oocytes. *Journal of Reproduction and Fertility, 113*(2), 167–172.

Isobe & Terada. (2001). Effect of the factor inhibiting germinal vesicle breakdown on the disruption of gap junctions and cumulus expansion of pig cumulus-oocyte complexes cultured in vitro. *Reproduction, 121*(2), 249–257.

Itahana, Morikazu, & Takeya. (1996) Differential expression of four connexin genes, Cx-26, Cx-30.3, Cx-32, and Cx-43, in the porcine ovarian follicle. *Endocrinology, 137*(11), 5036–5044.

Itahana, Tanaka, Morikazu, Komatu, Ishida, & Takeya. (1998). Isolation and characterization of a novel connexin gene, Cx-60, in porcine ovarian follicles. *Endocrinology, 139*(1), 320–329.

Jackson, Qiu, Sunnarborg, Chang, Zhang, Patterson, & Lee. (2003). Defective valvulogenesis in HB-EGF and TACE-null mice is associated with aberrant BMP signaling. *EMBO Journal, 22*(11), 2704–2716

Kawashima, Okazaki, Noma, Nishibori, Yamashita, & Shimada. (2008). Sequential exposure of porcine cumulus cells to FSH and/or LH is critical for appropriate expression of steroidogenic and ovulation-related genes that impact oocyte maturation in vivo and in vitro. *Reproduction, 136*(1), 9–21.

Kumagai & Dunphy. (1992). Regulation of the cdc25 protein during the cell cycle in Xenopus extracts. *Cell, 70*(1), 139–151.

Larsen, Wert, & Brunner. (1986). A dramatic loss of cumulus cell gap junctions is correlated with germinal vesicle breakdown in rat oocytes. *Developmental Biology, 113*(2), 517–521.

Larsen, Wert, & Brunner. (1987). Differential modulation of rat follicle cell gap junction populations at ovulation. *Developmental Biology, 122*(1), 61–71.

Lau, Kurata, Kanemitsu, Loo, Warn-Cramer, Eckhart, & Lampe. (1996). Regulation of connexin 43 function by activated tyrosine protein kinases. *Journal of Bioenergetics and Biomembranes, 28*(4), 359–367.

Lee, Sunnarborg, Hinkle, Myers, Stevenson et al. (2003). TACE/ADAM17 processing of EGFR ligands indicates a role as a physiological convertase. *Annals of New York Academy of Sciences, 995*, 22–38.

Leise III & Mueller. (2002). Multiple Cdk1 inhibitory kinases regulate the cell cycle during development. *Developmental Biology, 249*(1), 156–173.

Lohka, Hayes, & Maller. (1988). Purification of maturation-promoting factor, an intercellular regulator of early events. *Proceeding of the National Academy of Science USA, 85*(9), 3009–3013.

Luciano, Pocar, Milanesi, Modina, Rieger, Lauria, & Gandolfi. (1999). Effect of different levels of intracellular cAMP on the in vitro maturation of cattle oocytes and their subsequent development following in vitro fertilization. *Molecular Reproduction and Development, 54*(1), 86–91.

Masui & Markert. (1971). Cytoplasmic control of nuclear behavior during meiotic maturation of frog oocytes. *Journal of Experimental Zoology, 177*(2), 129–146.

Matten, Daar, & Vande Woude. (1994). Protein kinase A acts at multiple points to inhibit Xenopus oocyte maturation. *Molecular and Cellular Biology, 14*(7), 4419–4426.

Mattioli, Galeati, Barboni, & Seren. (1994). Concentration of cyclic AMP during the maturation of pig oocytes in vivo and in vitro. *Journal of Reproduction and Fertility, 100*(2), 403–409.

Mehlmann, Saeki, Tanaka, Brennan, Evsikov et al. (2004). The Gs-linked receptor GPR3 maintains meiotic arrest in mammalian oocytes. *Science, 306*(5703), 1947–1950.

Meyer, Yamaai, Garratt, Riethmacher-Sonnenberg, Kane, Theill, & Birchmeier. (1997). Isoform-specific expression and function of neuregulin. *Development, 124*(18), 3575–3586.

Morgan. (1995). Principles of CDK regulation. *Nature, 374*(6518), 131–134.

Musil, Cunningham, Edelman, & Goodenough. (1990). Differential phosphorylation of the gap junction protein connexin 43 in junctional communication-competent and -deficient cell lines. *Journal of Cell Biology, 111*(5), 2077–2088.

Muslin, Klippel, & Williams. (1993). Phosphatidylinositol 3-kinase activity is important for progesterone-induced Xenopus oocyte maturation. *Molecular and Cellular Biology, 13*(11), 6661–6666.

Naito & Toyoda. (1991). Fluctuation of histone H1 kinase activity during meiotic maturation in porcine oocytes. *Journal of Reproduction and Fertility, 93*(2), 467–473.

Naito, Hawkins, Yamashita, Nagahama, Aoki, Kohmoto, Toyoda, & Moor. (1995). Association of p34cdc2 and cyclin B1 during meiotic maturation in porcine oocytes. *Developmental Biology, 168*(2), 627–634.

Nebreda, Gannon, & Hunt. (1995). Newly synthesized protein(s) must associate with p34cdc2 to activate MAP kinase and MPF during progesterone-induced maturation of Xenopus oocytes. *EMBO Journal, 14*(22), 5597–5607.

Noma, Kawashima, Fan, Fujita, Kawai et al. (2010). LH-induced Neuregulin 1 (NRG1) type III transcripts control granulosa cell differentiation and oocyte maturation. *Molecular Endocrinology* [Epub].

Nuttinck, Peynot, Humblot, Massip, Dessy, & Fléchon. (2000). Comparative immunohistochemical distribution of connexin 37 and connexin 43 throughout folliculogenesis in the bovine ovary. *Molecular Reproduction and Development, 57*(1), 60–66.

Oh, Han, & Conti. (2010). Wee1B, Myt1, and Cdc25 function in distinct compartments of the mouse oocyte to control meiotic resumption. *Journal of Cell Biology, 188*(2), 199–207.

Okamoto, Nakajo, & Sagata. (2002). The existence of two distinct Wee1 isoforms in Xenopus: Implications for the developmental regulation of the cell cycle. *EMBO Journal, 21*(10), 2472–2484.

Park, Su, Ariga, Law, Jin, & Conti. (2004). EGF-like growth factors as mediators of LH action in the ovulatory follicle. *Science, 303*(5658), 682–684.

Peng, Hsueh, LaPolt, Bjersing, & Ny. (1991). Localization of luteinizing hormone receptor messenger ribonucleic acid expression in ovarian cell types during follicle development and ovulation. *Endocrinology, 129*(6), 3200–3207.

Peschon, Slack, Reddy, Stocking, Sunnarborg et al. (1998) An essential role for ectodomain shedding in mammalian development. *Science, 282*(5392), 1281–1284.

Racowsky. (1985a). Antagonistic actions of estradiol and tamoxifen upon forskolin-dependent meiotic arrest, intercellular coupling, and the cyclic AMP content of hamster oocyte-cumulus complexes. *Journal of Experimental Zoology, 234*(2), 251–260.

Racowsky. (1985b). Effect of forskolin on maintenance of meiotic arrest and stimulation of cumulus expansion, progesterone and cyclic AMP production by pig oocyte-cumulus complexes. *Journal of Reproduction and Fertility, 74*(1), 9–21.

Richard, Tsafriri, & Conti. (2001). Role of phosphodiesterase type 3A in rat oocyte maturation. *Biology of Reproduction, 65*(5), 1444–1451.

Richards. (1994). Hormonal control of gene expression in the ovary. *Endocrine Review, 15*(6), 725–751.

Rose-Hellekant & Bavister. (1996). Roles of protein kinase A and C in spontaneous maturation and in forskolin or 3-isobutyl-1-methylxanthine maintained meiotic arrest of bovine oocytes. *Molecular Reproduction and Development, 44*(2), 241–249.

Sahin, Weskamp, Kelly, Zhou, Higashiyama et al. (2004). Distinct roles for ADAM10 and ADAM17 in ectodomain shedding of six EGFR ligands. *Journal of Cell Biology, 164*(5), 769–779.

Sasseville, Côté, Vigneault, Guillemette, & Richard. (2007). 3′5′-cyclic adenosine monophosphate-dependent up-regulation of phosphodiesterase type 3A in porcine cumulus cells. *Endocrinology, 148*(4), 1858–1867.

Sasseville, Côté, Gagnon, & Richard. (2008). Up-regulation of 3′5′-cyclic guanosine monophosphate-specific phosphodiesterase in the porcine cumulus-oocyte complex affects steroidogenesis during in vitro maturation. *Endocrinology, 149*(11), 5568–5576.

Sasseville, Albuz, Côté, Guillemette, Gilchrist, & Richard. (2009a). Characterization of novel phosphodiesterases in the bovine ovarian follicle. *Biology of Reproduction, 81*(2), 415–425.

Sasseville, Gagnon, Guillemette, Sullivan, Gilchrist, & Richard. (2009b). Regulation of gap junctions in porcine cumulus-oocyte complexes: Contributions of granulosa cell contact, gonadotropins, and lipid rafts. *Molecular Endocrinology, 23*(5), 700–710.

Schultz, Montgomery, & Belanoff. (1983a). Regulation of mouse oocyte meiotic maturation: Implication of a decrease in oocyte cAMP and protein dephosphorylation in commitment to resume meiosis. *Developmental Biology, 97*(2), 264–273.

Schultz, Montgomery, Ward-Bailey, & Eppig. (1983b). Regulation of oocyte maturation in the mouse: Possible roles of intercellular communication, cAMP, and testosterone. *Developmental Biology, 95*(2), 294–304.

Sela-Abramovich, Chorev, Galiani, & Dekel. (2005). Mitogen-activated protein kinase mediates luteinizing hormone-induced breakdown of communication and oocyte maturation in rat ovarian follicles. *Endocrinology, 146*(3), 1236–1244.

Shao, Evers, & Sheng. (2004). Prostaglandin E2 synergistically enhances receptor tyrosine kinase-dependent signaling system in colon cancer cells. *Journal of Biological Chemistry, 279*(14), 14287–14293.

Shimada, Hernandez-Gonzalez, Gonzalez-Robayna, & Richards. (2006). Paracrine and autocrine regulation of epidermal growth factor-like factors in cumulus oocyte complexes and granulosa cells: key roles for prostaglandin synthase 2 and progesterone receptor. *Molecular Endocrinology, 20*(6), 1352–1365.

Shimada, Ito, Yamashita, Okazaki, & Isobe. (2003b). Phosphatidylinositol 3-kinase in cumulus cells is responsible for both suppression of spontaneous maturation and induction of gonadotropin-stimulated maturation of porcine oocytes. *Journal of Endocrinology, 179*(1), 25–34.

Shimada, Maeda, & Terada. (2001a). Dynamic changes of connexin-43, gap junctional protein, in outer layers of cumulus cells are regulated by PKC and PI 3-kinase during meiotic resumption in porcine oocytes. *Biology of Reproduction, 64*(4), 1255–1263.

Shimada, Nishibori, Isobe, Kawano, & Terada. (2003a). Luteinizing hormone receptor formation in cumulus cells surrounding porcine oocytes and its role during meiotic maturation of porcine oocytes. *Biology of Reproduction, 68*(4), 1142–1149.

Shimada, Nishibori, Yamashita, Ito, Mori, & Richards. (2004). Down-regulated expression of A disintegrin and metalloproteinase with thrombospondin-like repeats-1 by progesterone receptor antagonist is associated with impaired expansion of porcine cumulus-oocyte complexes. *Endocrinology, 145*(10), 4603–4614.

Shimada, Samizo, Yamashita, Matsuo, & Terada. (2002). Both Ca^{2+}-PKC pathway and cAMP-PKA pathway require for progesterone production in FSH- and LH-stimulated cumulus cells during in vitro maturation of porcine oocytes. *Journal of Mammalian Ova Research, 19*(3), 81–88.

Shimada & Terada. (2001). Phosphatidylinositol 3-kinase in cumulus cells and oocytes is responsible for activation of oocyte mitogen-activated protein kinase during meiotic progression beyond the meiosis I stage in pigs. *Biology of Reproduction, 64*(4), 1106–1114.

Shimada & Terada. (2002a). Roles of cAMP in regulation of both MAP kinase and p34(cdc2) kinase activity during meiotic progression, especially beyond the MI stage. *Molecular Reproduction and Development, 62*(1), 124–131.

Shimada & Terada. (2002b). FSH and LH induce progesterone production and progesterone receptor synthesis in cumulus cells: A requirement for meiotic resumption in porcine oocytes. *Molecular Human Reproduction, 8*(7), 612–618.

Shimada, Zeng, & Terada. (2001b). Inhibition of phosphatidylinositol 3-kinase or mitogen-activated protein kinase kinase leads to suppression of p34(cdc2) kinase activity and meiotic progression beyond the meiosis I stage in porcine oocytes surrounded with cumulus cells. *Biology of Reproduction, 65*(2), 442–448.

Shimaoka, Nishimura, Kano, & Naito. (2009). Critical effect of pigWee1B on the regulation of meiotic resumption in porcine immature oocytes. *Cell Cycle, 8*(15), 2375–2384.

Simon, Goodenough, Li, & Paul. (1997). Female infertility in mice lacking connexin 37. *Nature, 385*(6616), 525–529.

Sirois, Levy, Simmons, & Richard. (1993). Characterization and hormonal regulation of the promoter of the rat prostaglandin endoperoxide synthase 2 gene in granulosa cells. Identification of functional and protein-binding regions. *Journal of Biological Chemistry, 268*(16), 12199–12206.

Sirois & Richards. (1992). Purification and characterization of a novel, distinct isoform of prostaglandin endoperoxide synthase induced by human chorionic gonadotropin in granulosa cells of rat preovulatory follicles. *Journal of Biological Chemistry, 267*(9), 6382–6388.

Sirois, Simmons, & Richards. (1992). Hormonal regulation of messenger ribonucleic acid encoding a novel isoform of prostaglandin endoperoxide H synthase in rat preovulatory follicles. Induction in vivo and in vitro. *Journal of Biological Chemistry, 267*(16), 11586–11592.

Su, Wigglesworth, Pendola, O'Brien, & Eppig. (2002). Mitogen-activated protein kinase activity in cumulus cells is essential for gonadotropin-induced oocyte meiotic resumption and cumulus expansion in the mouse. *Endocrinology, 143*(6), 2221–2232.

Swenson, Farrell, & Ruderman. (1986). The clam embryo protein cyclin A induces entry into M phase and the resumption of meiosis in Xenopus oocytes. *Cell, 47*(6), 861–870.

Tatemoto & Horiuchi. (1995). Requirement for protein synthesis during the onset of meiosis in bovine oocytes and its involvement in maturation-promoting factor. *Molecular Reproduction and Development, 41*(1), 47–53.

Tatemoto & Terada. (1995). Time-dependent effects of cycloheximide and alpha-amanitin on meiotic resumption and progression in bovine follicular oocytes. *Theriogenology, 43*(6), 1107–1113.

Tatemoto & Terada. (1996). Involvement of cyclic-dependent protein kinase in chromatin condensation before germinal vesicle breakdown in bovine oocytes. *Animal Reproduction Science, 44*(2), 99–110.

Warn-Cramer, Lampe, Kurata, Kanemitsu, Loo et al. (1996). Characterization of the mitogen-activated protein kinase phosphorylation sites on the connexin-43 gap junction protein. *Journal of Biological Chemistry, 271*(7), 3779–3786.

Wu & Wolgemuth. (1995). The distinct and developmentally regulated patterns of expression of members of the mouse Cdc25 gene family suggest differential functions during gametogenesis. *Developmental Biology, 170*(1), 195–206.

Yamashita, Hishinuma, & Shimada. (2009). Activation of PKA, p38 MAPK and ERK1/2 by gonadotropins in cumulus cells is critical for induction of EGF-like factor and TACE/ADAM17 gene expression during in vitro maturation of porcine COCs. *Journal of Ovarian Research, 24*(2), 20.

Yamashita, Kawashima, Gunji, Hishinuma, & Shimada. (2010). Progesterone is essential for maintenance of Tace/Adam17 mRNA expression, but not EGF-like factor, in cumulus cells, which enhances the EGF receptor signaling pathway during in vitro maturation of porcine COCs. *Journal of Reproduction and Development, 56*(3), 315–323.

Yamashita, Kawashima, Yanai, Nishibori, Richards, & Shimada. (2007). Hormone-induced expression of tumor necrosis factor alpha-converting enzyme/A disintegrin and metalloprotease-17 impacts porcine cumulus cell oocyte complex expansion and meiotic maturation via ligand activation of the epidermal growth factor receptor. *Endocrinology, 148*(12), 6164–6175.

Yamashita, Nishibori, Terada, Isobe, & Shimada. (2005). Gonadotropin-induced delta14-reductase and delta7-reductase gene expression in cumulus cells during meiotic resumption of porcine oocytes. *Endocrinology, 146*(1), 186–194.

Yamashita, Shimada, Okazaki, Maeda, & Terada. (2003). Production of progesterone from de novo-synthesized cholesterol in cumulus cells and its physiological role during meiotic resumption of porcine oocytes. *Biology of Reproduction, 68*(4), 1193–1198.

Zhang, Su, Sugiura, Xia, & Eppig. (2010). Granulosa cell ligand NPPC and its receptor NPR2 maintain meiotic arrest in mouse oocytes. *Science, 330*(6002), 366–369.

4 Oocyte-Secreted Factors in Domestic Animals

Jeremy G. Thompson, David G. Mottershead, and
Robert B. Gilchrist

4.1 Introduction

The concept that the oocyte regulates follicular cell function and therefore plays a major role in regulating follicular development, ovulation, and fecundity is a relatively new concept that emerged during the 1990s. Studies of domestic animals, especially sheep, have played a significant part in this new paradigm. There can be no review of oocyte-secreted factors within domestic animals without acknowledging the enormous contribution by New Zealand researchers to this area, in particular, Ken McNatty. Much of the key literature on this topic has been generated by him and his colleagues, including Jenny Juengel, Grant Montgomery, and George Davis. A combination of interests came together to make this possible. First, a practical interest in sheep prolificacy for commercial production of lambs was matched by a desire to understand basic mechanisms of follicular growth and ovulation rate. This was also coupled to the recognition that within some sheep breeds selected for prolificacy, single gene mutants appeared to exist that significantly altered prolificacy. This combination of factors led these researchers within the New Zealand Ministry of Agriculture and Fisheries (now "AgResearch") to search for the genetic regulators of sheep prolificacy. Through their work, we now recognize these unique genotypes are a consequence of mutations in specific-oocyte and granulosa cell genes that are members of the transforming growth factor β superfamily, especially oocyte-specific growth differentiation factor 9 (GDF9) and bone morphogenetic protein 15 (BMP15, sometimes also referred to as GDF9b), and their signalling cascades.

This chapter will first discuss briefly the historical context of the influence of sheep genetic models and the role they played in establishing the importance of oocyte-secreted factors (OSFs) during follicular development, followed by an examination of the genetic diversity in the two most prominent OSFs, GDF9 and BMP15. We will then touch upon biological roles of OSFs within the ovarian follicle and look at their potential applications in advanced animal breeding. As another chapter of this book is on folliculogenesis, much of this chapter will focus on the roles of OSFs as occurs in antral follicles in ruminants and pigs, but will acknowledge the important work performed using rodent models.

Oocyte Physiology and Development in Domestic Animals, First Edition. Edited by Rebecca L. Krisher.

4.2 Historical Background

Although the concept of "oocyte-secreted factors" (OSFs) was first proposed in the early 1970's by Nalbandov and colleagues from their work in rats and rabbits (el-Fouly et al., 1970; Nekola & Nalbandov, 1971), one could argue that the story of OSFs in domestic animals, especially in ruminants, can be traced to the discovery of the FecB gene in Booroola merinos in 1980 by Bernie Bindon and Laurie Piper (CSIRO Livestock Division, Armidale, Australia) (Piper & Bindon, 1982). This was the first gene associated with high prolificacy and specifically ovulation rate (Davis et al., 1982) and although it created much interest, determining the molecular mechanism took a further two decades. Progress with mapping the gene was significantly hampered by the relatively poor molecular tools and genomic information available to sheep geneticists at the time. From its discovery, it took a decade of work to map FecB to an area on ovine chromosome 6 (Montgomery et al., 1994). FecB was finally identified simultaneously by three independent groups in 2001 (Mulsant et al., 2001; Souza et al., 2001; Wilson et al., 2001) as a point mutation within the bone morphogenetic protein receptor type 1, also known as activin receptor-like kinase type 6 (ALK6, the BMP15 type-I receptor).

In 1991, George Davis and team had discovered a further genetic mutation in a line of Romney breed sheep selected for prolificacy, known as the "Inverdale" (Davis et al., 1991). This line was established in 1979 at the Invermay Research Centre, New Zealand. Interestingly, this mutation was found to be X-linked, so that ewes homozygous for the Inverdale gene, $FecX^I$, are sterile, whereas intriguingly the heterozygote is more prolific than wild-type sheep, with an average ovulation rate of 2.9 (Davis et al., 2001). Being X-linked allowed the $FecX^I$ to be more rapidly identified as a mutation within the BMP15 gene (Galloway et al., 2000), which was a significant breakthrough for the field. Later, the first functional GDF9 mutant was identified in Irish flocks with high prolifacy (Hanrahan et al., 2004), along with two new BMP15 mutations – all three lead to sterility as homozygotes and increased ovulation rate as heterozygotes. Of note was the observation that animals that are heterozygous for mutations in both the GDF9 and BMP15 genes have additive effects on ovulation rate (Hanrahan et al., 2004), suggesting some form of GDF9/BMP15 interaction. These are on-going areas of research, and to date we have found within the literature at least 15 different known point mutations described in the GDF9, BMP15, and ALK6 genes in sheep that affect ovulation rate (Scaramuzzi et al., 2011) (see Figure 4.1).

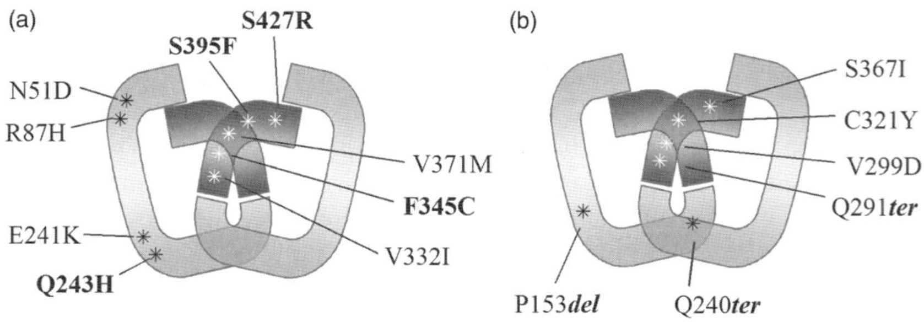

Figure 4.1 GDF9 and BMP15 mutations reported in sheep. A schematic of a pro-mature dimer complex (Walton et al., 2010) is shown with mutant amino acids shown on one half of the dimer. **(a)** GDF9, residues in **bold** exhibit a fertility phenotype. **(b)** BMP15, all mutants exhibit a fertility phenotype. GDF9 references: N51D (Li et al., 2003); R87H, E241K, V332I, V371M, S395F (Hanrahan et al., 2004); Q243H (**Chu et al., 2010**); F345C (**Silva et al., 2011**); S427R (**Nicol et al., 2009**). BMP15 references: P153*del* (Martinez-Royo et al., 2008) & (Monteagudo et al., 2009); Q240*ter* & S367I (Hanrahan et al., 2004); Q291*ter* & V299D (Galloway et al., 2000); C321Y (Bodin et al., 2007).

The determination that mutations in single genes affect fecundity was undeniably important information, but did not directly indentify OSFs as mediators of folliculogenesis. Much of this discovery work occurred in the mouse. For two decades the work of Nalbandov and colleagues remained unrecognized. It was not until 1990 that John Eppig reignited the interest in OSFs when his laboratory demonstrated that mouse cumulus expansion was an oocyte-regulated function (Buccione et al., 1990). This initiated the search for the oocyte-secreted "Cumulus Expansion Enabling Factor" (CEEF). In unrelated work, McPherron and Lee (1993) and McGrath et al. (1995) identified GDF9 as an ovary (oocyte)-specific protein. However, it was the landmark study of Matzuk and colleagues who generated GDF9 knockout mice and revealed that they are infertile, with ovaries similar to the streak ovaries identified in the homozygous carriers of the Inverdale mutation (Dong et al., 1996). Thus a link between the sheep mutations and a TGFβ superfamily oocyte-specific growth factor was made. Murine BMP15 (also known briefly as GDF9b) was then identified by two independent laboratories (Dube et al., 1998; Laitinen et al., 1998); within the ovary, it is also oocyte-specific. However, the BMP15 mouse knockout has no major loss of fertility (Yan et al., 2001), as opposed to sheep genetically deficient in BMP15, which are sterile (Galloway et al., 2000). This demonstrates a major difference between mice and other species studied to date, where in the latter, BMP15 plays a major role in folliculogenesis and fecundity. This led to the hypothesis that oocyte-secreted GDF9 and BMP15 in some manner are major determinants of mammalian ovulation rate and fecundity (McNatty et al., 2003). Recent evidence provides strong support for this notion by demonstrating that the ratio of *GDF:BMP15* mRNA expression is species-specific and that GDF9 is the predominate OSF in polyovular species whereas BMP15 is predominate in monovular species, including in cattle and deer (Crawford & McNatty, 2012).

4.3 Localization and Specificity

4.3.1 TGFβ Superfamily

GDF9 and BMP15

Given their significant effects on reproduction, both GDF9 and BMP15 remain the most recognized and researched of the OSFs. Their expression is largely restricted to the gametes, and localization outside of the ovary (or testes in the male) remains equivocal. Within ruminants, their localization within the ovary is highly restricted to the oocyte (McNatty et al., 2003), whereas for GDF9 at least, porcine cumulus and granulosa cells express GDF9 mRNA (Prochazka et al., 2004). In contrast to rodents, in sheep and cattle, GDF9 is expressed in oocytes of primordial follicles and then throughout folliculogenesis, whereas BMP15 is expressed from primary follicles onward (Juengel & McNatty, 2005).

Other TGFβ Superfamily Members

Although many members of the large TGFβ superfamily of growth factors play a critical role in the fine control regulation of folliculogenesis, no other superfamily members other than GDF and BMP15 have so far been attributed as important OSFs. BMP6 may be an oocyte-secreted factor in several species, including sheep (Juengel & McNatty, 2005). Unlike BMP15 and GDF9, expression of BMP6 is widespread within the ovary (as well as other tissues), although the oocyte is thought of as an important follicular source. Unlike GDF9 and BMP15, BMP6 has the potential to bind to multiple TGFβ superfamily receptors, but activates the same intracellular signalling cascade as BMP15 (see below). The role of BMP6 during folliculogenesis is well reviewed by

Juengel and McNatty (2005). BMP6 appears to be both mitogenic and anti-apoptotic (Glister et al., 2004; Hussein et al., 2005) and can also regulate steroid production by granulosa cells (Juengel & McNatty, 2005).

TGF-β1, β2, and β3 are expressed within the mammalian ovary; however there is equivocal evidence for the mRNA expression and production of TGFβ1-3 within domestic animal oocytes (reviewed by Juengel & McNatty, 2005). More important, there appears to be no biological activity associated with TGFβs derived from bovine oocytes (Gilchrist et al., 2003a), even though exogenous TGFβ1 and β2 can replicate the proliferative effects of OSFs on mural granulosa cells (Gilchrist et al., 2003a).

4.3.2 Fibroblast Growth Factors

The Fibroblast Growth Factors (FGFs) make up a complex family of proteins, consisting of 22 members grouped into 7 subfamilies. In rodents, FGF8 appears to be an oocyte-specific growth factor within the ovary (Valve et al., 1997), and importantly has been found to synergize with BMP15 in the regulation of (at least) glycolytic enzyme expression in murine cumulus cells (Sugiura et al., 2007). Compared to the TGFβ superfamily, the FGF family has not yet been extensively investigated in domestic species. FGF2 is expressed in bovine oocytes from primordial and primary follicles and may play a role in early folliculogenesis (van Wezel et al., 1995). Buratini and colleagues have since undertaken a significant level of FGF characterisation in bovine oocytes and ovarian tissues. To date they have described FGF8, 10 and 17 within bovine oocytes and have deduced that FGF18 is not in oocytes (Buratini et al., 2005; Buratini et al., 2007; Machado et al., 2009; Portela et al., 2010). Interestingly, the expression of FGF8 appears not specific to the oocyte in bovine, but was observed in granulosa cells as well.

4.4 Structure and Genetic Diversity of Gdf9 and Bmp15

GDF9 and BMP15 share a similar overall structure, as do all members of the TGFβ superfamily of proteins. All these family members are produced as pre-proproteins, with an amino terminal signal sequence (cleaved within the endoplasmic reticulum) followed by the pro-region, and finally the carboxy-terminal mature region. To obtain the fully biologically active protein the pro-mature precursor needs to be proteolytically processed, and this may represent an important event regulating the bioavailability of GDF9 and BMP15. This is the subject of ongoing research; however it is interesting to note that the major forms detected in sheep follicular fluid are the unprocessed pro-mature forms of GDF9 and BMP15 (McNatty et al., 2006). Unlike the majority of TGFβ superfamily members, which have a total of 7 conserved cysteine amino acids within their mature region, both GDF9 and BMP15 lack the 4th cysteine residue (McPherron & Lee, 1993; Dube et al., 1998; Laitinen et al., 1998) that normally forms a covalent disulphide bond binding the two mature regions together forming an active homo- or heterodimer. Dimerization of GDF9 and BMP15 is therefore thought to depend upon a non-covalent configuration of mature regions. This arrangement may allow for, and indeed facilitate, the formation of a GDF9/BMP15 heterodimer, a possibility that remains an area of active research.

The amino acid sequences of GDF9 and BMP15 for quite a number of different species are known, and in Figure 4.2 we show an alignment of the carboxy-terminal mature regions (an

(a)

```
sh   DQESASSELK KPLVPASVNL SEYFKQFLFP QNECELHDFR LSFSQLKWDN WIVAPHKYNP RYCKGDCPRA
ca   DQESVSSELK KPLVPASFNL SEYFKQFLFP QNECELHDFR LSFSQLKWDN WIVAPHKYNP RYCKGDCPRA
go   DQESVSSELK KPLVPASVNL SEYFKQFLFP QNECELHDFR LSFSQLKWDN WIVAPHKYNP RYCKGDCPRA
pi   AQDTVSSELK KPLVPASFNL SEYFKQFLFP QNECELHDFR LSFSQLKWDN WIVAPHKYNP RYCKGDCPRA

hu   GQETVSSELK KPLGPASFNL SEYFRQFLLP QNECELHDFR LSFSQLKWDN WIVAPHRYNP RYCKGDCPRA
mo   GQKAIRSEAK GPLLTASFNL SEYFKQFLFP QNECELHDFR LSFSQLKWDN WIVAPHRYNP RYCKGDCPRA
ra   GQKTLSSETK KPL-TASFNL SEYFRQFLFP QNECELHDFR LSFSQLKWDN WIVAPHRYNP RYCKGDCPRA

sh   VGHRYGSPVH TMVQNIIHEK LDSSVPRPSC VPAKYSPLSV LAIEPDGSIA YKEYEDMIAT KCTCR
ca   VGHRYGSPVH TMVMNIIHEK LDSSVPRPSC VPAKYSPLSV LAIEPDGSIA YKEYEDMIAT KCTCR
go   VGHRYGSPVH TMVQNIIHEK LDSSVPRPSC VPAKYSPLSV LAIEPDGSIA YKEYEDMIAT KCTCR
pi   VGHRYGSPVH TMVQNIIHEK LDSSVPRPSC VPAKYSPLSV LAIEPDGSIA YKEYEDMIAT KCTCR

hu   VGHRYGSPVH TMVQNIIYEK LDSSVPRPSC VPAKYSPLSV LTIEPDGSIA YKEYEDMIAT KCTCR
mo   VRHRYGSPVH TMVQNIIYEK LDPSVPRPSC VPGKYSPLSV LTIEPDGSIA YKEYEDMIAT RCTCR
ra   VRHRYGSPVH TMVQNIIYEK LDPSVPRPSC VPGKYSPLSV LTIEPDGSIA YKEYEDMIAT RCTCR
```

(b)

```
sh   QAGSIASEVP GPSREHDGPE SNQCSLHPFQ VSFQQLGWDH WIIAPHLYTP NYCKGVCPRV LHYGLNSPNH
ca   QAGSIASEVP GPSREHDGPE SNLCSLHPFQ VSFQQLGWDH WIIAPHLYTP NYCKGVCPRV LHYGLNSPNH
go   QAGSIASEVP GPSREHDGPE SNQCSLHPFQ VSFQQLGWDH WIIAPHLYTP NYCKGVCPRV LYYGLNSPNH
pi   QAGSIASEVL GPSREHDGPE SNQCSLHPFQ VSFHQLGWDH WIIAPHFYTP NYCKGVCPRV LHYGLNSPNH

hu   QADGISAEVT ASSSKHSGPE NNQCSLHPFQ ISFRQLGWDH WIIAPPFYTP NYCKGTCLRV LRDGLNSPNH
mo   QACSIESDAS CPSQEHDGSV NNQCSLHPYK VSFHQLGWDH WIIAPRLYTP NYCKGICTRV LPYGLNSPNH
ra   QTCSIASDVP CPSQEQDRSV NNQCSLHPYK VSFHQLGWDH WIIAPRLYTP NYCKGICTGV LPYGLNSPNH

sh   AIIQNLVSEL VDQNVPQPSC VPYKYVPISI LLIEANGSIL YKEYEGMIAQ SCTCR
ca   AIIQNLVNEL VDQSVPQPSC VPYKYVPISI LLIEANGSIL YKEYEGMIAQ SCTCR
go   AIIQNLVNEL VDQNVPQPSC VPYKYVPISI LLIEANGSIL YKEYEGMIAQ SCTCR
pi   AIIQNLVNEL VDQSVPQPSC VPYKYVPISI LLIEANGSIL YKEYEDMIAQ PCTCR

hu   AIIQNLINQL VDQSVPRPSC VPYKYVPISV LMIEANGSIL YKEYEGMIAE SCTCR
mo   AIIQSLVNEL VNHSVPQPSC VPYNFLPMSI LLIETNGSIL YKEYEGMIAQ SCTCR
ra   AIIQSLVNEL VNRSVPQLSC VPYKFLPMSI LLIEANGSIL YKEYEGMIAQ SCTCR
```

Figure 4.2 Alignment of GDF9 and BMP15 mature region amino acid sequences. Multiple sequence alignment was carried out using the T-Coffee software (www.tcoffee.org). **(a)** GDF9 sequences. **(b)** BMP15 sequences. sh: sheep, ca: cattle, go: goat, pi: pig, hu: human, mo: mouse, ra: rat. The six conserved cysteines that make up the "cysteine knot" are shown in yellow. Sequence differences between the four domestic species are highlighted in blue. Sequence differences between the domestic species and the human/rodent sequences are highlighted in green. (For color detail, see color plate section.)

analysis of the GDF9 & BMP15 pro-region sequences is beyond the scope of this review) for four domestic species (sheep, cattle, goat, and pig) in comparison with the human and rodent (mouse and rat) sequences. Comparison of these sequences makes it clear that all the domestic species sequences are very similar. The focus of the amino acid variation observed between the domestic species and the human/rodent sequences (shown in green, Figure 4.2) is the amino terminus of the mature region, prior to the first cysteine residue. This is true for both GDF9 and BMP15; however in the case of BMP15 there are also variant amino acids within the "cysteine knot" region of the protein (i.e., that portion of GDF9 & BMP15 bordered by the conserved cysteine residues that stabilize the tertiary fold of the mature region). A novel structural feature reported for recombinantly produced GDF9 and BMP15 (using HEK-293 cells as expression hosts) is the phosphorylation of a specific serine residue (Ser-6) at the amino terminus of the mature region of the protein (Tibaldi et al., 2010). This post-translational modification of GDF9 and BMP15 was reported to be essential for bioactivity (McMahon et al., 2008). Interestingly, although this amino acid residue is conserved across all species of GDF9 (Figure 4.2), with the exception of the mouse, in the case of BMP15, the human is the only species with a Ser at position 6. In the other species BMP15 has a Ser at position 7 of the mature region, but it is unknown if this residue is post-translationally modified, indeed, it is unknown if any of these post-translational modifications occur *in vivo*.

A number of mutations have been reported in sheep, which are located within the coding sequences of the GDF9 or BMP15 proteins (Figure 4.1), and these have often been associated with a fertility phenotype. In the case of BMP15, all the reported mutations are associated with a fertility phenotype; that is, animals homozygous for the mutation have streak ovaries and do not ovulate, whereas those that are heterozygous have a higher ovulation rate than wild-type ewes (see references in Figure 4.1). These BMP15 mutations consist of either point mutations or a more drastic change resulting in a premature termination of the protein, and are predominantly located within the mature region of the protein. In the case of GDF9, the majority of the mutations associated with a fertility phenotype are also localized to the biologically active mature region of the protein. Interestingly, a similar phenotype to the GDF9 or BMP15 mutations can be reproduced by immunization of ewes with GDF9 or BMP15 peptides (Juengel et al., 2004; McNatty et al., 2007).

4.5 Signalling Mechanisms of Gdf9 and Bmp15

It has only been in the last decade or so that the signalling pathways of GDF9 and BMP15 have been elucidated, most of which work has primarily been conducted in cell culture or rodent models (reviewed by Gilchrist, 2011). GDF9 and BMP15 signal using TGFβ superfamily receptors and intracellular transducer molecules. Both proteins signal through the BMP receptor Type 2 (BMPR2) on granulosa and cumulus cells, but recruit different co-receptors, the activin receptor-like kinases (ALKs). *In vitro* evidence indicates that GDF9 signals via a combination of BMPR2 and ALK5 (Vitt et al., 2002; Mazerbourg et al., 2004; Kaivo-Oja et al., 2005) (also known as the TGFβ receptor type 1), a unique situation within the superfamily. However recent data suggests that ALK5 is dispensable for GDF9 signaling *in vivo* (Li et al., 2011); hence the identity of the functional *in vivo* GDF9 receptors remains an open question. In the case of BMP15, studies with receptor ecto-domains indicate that ALK6 would be the type I receptor recruited by this ligand (Moore et al., 2003). This has very recently been confirmed with the use of a novel mutant of BMP15 that forms a covalent dimer enabling the use of this reagent to identify the cell surface receptor complex to which BMP15 binds on a human granulosa cell line (Pulkki et al., 2012). This study confirms that ALK6 and BMPR2 are indeed the cell surface receptors for BMP15. Activation of the ALK receptor kinases causes the immediate and transient phosphorylation of *Sma*- and *Mad*-related (SMAD) intracellular signals. GDF9 signals through the SMAD 2/3 pathway (Kaivo-Oja et al., 2003; Mazerbourg et al., 2004; Kaivo-Oja et al., 2005) whereas BMP15 activates the SMAD 1/5/8 pathway (Moore et al., 2003; Pulkki et al., 2011). Inhibition of these pathways using a variety of agents has been invaluable in determining the roles of the various OSFs and respective signal transduction pathways (Table 4.1).

Table 4.1 A selection of physiological or pharmacological agents that reduce GDF9/BMP15 signaling.

Agent/Name	Description	Relevant Actions	References
MAb53	Monoclonal antibody	GDF9 neutralizing antibody	(Gilchrist et al., 2004); (Dragovic et al., 2005)
a/b#1	Polyclonal antibody	GDF9 neutralizing antibody	(Wang et al., 2009)
Follistatin	Ovarian peptide	Weak BMP15 antagonist	(Otsuka et al., 2001); (Hussein et al., 2005); (Hussein et al., 2006)
SB-431542	ALK4/5/7 inhibitor	GDF9 antagonist	(Gilchrist et al., 2006); (Hussein et al., 2006); (Dragovic et al., 2007); (Diaz et al., 2007)
SIS3	SMAD3 inhibitor	GDF9 antagonist	(Diaz et al., 2006); (Diaz et al., 2007)
BMPR2-ECD	Soluble receptor fragment	GDF9/BMP15 antagonist	(Dragovic et al., 2005); (Gilchrist et al., 2006); (Myllymaa et al., 2010).

4.6 Roles of Oocyte-Secreted Factors

A fundamental concept that has developed over the past two decades is that oocyte paracrine signalling is critical for normal preantral granulosa cell and cumulus cell function and for establishing and maintaining the cumulus cell phenotype in antral follicles (reviewed by Gilchrist et al. (2008); see Figure 4.3). OSF signalling is therefore an integral component of the bidirectional communication loop between the oocyte and granulosa/cumulus cells, which also includes communication through the gap-junctions between the oocyte and the surrounding granulosa and cumulus cells. This communication loop is imperative for acquisition of full developmental capacity by the oocyte during antral development leading to ovulation. Native OSFs, a term used to describe the uncharacterised mix of growth factors secreted by an oocyte (Hussein et al., 2006), have been attributed a large array of roles during antral folliculogenesis, all of which relate to the differentiation of the cumulus cell phenotype from the surrounding mural granulosa cells (Gilchrist et al., 2008; Gilchrist, 2011; plus see Table 4.2, modified from review by Gilchrist et al., 2008). However, much of this information has been determined using rodent models. As discussed by Juengel and McNatty (2005), a problem within the field is that when treated with either homologous or heterologous native OSFs or recombinant proteins, ovarian cells from different species yield varied results, depending on the species of native or recombinant OSF. In support of this, it has recently been demonstrated that human GDF9 is secreted in a latent form as a complex of pro and mature regions while mouse GDF9 is not latent (Mottershead et al., 2008; Simpson et al., 2012). Again this appears to be a peculiarity of GDF9 (and potentially BMP15), as most other proteins in the TGFβ superfamily do not generally show a high degree of species-specific variability in response of the target cell.

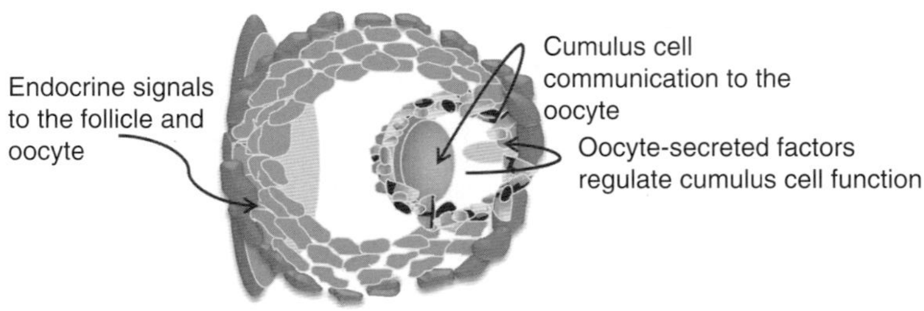

Figure 4.3 Diagrammatic representation of an antral follicle showing the communication pathways that operate between cell types.

Table 4.2 Known effects of oocyte secreted factors on cumulus cells.

Effect of native OSFs on GC, CC or oocyte function		References*
Signaling cascades	Activation of GC/CC SMAD signaling	(Gilchrist et al., 2006)
	Activation of MAPK signaling	(Su et al., 2003)
Oocyte growth	Stimulation/suppression *KitL*	(Joyce et al., 1999)
CC/MGC proliferation	Stimulation of *Ccnd2*	(Gilchrist et al., 2006)
	Stimulation of GC/CC DNA synthesis, cell number or follicle growth	(Vanderhyden et al., 1992; Lanuza et al., 1998; Li et al., 2000; Gilchrist et al., 2001; Eppig et al., 2002; Brankin et al., 2003; Gilchrist et al., 2003b; Glister et al., 2003; Gilchrist et al., 2004; Hickey et al., 2005; Gilchrist et al., 2006)
	Interaction of OSFs with IGF-I	(Lanuza et al., 1998; Li et al., 2000; Brankin et al., 2003; Gilchrist et al., 2003b; Hickey et al., 2005)
	Stimulation of CC *Ar*	(Diaz et al., 2007)
CC apoptosis	Prevention of CC apoptosis	(Hussein et al., 2005)
CC/MGC luteinization	Regulation of MGC/CC progesterone or oestradiol production	(Nekola and Nalbandov, 1971; Vanderhyden et al., 1993; Coskun et al., 1995; Vanderhyden and Tonary, 1995; Vanderhyden and Macdonald, 1998; Lanuza et al., 1999; Li et al., 2000; Glister et al., 2003)
	Suppression of CC *Cyp11a1*	(Diaz et al., 2007)
	Suppression of FSH-induced *Lhcgr*	(Eppig et al., 1997)
	Regulation of MGC inhibin-follistatin-activin production	(Lanuza et al., 1999; Glister et al., 2003)
	Stimulation of CC *Amh*	(Salmon et al., 2004; Diaz et al., 2007)
	Suppression of CC *Cd44*	(Diaz et al., 2007)
CC metabolism	Stimulation of CC glycolysis	(Su et al., 2004; Sugiura et al., 2005)
	Stimulation of CC AA transport	(Eppig et al., 2005)
CC expansion	Enabling FSH/EGF-stimulated CC expansion (CEEF – murine)	(Buccione et al., 1990; Salustri et al., 1990b; Vanderhyden et al., 1990; Dragovic et al., 2005; Dragovic et al., 2007)
	Production of CEEF (non-murine)	(Prochazka et al., 1991; Singh et al., 1993; Vanderhyden, 1993; Ralph et al., 1995; Prochazka et al., 1998)
	Enabling FSH/EGF-induction of *Has2, Ptgs2, Ptx3, Tnfaip6* and secretion of hyaluronic acid	(Salustri et al., 1990a; Salustri et al., 1990b; Joyce et al., 2001; Dragovic et al., 2005; Diaz et al., 2006; Dragovic et al., 2007)
	Regulation of plasminogen activator	(Canipari et al., 1995; D'Alessandris et al., 2001)
Oocyte quality	IVM additive increasing blastocyst development	(Hussein et al., 2006; Yeo et al., 2008; Romaguera et al., 2010; Gomez et al., 2011; Hussein et al., 2011)
	IVM additive increasing fetal survival	(Yeo et al., 2008)

*Studies using native OSFs. Studies on domestic species are in bold.

GC = granulosa cell, CC = cumulus cell, MAPK = mitogen-activated protein kinase, OSF = oocyte-secreted factors, MGC = mural granulosa cell, AA = amino acid, CEEF = cumulus-expansion enabling factors, EGF = epidermal growth factor, IVM = in vitro maturation.

Access to high purity recombinant GDF9/BMP15 and effective antibodies against GDF9 and BMP15 also remain technical problems. There are a number of reasons for this, including their relatively recent discovery: the fact that their expression and therefore commercial potential is largely restricted to reproductive tissues and the species-specific peculiarities of the proteins as discussed above. These issues are particularly evident in the domestic animal species, where very few species-specific reagents have been generated. With the recent advances in the understanding of species-specific potency of the proteins (Simpson et al., 2012) and the relative importance of BMP15 versus GDF9 in mono- versus poly-ovular species, respectively (Crawford & McNatty, 2012), we can expect an improvement in the availability of high quality recombinant proteins in coming years. In particular, the importance of the role for BMP15 in many non-murine species remains a major difference, which will emerge during the remainder of this chapter.

4.6.1 Follicle Growth, Cell Proliferation, and Apoptosis

GDF9 and BMP15 are clearly essentially for progression of the earliest stages of ovine folliculogenesis. Sheep genetically deficient in GDF9 or BMP15 exhibit arrested follicular growth at the primary stage of folliculogenesis (Galloway et al., 2000; Hanrahan et al., 2004). This suggests that GDF9 and/or BMP15 secreted by oocytes in primary follicles may directly promote proliferation of granulosa cells. Alternatively GDF9/BMP15 may counter the actions of a follicle growth inhibiting factor(s).

Use of a bioassay where denuded oocytes (producing native OSFs) from large antral follicles are placed together with either cultures of mural granulosa cells or with oocytectomized COCs has revealed that OSFs in all species studied to date have a pronounced effect on proliferation of granulosa and cumulus cells (Lanuza et al., 1998; Li et al., 2000; Gilchrist et al., 2003b). More recently, these observations have been extended to specific recombinant factors (McNatty et al., 2005; Spicer et al., 2006). However, unlike in the mouse where GDF9 alone is potently mitogenic, both recombinant ovine GDF9 and BMP15 were required together to stimulate rat granulosa cell proliferation (McNatty et al., 2005). Furthermore, recombinant murine GDF9 alone stimulates ovine granulosa cell proliferation, highlighting the difficulty of interpretation of results when using reagents developed in heterologous species. Moreover, there are notable differences in the actions of native OSFs between species. For example, native OSFs or recombinant GDF9 in rodents significantly increase granulosa cell proliferation without further hormone stimulation (Gilchrist et al., 2001; Gilchrist et al., 2006). In contrast, the proliferating effect of OSFs on cattle and pig granulosa and cumulus cells is relatively minor, unless IGF-1 is present (Li et al., 2000; Hickey et al., 2005). In addition, for the pig at least, a further enhancement of proliferation is observed with stimulation of the androgen receptor in the presence of OSFs and IGF-1, suggesting important interactions between endocrine signalling and the function of OSFs (Hickey et al., 2005).

The effect of OSFs on apoptosis of ruminant follicular cells has been restricted to the observations of Hussein et al. (2005). This study revealed that apoptosis was prevented by native OSFs in cattle COCs, as well as recombinant BMP15 treatment. Surprisingly, recombinant GDF9 had little influence on apoptosis levels, whereas its proliferative effect is well documented in many species.

4.6.2 Regulation of Steroidogenesis and Luteinization

OSFs play an important role in regulating cumulus and granulosa cell steroid metabolism. Indeed, this was the first attribute accredited to OSFs from the early work of Nalbandov and colleagues in the early 1970s (el-Fouly et al., 1970; Nekola & Nalbandov, 1971). Across many species, OSFs have been reported to inhibit the production of progesterone from cumulus and granulosa cells (pig: Coskun et al., 1995; Brankin et al., 2003; cattle: Li et al., 2000). These effects have also been observed with recombinant GDF9. Nevertheless, as observed for cellular proliferation (McNatty et al., 2005), inhibition of progesterone production by ovine granulosa cells occurred most notably when treated with both oGDF9 and BMP. Eppig and colleagues (1997) first demonstrated in the mouse that LH receptors in cumulus cells are repressed by the actions of OSFs, and have since gone on to catalog a large array of genes associated with granulosa cell luteinization that are suppressed by native OSFs and recombinant GDF9 and BMP15.

4.6.3 Cumulus Cell Expansion

In mice, there is an absolute need for a cumulus expansion enabling factor, secreted from the oocyte (Buccione et al., 1990; Eppig et al., 1993a; Eppig et al., 1993b). In particular, oocyte-activation of SMAD 2/3 signalling for cumulus expansion is an absolute requirement (Dragovic et al., 2007), but this can only occur with the co-participation of extracellular signal regulated kinases 1/2 (ERK1/2) (Sasseville et al., 2010). However, in most domestic species studied to date (pig and cattle COCs), this interactive signalling between SMADs and ERKs appears significantly less important, as bovine and porcine cumulus cells in oocytectomized complexes (i.e., lacking OSFs) will still expand *in vitro* when either FSH and/or EGF treatments are applied (Prochazka et al., 1991; Singh et al., 1993; Ralph et al., 1995).

4.6.4 Cumulus Cell Metabolism

There is also some disparity in results concerning the influence of OSFs on cumulus cell metabolism. Eppig has provided substantial evidence in the mouse that within COCs from antral follicles, cumulus cell glycolysis and amino acid uptake is regulated by native OSFs (Sugiura & Eppig, 2005; Sugiura et al., 2005; Sugiura et al., 2007; Su et al., 2008) and, at least for the regulation of glycolysis, by BMP15 and FGF8 in a synergistic manner (Sugiura et al., 2007). In contrast, Sutton et al failed to find any difference in glucose uptake or lactate production in oocytectomized cattle COCs (Sutton et al., 2003) Nevertheless, this latter work was conducted *in vitro* in the presence of FSH and hence during cumulus expansion, raising issues as to whether effects on cumulus cell metabolism by OSFs are masked under such conditions.

4.7 Manipulation and Use in Reproductive Technologies

4.7.1 Altering Ovulation Rate—Antibodies

Altering levels of endogenous GDF9 and BMP15 using antibody therapy has been found to either annul or enhance prolificacy, as both scenarios have practical uses in domestic animal breeding including in companion animals and in conservation biology. Partial immunization against either GDF9 or BMP15 or against both in sheep, by short-term passive antibody treatment, increases ovulation and lambing rates (Juengel et al., 2004; McNatty et al., 2006). In contrast, prolonged immunisation shuts down folliculogenesis at the primary or primordial stages (Juengel et al., 2002;

McNatty et al., 2006), with antisera generated from peptide sequences from the N-terminal region having the most potency (McNatty et al., 2007). Immunization against GDF9 or BMP15 proteins decreased cattle ovulation rates, but caused variable responses when the N-terminal peptide of BMP-15 was targeted (alone or with peptide of GDF-9), such that some cattle had up to 4 CL and others had none (Juengel et al., 2009).

4.7.2 **In Vitro** ***Embryo Production***

Our own laboratory pioneered the idea that OSFs have an application in embryo production technologies, and this has since been confirmed by two other laboratories. We have investigated exogenous OSF treatment during oocyte *in vitro* maturation (IVM), based on the hypothesis that IVM poorly replicates the follicular environment and there is disrupted signalling between the cumulus cells and oocyte, leading to reduced oocyte quality (Gilchrist, 2011). We have demonstrated that supplementing cattle or mouse IVM medium with native OSFs (cattle), recombinant GDF9 (cattle and mouse), or BMP15 (cattle) increases subsequent blastocyst production and improved post-transfer outcomes in terms of fetal development (Hussein et al., 2006; Yeo et al., 2008; Hussein et al., 2011). More recently, similar improvements have been observed in pre-pubertal goat oocytes and pig oocytes by co-incubation during IVM with denuded oocytes as a source of native OSFs (Romaguera et al., 2010; Gomez et al., 2011). To date there are no reports on the applications of OSFs in IVM in other species, including in humans. The relative utility of GDF9 versus BMP15 as IVM additives is likely to differ between species, and this is an ongoing area of investigation in our laboratory.

The mechanism by which exogenous OSFs added during IVM improve oocyte quality and developmental outcomes is currently unclear. It may be that the artefactual nature of IVM perturbs either the oocyte's capacity to secrete appropriate levels or forms of endogenous GDF9/BMP15, or the cumulus cells' capacity to receive these paracrine signals. In a recent publication, we have shown that the capacity for co-incubation with denuded oocytes to improve the developmental competence of intact bovine COCs is temporally regulated (Hussein et al., 2011). We found that culturing the oocytes that contribute the native OSFs as intact COCs for 9 hours prior to denuding, and then co-incubating the subsequent denuded oocytes with COCs, led to a marked increase in developmental competence of the COCs. This suggests there are temporal changes to either the "mixture" or quantities of OSFs during maturation. In contrast, we found that during IVM, both supplemental recombinant GDF9 and BMP15 had their beneficial effects on oocyte quality within the first 9h of maturation (Hussein et al., 2011).

Recent work has also examined the benefit of FGF10 to IVM medium and has been observed to increase cumulus expansion, accelerate meiosis, and yield more blastocysts. These affects could be reversed using a specific FGF10 antibody (Zhang et al., 2010).

4.8 Concluding Remarks

Over the past decade, our understanding of the importance and roles of oocyte-secreted factors in regulating cumulus cell function has been a major advancement in ovarian and oocyte biology. Studies of domestic species, especially sheep, have played an important contributing role in this understanding, especially the genetic studies first identifying that ovarian function was highly regulated by single genes.

Nevertheless, our understanding of specific roles that domestic animal oocyte-secreted factors play, especially in antral follicle selection and ovulation, is at best, rudimentary. The lack of necessary tools is still limiting research capacity, and our knowledge is significantly lagging behind the rapid advances in knowledge that are being achieved with the mouse. Furthermore, what has been investigated does not necessarily fit with the mouse model, especially the importance of the role of BMP-15 in mono-ovulatory animals. Most important, manipulation of OSF levels during *in vitro* or *in vivo* oocyte maturation has a great untapped capacity to modulate reproductive performance, which should be explored further.

References

Bodin, L., Di Pasquale, E., Fabre, S., Bontoux, M., Monget, P., et al. (2007). A novel mutation in the bone morphogenetic protein 15 gene causing defective protein secretion is associated with both increased ovulation rate and sterility in Lacaune sheep. *Endocrinology, 148*, 393–400.

Brankin, V., Mitchell, M. R., Webb, B., & Hunter, M. G. (2003). Paracrine effects of oocyte secreted factors and stem cell factor on porcine granulosa and theca cells *in vitro*. *Reproductive Biology and Endocrinology, 1*, 55.

Buccione, R., Vanderhyden, B. C., Caron, P. J., Eppig, J. J. (1990). FSH-induced expansion of the mouse cumulus oophorus *in vitro* is dependent upon a specific factor(s) secreted by the oocyte. *Developmental Biology, 138*, 16–25.

Buratini, Jr., J., Pinto, M. G., Castilho, A. C., Amorim, R. L., Giometti, I. C. et al. (2007). Expression and function of fibroblast growth factor 10 and its receptor, fibroblast growth factor receptor 2B, in bovine follicles. *Biology of Reproduction, 77*, 743–750.

Buratini, Jr., J., Teixeira, A. B., Costa, I. B., Glapinski, V. F., Pinto, M. G. et al. (2005). Expression of fibroblast growth factor-8 and regulation of cognate receptors, fibroblast growth factor receptor-3c and -4, in bovine antral follicles. *Reproduction, 130*, 343–350.

Canipari, R., Epifano, O., Siracusa, G., & Salustri, A. (1995). Mouse oocytes inhibit plasminogen activator production by ovarian cumulus and granulosa cells. *Developmental Biology, 167*, 371–378.

Chu, M. X., Yang, J., Feng, T., Cao, G. L., Fang, L. et al. (2010). GDF9 as a candidate gene for prolificacy of Small Tail Han sheep. *Molecular Biology and Reproduction*.

Coskun, S., Uzumcu, M., Lin, Y. C., Friedman, C. I., & Alak, B. M. (1995). Regulation of cumulus cell steroidogenesis by the porcine oocyte and preliminary characterization of oocyte-produced factor(s). *Biol Reprod, 53*, 670–675.

Crawford, J. L. & McNatty, K. P. (2012). The ratio of growth differentiation factor 9: bone morphogenetic protein 15 mRNA expression is tightly co-regulated and differs between species over a wide range of ovulation rates. *Mollecular and Cellular Endocrinology, 348*, 339–343.

D'Alessandris, C., Canipari, R., Di Giacomo, M., Epifano, O., Camaioni, A., et al., (2001). Control of mouse cumulus cell-oocyte complex integrity before and after ovulation: plasminogen activator synthesis and matrix degradation. *Endocrinology, 142*, 3033–3040.

Davis, G. H., Bruce, G. D., & Dodds, K. G. (2001). Ovulation rate and litter size of prolific Inverdale (FecX I) and Hanna (FecX H) sheep. *Proceedings of the Association for the Advancement of Animal Breeding and Genetics, 14*, 175–178.

Davis, G. H., McEwan, J. C., Fennessy, P. F., Dodds, K. G., & Farquhar, P. A. (1991). Evidence for the presence of a major gene influencing ovulation rate on the X chromosome of sheep. *Biology of Reproduction, 44*, 620–624.

Davis, G. H., Montgomery, G. W., Allison, A. J., Kelly, R. W., & Bray, A. R. (1982). Segregation of a major gene influencing fecundity in progeny of Booroola sheep. *New Zealand Journal of Agricultural Research, 25*, 525–529.

Diaz, F. J., O'Brien, M. J., Wigglesworth, K., & Eppig, J. J. (2006). The preantral granulosa cell to cumulus cell transition in the mouse ovary: Development of competence to undergo expansion. *Developmental Biology, 299*, 91–104.

Diaz, F. J., Wigglesworth, K., & Eppig, J. J. (2007). Oocytes determine cumulus cell lineage in mouse ovarian follicles. *Journal of Cell Science, 120*, 1330–1340.

Dong, J., Albertini, D. F., Nishimori, K., Kumar, T. R., Lu, N., & Matzuk, M. M. (1996). Growth differentiation factor-9 is required during early ovarian folliculogenesis. *Nature, 383*, 531–535.

Dragovic, R. A., Ritter, L. J., Schulz, S. J., Amato, F., Armstrong, D. T., & Gilchrist, R. B. (2005). Role of oocyte-secreted growth differentiation factor 9 in the regulation of mouse cumulus expansion. *Endocrinology, 146*, 2798–2806.

Dragovic, R. A., Ritter, L. J., Schulz, S. J., Amato, F., Thompson, J. G., et al. (2007). Oocyte-secreted factor activation of SMAD 2/3 signaling enables initiation of mouse cumulus cell expansion. *Biology of Reproduction, 76*, 848–857.

Dube, J. L., Wang, P., Elvin, J., Lyons, K. M., Celeste, A. J., & Matzuk, M. M. (1998). The bone morphogenetic protein 15 gene is X-linked and expressed in oocytes. *Molecular Endocrinology, 12*, 1809–1817.

el-Fouly, M. A., Cook, B., Nekola, M., & Nalbandov, A. V. (1970). Role of the ovum in follicular luteinization. *Endocrinology, 87*, 286–293.

Eppig, J. J., Pendola, F. L., Wigglesworth, K., & Pendola, J. K. (2005). Mouse oocytes regulate metabolic cooperativity between granulosa cells and oocytes: Amino acid transport. *Biology of Reproduction, 73*, 351–357.

Eppig, J. J., Peters, A. H., Telfer, E. E., & Wigglesworth, K. (1993a). Production of cumulus expansion enabling factor by mouse oocytes grown *in vitro*: Preliminary characterization of the factor. *Molecular Reproduction and Development, 34*, 450–456.

Eppig, J. J., Wigglesworth, K., & Chesnel, F. (1993b). Secretion of cumulus expansion enabling factor by mouse oocytes: relationship to oocyte growth and competence to resume meiosis. *Developmental Biology, 158*, 400–409.

Eppig, J. J., Wigglesworth, K., Pendola, F., & Hirao, Y. (1997). Murine oocytes suppress expression of luteinizing hormone receptor messenger ribonucleic acid by granulosa cells. *Biology of Reproduction, 56*, 976–984.

Eppig, J. J., Wigglesworth, K., & Pendola, F. L. (2002). The mammalian oocyte orchestrates the rate of ovarian follicular development. *Proceedings of the National Academy of Sciences USA, 99*, 2890–2894.

Galloway, S. M., McNatty, K. P., Cambridge, L. M., Laitinen, M. P., Juengel, J. L., et al. (2000). Mutations in an oocyte-derived growth factor gene (BMP15) cause increased ovulation rate and infertility in a dosage-sensitive manner. *Nature Genetics, 25*, 279–283.

Gilchrist, R. B. (2011). Recent insights into oocyte-follicle cell interactions provide opportunities for the development of new approaches to *in vitro* maturation. *Reproduction Fertility and Development, 23*, 23–31.

Gilchrist, R. B., Lane, M., & Thompson, J. G. (2008). Oocyte-secreted factors: regulators of cumulus cell function and oocyte quality. *Human Reproduction Update, 14*, 159–177.

Gilchrist, R. B., Morrissey, M. P., Ritter, L. J., & Armstrong, D. T. (2003). Comparison of oocyte factors and transforming growth factor-beta in the regulation of DNA synthesis in bovine granulosa cells. *Molecular and Cellular Endocrinology, 201*, 87–95.

Gilchrist, R. B., Ritter, L. J., & Armstrong, D. T. (2001). Mouse oocyte mitogenic activity is developmentally coordinated throughout folliculogenesis and meiotic maturation. *Developmental Biology, 240*, 289–298.

Gilchrist, R.B., Ritter, L.J., Cranfield, M., Jeffery, L.A., Amato, F., et al. (2004). Immunoneutralization of growth differentiation factor 9 reveals it partially accounts for mouse oocyte mitogenic activity. *Biology of Reproduction, 71*, 732–739.

Gilchrist, R. B., Ritter, L. J., Myllymaa, S., Kaivo-Oja, N., Dragovic, R. A., et al. (2006). Molecular basis of oocyte-paracrine signalling that promotes granulosa cell proliferation. *Journal of Cell Science, 119*, 3811–3821.

Glister, C., Groome, N. P., & Knight, P. G. (2003). Oocyte-mediated suppression of follicle-stimulating hormone- and insulin-like growth factor-induced secretion of steroids and inhibin- related proteins by bovine granulosa cells *in vitro*: Possible role of transforming growth factor alpha. *Biology of Reproduction, 68*, 758–765.

Glister, C., Kemp, C. F., & Knight, P. G. (2004). Bone morphogenetic protein (BMP) ligands and receptors in bovine ovarian follicle cells: actions of BMP-4, -6 and -7 on granulosa cells and differential modulation of Smad-1 phosphorylation by follistatin. *Reproduction, 127*, 239–254.

Gomez, M. N., Kang, J. T., Koo, O. J., Kim, S. J., Kwon, D. K., et al. (2011). Effect of oocyte-secreted factors on porcine *in vitro* maturation, cumulus expansion and developmental competence of parthenotes. *Zygote*, 1–11.

Hanrahan, J. P., Gregan, S. M., Mulsant, P., Mullen, M., Davis, G.H., et al. (2004). Mutations in the genes for oocyte-derived growth factors GDF9 and BMP15 are associated with both increased ovulation rate and sterility in Cambridge and Belclare sheep (Ovis aries). *Biology of Reproduction, 70*, 900–909.

Hickey, T. E., Marrocco, D. L., Amato, F., Ritter, L. J., Norman, R. J., et al. (2005). Androgens augment the mitogenic effects of oocyte-secreted factors and growth differentiation factor 9 on porcine granulosa cells. *Biology of Reproduction, 73*, 825–832.

Hussein, T. S., Froiland, D. A., Amato, F., Thompson, J. G., & Gilchrist, R. B. (2005). Oocytes prevent cumulus cell apoptosis by maintaining a morphogenic paracrine gradient of bone morphogenetic proteins. *Journal of Cell Science, 118*, 5257–5268.

Hussein, T. S., Sutton-McDowall, M. L., Gilchrist, R. B., & Thompson, J. G. (2011). Temporal effects of exogenous oocyte-secreted factors on bovine oocyte developmental competence during IVM. *Reproduction Fertility and Development, RD10323* (in press).

Hussein, T. S., Thompson, J. G., & Gilchrist, R. B. (2006). Oocyte-secreted factors enhance oocyte developmental competence. *Developmental Biology, 296*, 514–521.

Joyce, I. M., Pendola, F. L., O'Brien, M., & Eppig, J. J. (2001). Regulation of prostaglandin-endoperoxide synthase 2 messenger ribonucleic acid expression in mouse granulosa cells during ovulation. *Endocrinology, 142*, 3187–3197.

Joyce, I. M., Pendola, F. L., Wigglesworth, K., & Eppig, J. J. (1999). Oocyte regulation of kit ligand expression in mouse ovarian follicles. *Developmental Biology, 214*, 342–353.

Juengel, J. L., Hudson, N. L., Berg, M., Hamel, K., Smith, P., et al. (2009). Effects of active immunization against growth differentiating factor 9 and/or bone morphogenetic protein 15 on ovarian function in cattle. *Reproduction, 138*, 107–114.

Juengel, J. L., Hudson, N. L., Heath, D. A., Smith, P., Reader, K. L., et al. (2002). Growth differentiation factor 9 and bone morphogenetic protein 15 are essential for ovarian follicular development in sheep. *Biology of Reproduction, 67*, 1777–1789.

Juengel, J. L., Hudson, N. L., Whiting, L., & McNatty, K. P. (2004). Effects of immunization against bone morphogenetic protein 15 and growth differentiation factor 9 on ovulation rate, fertilization, and pregnancy in ewes. *Biology of Reproduction, 70*, 557–561.

Juengel, J. L. & McNatty, K. P. (2005). The role of proteins of the transforming growth factor-beta superfamily in the intraovarian regulation of follicular development. *Human Reproduction Update, 11*, 143–160.

Kaivo-Oja, N., Bondestam, J., Kamarainen, M., Koskimies, J., Vitt, U., et al. (2003). Growth differentiation factor-9 induces Smad2 activation and inhibin B production in cultured human granulosa-luteal cells. *The Journal of Clinical Endocrinology & Metabolism, 88*, 755–762.

Kaivo-Oja, N., Mottershead, D. G., Mazerbourg, S., Myllymaa, S., Duprat, S., et al. (2005). Adenoviral gene transfer allows Smad-responsive gene promoter analyses and delineation of type I receptor usage of transforming growth factor-beta family ligands in cultured human granulosa luteal cells. *The Journal of Clinical Endocrinology & Metabolism, 90*, 271–278.

Laitinen, M., Vuojolainen, K., Jaatinen, R., Ketola, I., Aaltonen, J., et al. (1998). A novel growth differentiation factor-9 (GDF-9) related factor is co- expressed with GDF-9 in mouse oocytes during folliculogenesis. *Mechanics of Development, 78*, 135–140.

Lanuza, G. M., Fischman, M. L., & Baranao, J. L. (1998). Growth promoting activity of oocytes on granulosa cells is decreased upon meiotic maturation. *Developmental Biology, 197*, 129–139.

Lanuza, G. M., Groome, N. P., Baranao, J. L., & Campo, S. (1999). Dimeric inhibin A and B production are differentially regulated by hormones and local factors in rat granulosa cells. *Endocrinology, 140*, 2549–2554.

Li, B. X., Chu, M. X., & Wang, J. Y. (2003). [PCR-SSCP analysis on growth differentiation factor 9 gene in sheep]. *Yi Chuan Xue Bao, 30*, 307–310.

Li, Q., Agno, J. E., Edson, M. A., Nagaraja, A. K., Nagashima, T., & Matzuk, M. M. (2011). Transforming growth factor beta receptor type 1 is essential for female reproductive tract integrity and function. *PLOS Genetics, 7*, e1002320.

Li, R., Norman, R. J., Armstrong, D. T., & Gilchrist, R. B. (2000). Oocyte-secreted factor(s) determine functional differences between bovine mural granulosa cells and cumulus cells. *Biology of Reproduction, 63*, 839–845.

Machado, M. F., Portela, V. M., Price, C. A., Costa, I. B., Ripamonte, P., et al. (2009). Regulation and action of fibroblast growth factor 17 in bovine follicles. *Journal of Endocrinology, 202*, 347–353.

Martinez-Royo, A., Jurado, J. J., Smulders, J. P., Marti, J. I., Alabart, J. L., et al. (2008). A deletion in the bone morphogenetic protein 15 gene causes sterility and increased prolificacy in Rasa Aragonesa sheep. *Animal Genetics, 39*, 294–297.

Mazerbourg, S., Klein, C., Roh, J., Kaivo-Oja, N., Mottershead, D. G., et al. (2004). Growth differentiation factor-9 (GDF-9) signaling is mediated by the type I receptor, activin re3ceptor-like kinase 5. *Molecular Endocrinology, 18*, 653–665.

McGrath, S. A., Esquela, A. F., & Lee, S. J. (1995). Oocyte-specific expression of growth/differentiation factor-9. *Molecular Endocrinology, 9*, 131–136.

McMahon, H. E., Hashimoto, O., Mellon, P. L., & Shimasaki, S. (2008). Oocyte-specific overexpression of mouse bone morphogenetic protein-15 leads to accelerated folliculogenesis and an early onset of acyclicity in transgenic mice. *Endocrinology, 149*, 2807–2815.

McNatty, K. P., Hudson, N. L., Whiting, L., Reader, K. L., Lun, S., et al. (2007). The effects of immunizing sheep with different BMP15 or GDF9 peptide sequences on ovarian follicular activity and ovulation rate. *Biology of Reproduction, 76*, 552–560.

McNatty, K. P., Juengel, J. L., Reader, K. L., Lun, S., Myllymaa, S., et al. (2005). Bone morphogenetic protein 15 and growth differentiation factor 9 co-operate to regulate granulosa cell function. *Reproduction, 129*, 473–480.

McNatty, K. P., Juengel, J. L., Wilson, T., Galloway, S. M., Davis, G. H., et al. (2003). Oocyte-derived growth factors and ovulation rate in sheep. *Reproduction Supplement, 61*, 339–351.

McNatty, K. P., Lawrence, S., Groome, N. P., Meerasahib, M. F., Hudson, N. L., et al. (2006). Oocyte signalling molecules and their effects on reproduction in ruminants. *Reproduction, Fertility and Development, 18*, 403–412.

McPherron, A. C. & Lee, S. J. (1993). GDF-3 and GDF-9: two new members of the transforming growth factor-beta superfamily containing a novel pattern of cysteines. *The Journal of Biological Chemistry, 268*, 3444–3449.

Monteagudo, L. V., Ponz, R., Tejedor, M. T., Lavina, A., & Sierra, I. (2009). A 17 bp deletion in the Bone Morphogenetic Protein 15 (BMP15) gene is associated to increased prolificacy in the Rasa Aragonesa sheep breed. *Animal Reproduction Science, 110*, 139–146.

Montgomery, G. W., Lord, E. A., Penty, J. M., Dodds, K. G., Broad, T. E., et al. (1994). The Booroola fecundity (FecB) gene maps to sheep chromosome 6. *Genomics, 22*, 148–153.

Moore, R. K., Otsuka, F., & Shimasaki, S. (2003). Molecular basis of bone morphogenetic protein-15 signaling in granulosa cells. *The Journal of Biological Chemistry, 278*, 304–310.

Mottershead, D. G., Pulkki, M. M., Muggalla, P., Pasternack, A., Tolonen, M., et al. (2008). Characterization of recombinant human growth differentiation factor-9 signaling in ovarian granulosa cells. *Molecular and Cellular Endocrinology, 283*, 58–67.

Mulsant, P., Lecerf, F., Fabre, S., Schibler, L., Monget, P., et al. (2001). Mutation in bone morphogenetic protein receptor-IB is associated with increased ovulation rate in Booroola Merino ewes. *Proceedings of the National Academy of Sciences USA, 98*, 5104–5109.

Myllymaa, S., Pasternack, A., Mottershead, D. G., Poutanen, M., Pulkki, M. M., et al. (2010). Inhibition of oocyte growth factors *in vivo* modulates ovarian folliculogenesis in neonatal and immature mice. *Reproduction, 139*, 587–598.

Nekola, M. V. & Nalbandov, A. V. (1971). Morphological changes of rat follicular cells as influenced by oocytes. *Biology of Reproduction, 4*, 154–160.

Nicol, L., Bishop, S. C., Pong-Wong, R., Bendixen, C., Holm, L. E., et al. (2009). Homozygosity for a single base-pair mutation in the oocyte-specific GDF9 gene results in sterility in Thoka sheep. *Reproduction, 138*, 921–933.

Otsuka, F., Moore, R. K., Iemura, S., Ueno, N., & Shimasaki, S. (2001). Follistatin inhibits the function of the oocyte-derived factor BMP-15. *Biochemical and Biophysical Research Communications, 289*, 961–966.

Piper, L. R. & Bindon, B. M. (1982). The Booroola Merino and the performance of medium non-peppin crosses at Armidale. In L. R. Piper (ed.), *The Booroola Merino: Proceedings of a workshop held in Armidale, New SouthWales, 24–25 August 1980* (pp. 9–19). Armidale: CSIRO.

Portela, V. M., Machado, M., Buratini, J., Jr., Zamberlam, G., Amorim, R. L., et al. (2010). Expression and function of fibroblast growth factor 18 in the ovarian follicle in cattle. *Biology of Reproduction, 83*, 339–346.

Prochazka, R., Nagyova, E., Brem, G., Schellander, K., & Motlik, J. (1998). Secretion of cumulus expansion-enabling factor (CEEF) in porcine follicles. *Molecular Reproduction and Development, 49*, 141–159.

Prochazka, R., Nagyova, E., Rimkevicova, Z., Nagai, T., Kikuchi, K., & Motlik, J. (1991). Lack of effect of oocytectomy on expansion of the porcine cumulus. *Journal of Reproduction and Fertility, 93*, 569–576.

Prochazka, R., Nemcova, L., Nagyova, E., & Kanka, J. (2004). Expression of growth differentiation factor 9 messenger RNA in porcine growing and preovulatory ovarian follicles. *Biology of Reproduction, 71*, 1290–1295.

Pulkki, M. M., Mottershead, D. G., Pasternack, A. H., Muggalla, P., Ludlow, H., et al. (2012). A covalently dimerized recombinant human bone morphogenetic protein-15 variant identifies Bone Morphogenetic Protein Receptor Type 1B as a key cell surface receptor on ovarian granulosa cells. *Endocrinology* (in press).

Pulkki, M. M., Myllymaa, S., Pasternack, A., Lun, S., Ludlow, H., et al. (2011). The bioactivity of human bone morphogenetic protein-15 is sensitive to C-terminal modification: characterization of the purified untagged processed mature region. *Molecular and Cellular Endocrinology, 332*, 106–115.

Ralph, J. H., Telfer, E. E., & Wilmut, I. (1995). Bovine cumulus cell expansion does not depend on the presence of an oocyte secreted factor. *Molecular Reproduction and Development, 42*, 248–253.

Romaguera, R., Morato, R., Jimenez-Macedo, A. R., Catala, M., Roura, M., et al. (2010). Oocyte secreted factors improve embryo developmental competence of COCs from small follicles in prepubertal goats. *Theriogenology*.

Salmon, N. A., Handyside, A. H., & Joyce, I. M. (2004). Oocyte regulation of anti-Mullerian hormone expression in granulosa cells during ovarian follicle development in mice. *Developmental Biology, 266*, 201–208.

Salustri, A., Ulisse, S., Yanagishita, M., Hascall, V. C. (1990a). Hyaluronic acid synthesis by mural granulosa cells and cumulus cells *in vitro* is selectively stimulated by a factor produced by oocytes and by transforming growth factor-beta. *The Journal of Biological Chemistry, 265*, 19517–19523.

Salustri, A., Yanagishita, M., & Hascall, V. C. (1990b). Mouse oocytes regulate hyaluronic acid synthesis and mucification by FSH-stimulated cumulus cells. *Developmental Biology, 138*, 26–32.

Sasseville, M., Ritter, L. J., Nguyen, T. M., Liu, F., Mottershead, D. G., et al. (2010). Growth differentiation factor 9 signaling requires ERK1/2 activity in mouse granulosa and cumulus cells. *Journal of Cell Science, 123*, 3166–3176.

Scaramuzzi, R. J., Baird, D. T., Campbell, B. K., Draincourt, M. A., Dupont, J., et al. (2011). Regulation of folliculogenesis and the determination of ovulation rate in ruminants. *Reproduction, Fertility and Development, 23*, 444–467.

Silva, B. D., Castro, E. A., Souza, C. J., Paiva, S. R., Sartori, R., et al. (2011). A new polymorphism in the Growth and Differentiation Factor 9 (GDF9) gene is associated with increased ovulation rate and prolificacy in homozygous sheep. *Animal Genetics, 42*, 89–92.

Simpson, C. M., Stanton, P. G., Walton, K. L., Chan, K. L., Ritter, L. J., et al. (2012). Activation of Latent Human GDF9 by a Single Residue Change (Gly391Arg) in the Mature Domain. *Endocrinology* (Epub).

Singh, B., Zhang, X., & Armstrong, D. T. (1993). Porcine oocytes release cumulus expansion-enabling activity even though porcine cumulus expansion *in vitro* is independent of the oocyte. *Endocrinology, 132*, 1860–1862.

Souza, C. J., MacDougall, C., Campbell, B. K., McNeilly, A. S., & Baird, D. T. (2001). The Booroola (FecB) phenotype is associated with a mutation in the bone morphogenetic receptor type 1 B (BMPR1B) gene. *Journal of Endocrinology, 169*, R1–R6.

Spicer, L. J., Aad, P. Y., Allen, D., Mazerbourg, S., & Hsueh, A. J. (2006). Growth differentiation factor-9 has divergent effects on proliferation and steroidogenesis of bovine granulosa cells. *Journal of Endocrinology, 189*, 329–339.

Su, Y. Q., Denegre, J. M., Wigglesworth, K., Pendola, F. L., O'Brien, M. J., & Eppig, J. J. (2003). Oocyte-dependent activation of mitogen-activated protein kinase (ERK1/2) in cumulus cells is required for the maturation of the mouse oocyte-cumulus cell complex. *Developmental Biology, 263*, 126–138.

Su, Y. Q., Sugiura, K., Wigglesworth, K., O'Brien, M. J., Affourtit, J. P., et al. (2008). Oocyte regulation of metabolic cooperativity between mouse cumulus cells and oocytes: BMP15 and GDF9 control cholesterol biosynthesis in cumulus cells. *Development, 135*, 111–121.

Su, Y. Q., Wu, X., O'Brien, M. J., Pendola, F. L., Denegre, J. N., et al. (2004). Synergistic roles of BMP15 and GDF9 in the development and function of the oocyte-cumulus cell complex in mice: Genetic evidence for an oocyte-granulosa cell regulatory loop. *Developmental Biology, 276*, 64–73.

Sugiura, K. & Eppig, J. J. (2005). Society for Reproductive Biology Founders' Lecture 2005. Control of metabolic cooperativity between oocytes and their companion granulosa cells by mouse oocytes. *Reproduction, Fertility and Development, 17*, 667–674.

Sugiura, K., Pendola, F. L., & Eppig, J. J. (2005). Oocyte control of metabolic cooperativity between oocytes and companion granulosa cells: energy metabolism. *Developmental Biology, 279*, 20–30.

Sugiura, K., Su, Y. Q., Diaz, F. J., Pangas, S. A., Sharma, S., et al. (2007). Oocyte-derived BMP15 and FGFs cooperate to promote glycolysis in cumulus cells. *Development, 134*, 2593–2603.

Sutton, M. L., Cetica, P. D., Beconi, M. T., Kind, K. L., Gilchrist, R. B., & Thompson, J. G. (2003). Influence of oocyte-secreted factors and culture duration on the metabolic activity of bovine cumulus cell complexes. *Reproduction, 126*, 27–34.

Tibaldi, E., Arrigoni, G., Martinez, H. M., Inagaki, K., Shimasaki, S., & Pinna, L. A. (2010). Golgi apparatus casein kinase phosphorylates bioactive Ser-6 of bone morphogenetic protein 15 and growth and differentiation factor 9. *FEBS Letters, 584*, 801–805.

Valve, E., Penttila, T. L., Paranko, J., & Harkonen, P. (1997). FGF-8 is expressed during specific phases of rodent oocyte and spermatogonium development. *Biochemical and Biophysical Research Communications, 232*, 173–177.

van Wezel, I. L., Umapathysivam, K., Tilley, W. D., & Rodgers, R. J. (1995). Immunohistochemical localization of basic fibroblast growth factor in bovine ovarian follicles. *Molecular and Cellular Endocrinology, 115*, 133–140.

Vanderhyden, B. C. (1993). Species differences in the regulation of cumulus expansion by an oocyte-secreted factor(s). *Journal of Reproduction and Fertility, 98*, 219–227.

Vanderhyden, B. C., Caron, P. J., Buccione, R., & Eppig, J. J. (1990). Developmental pattern of the secretion of cumulus expansion-enabling factor by mouse oocytes and the role of oocytes in promoting granulosa cell differentiation. *Developmental Biology, 140*, 307–317.

Vanderhyden, B. C., Cohen, J. N., & Morley, P. (1993). Mouse oocytes regulate granulosa cell steroidogenesis. *Endocrinology, 133*, 423–426.

Vanderhyden, B. C. & Macdonald, E. A. (1998). Mouse oocytes regulate granulosa cell steroidogenesis throughout follicular development. *Biology of Reproduction, 59*, 1296–1301.

Vanderhyden, B. C., Telfer, E. E., & Eppig, J. J. (1992). Mouse oocytes promote proliferation of granulosa cells from preantral and antral follicles *in vitro*. *Biology of Reproduction, 46*, 1196–1204.

Vanderhyden, B. C. & Tonary, A. M. (1995). Differential regulation of progesterone and estradiol production by mouse cumulus and mural granulosa cells by A factor(s) secreted by the oocyte. *Biology of Reproduction, 53*, 1243–1250.

Vitt, U. A., Mazerbourg, S., Klein, C., & Hsueh, A. J. (2002). Bone morphogenetic protein receptor type II is a receptor for growth differentiation factor-9. *Biology of Reproduction, 67*, 473–480.

Walton, K. L., Makanji, Y., Chen, J., Wilce, M. C., Chan, K. L., et al. (2010). Two distinct regions of latency-associated peptide coordinate stability of the latent transforming growth factor-beta1 complex. *The Journal of Biological Chemistry, 285*, 17029–17037.

Wang, Y., Nicholls, P. K., Stanton, P. G., Harrison, C. A., & Sarraj, M., et al. (2009). Extra-ovarian expression and activity of growth differentiation factor 9. *Journal of Endocrinology, 202*, 419–430.

Wilson, T., Wu, X. Y., Juengel, J. L., Ross, I. K., & Lumsden, J. M., et al. (2001). Highly prolific Booroola sheep have a mutation in the intracellular kinase domain of bone morphogenetic protein IB receptor (ALK-6) that is expressed in both oocytes and granulosa cells. *Biology of Reproduction, 64*, 1225–1235.

Yan, C., Wang, P., DeMayo, J., DeMayo, F. J., Elvin, J. A., et al. (2001). Synergistic roles of bone morphogenetic protein 15 and growth differentiation factor 9 in ovarian function. *Molecular Endocrinology, 15*, 854–866.

Yeo, C. X., Gilchrist, R. B., Thompson, J. G., & Lane, M. (2008). Exogenous growth differentiation factor 9 in oocyte maturation media enhances subsequent embryo development and fetal viability in mice. *Human Reproduction, 23*, 67–73.

Zhang, K., Hansen, P. J., & Ealy, A. D. (2010). Fibroblast growth factor 10 enhances bovine oocyte maturation and developmental competence *in vitro*. *Reproduction, 140*, 815–826.

5 MicroRNAs in Oocyte Physiology and Development

Dawit Tesfaye, Md M. Hossain, and Karl Schellander

5.1 Introduction

Advanced genomic analyses have disclosed an enormous repertoire of functional noncoding RNA molecules (ncRNAs). Among the ncRNAs that have revolutionized our understanding of eukaryotic gene expression, microRNAs (miRNAs) have recently been emphasized extensively with massive potential for playing a pivotal role in disease, fertility, and development. They are found to be potentially involved in various aspects of physiological regulation of reproductive tissues and embryonic development in addition to other physiological systems. Molecular and cellular processes required for the gradual oocyte growth throughout folliculogenesis leading to the ovulation of a competent oocyte are dependent on temporal expression and interaction of multiple regulatory gene products. This translational reprogramming during oogenesis, oocyte growth, and maturation involves synthesis and storage of large quantities of dormant mRNAs, spatially restricted protein synthesis, repression of localized mRNAs, and finally activation in time for embryonic axis formation. The precise translational regulation of mRNAs are dictated and controlled by the combination of key elements on the untranslated regions and other post-transcriptional regulatory factors. By the nature of binding to 3′ untranslated region and silencing mRNAs at a post-transcriptional level, miRNAs could be involved in such translational programming, as has been evidenced already. However, this is just beginning to elucidate the role of miRNAs in the cellular mechanisms that impart subsequent oocyte growth and development. This chapter will deal with the recent works regarding miRNAs, focusing on their diverse roles in oocyte physiology and development together with the perspectives for further implications.

5.2 Biogenesis of miRNA

MiRNAs are a class of small non-coding RNAs comprised of 18–24 nucleotides. Most miRNA genes are transcribed by RNA polymerase II (Pol II) to generate a primary transcript called "primary miRNA" (pri-miRNA), which can range in size from several hundred nucleotides (nt) to tens of kilobases (kb) (Cai et al., 2004). Like mRNAs, Pol II transcribed pri-miRNAs contain 5′ cap structures, are polyadenylated, and may be spliced (Bracht et al., 2004; Cai et al., 2004). The tiny (18–24 nt in length) and single-stranded miRNA molecule is derived from this pri-miRNA, which has a RNA hairpin structure of 60–120 nt with a mature miRNA in one of the two strands

Oocyte Physiology and Development in Domestic Animals, First Edition. Edited by Rebecca L. Krisher.

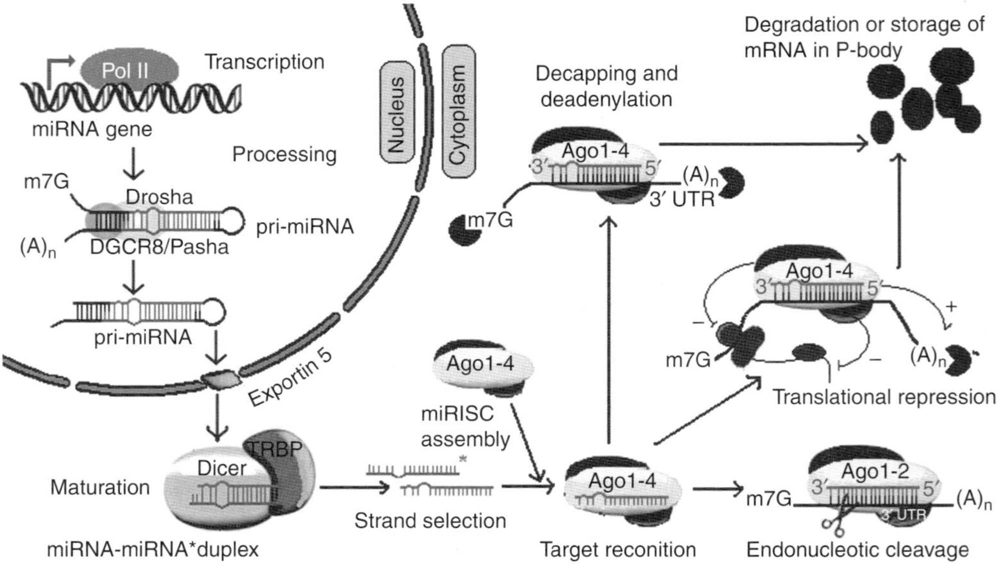

Figure 5.1 Biogenesis of miRNAs and their mechanism of gene regulation. MiRNAs are transcribed intragenic or intergenic chromosomal DNA regions by RNA polymerase II (Pol II) into primary transcripts of primary miRNA (pri-miRNA), which are then subjected to two RNase III endonucleases (Drosha and DGCR8/Pasha) to result in a stem-loop hairpin double stranded structure pre-miRNA. After being transported to the cytoplasm by the enzyme exportin 5, the pre-miRNA will be further processed to a 20–22 nt long miRNA duplex by the enzyme Dicer. Depnding on the sequence similarity with the target mRNA, the mature miRNA stand coupled with the RISC complex lead to the translational repression or endonucleatic cleavage of the target mRNA. Argonaute proteins 1-4 (Ago1-4), inhibition of initiation of translation (−), Promoting deadenylation (+), 7-methyl-G cap (m7G), the passenger strand (miRNA*) (For color detail, see color plate section.)

(Figure 5.1). The pri-miRNA is processed within the nucleus by a multi-core protein complex called the Microprocessor, which is composed of the RNAse III enzyme Drosha and the double-stranded RNA binding domain (dsRBD) protein DGCR8/Pasha (Denli et al., 2004; Gregory et al., 2004; Han et al., 2004; Landthaler, Yalcin, & Tuschl, 2004; Lee et al., 2003). This microprocessor cleaves the pri-miRNA in the nucleus to produce a stem-loop pairpin double strand structure called precursor miRNA (pre-miRNA) (Lee et al., 2002). The resulting pre-miRNA is transported to the cytoplasm via a process that involves binding of Exportin-5 to pre-miRNA through recognizing the 2-nt 3′ overhang produced by RNAse III mediated cleavage (Yi et al., 2003). This pre-miRNA undergoes a cleavage mediated by another RNAse III enzyme, Dicer, which interacts with the dsRBD proteins TRBP to generate the mature ∼22nt miRNA:miRNA* duplex (Chendrimada et al., 2005; Forstemann et al., 2005; Hutvagner et al., 2001; Jiang et al., 2005; Ketting et al., 2001; Lee et al., 2006; Saito et al., 2005; Lee et al., 2003). This duplex comprises the mature miRNA guide strand and the miRNA* passenger strand. The mature miRNA strand is loaded into the RNA-induced silencing complex (RISC) whereas the passenger strand is degraded (Figure 5.1). Subsequently, TRBP/Loquacious recruits the argonaute (Ago) protein and together with Dicer they form a complex that initiates the assembly of the RISC, a ribonucleoprotein (RNP) complex (Gregory et al., 2005; Maniataki & Mourelatos, 2005).

Recently, it has been proposed that an alternative pathway is also involved in microRNA biogenesis from the pre-miRNA hairpin to the mature functional miRNA (Diederichs & Haber 2007). In this pathway, Ago2 has been found as a highly specialized member of the argonaute family with an essential nonredundant Dicer-independent function within the mammalian miRNA pathway (O'Carroll et al., 2007). Ago2-mediated pre-miRNA cleavage has been observed in the processing of miRNAs derived from the 5′-arm of the pre-miRNA hairpin having no mismatches at the immediate cleavage site (Diederichs & Haber, 2007; Han et al., 2006). The product of Ago2-mediated pre-miRNA cleavage (ac-pre-miRNA) was found to be an intrinsic on-pathway intermediate during miRNA biogenesis and hence a substrate to Dicer, or it could be a by-product that cannot be further processed toward the mature miRNA. These data identify the ac-pre-miRNA as a pathway intermediate in miRNA biogenesis that is generated from the pre-miRNA by Ago2 and serves as a substrate for Dicer to mature into the active miRNA. Some microRNAs are reported to be processed without the Dicer enzyme in which the pre-miRNA will be loaded into Argonaute (Ago) and is cleaved by the Ago catalytic center to generate an intermediate 3′ end, which is then further trimmed and processed into mature miRNA (Cheloufi et al., 2010). So, Ago2-mediated cleavage of pre-miRNAs, followed by uridylation and trimming, generates functional miRNAs independently of Dicer (Cifuentes et al., 2010). Based on the relative stability of the two ends of the duplex, the miRNA strand, with relatively unstable base pairs at the 5′ end, remains incorporated into the RISC, whereas the passenger strand, or miRNA* strand, is degraded (Leuschner & Martinez, 2007; Matranga et al., 2005; Schwarz et al., 2003).

5.3 Recognition and Post-Transcriptional Regulation of Target mRNA by miRNA

miRNAs, which are incorporated within the RISC complex, direct RISC to downregulate expression of target mRNAs containing complementary sequences in the 3′ untranslated region to a 7- to 8-nt region of the 5′ end of miRNA called the seed sequence (Pillai et al., 2005) (Figure 5.1). Depending on the degree of complementarity between the miRNA and the target sequence, mRNAs are either cleaved or degraded (perfect or near perfect complementarity) or their translation is repressed (imprecise complementarity) (Hutvagner & Zamore, 2002; Martinez & Tuschl, 2004). The microRNA ribonucleoprotein (miRNP) complex that is loaded onto the target mRNA exhibits a direct or indirect effect in translational repression. Direct effects occur either through inhibition of initiation of translation through binding of Ago2 to m^{7}G (7-methyl-G cap) resulting in prevention of ribosome association with the target mRNA, or through inhibition of translation post-initiation, which includes premature ribosome drop-off, slowed or stalled elongation, and co-translational protein (Figure 5.1).

In addition to direct effects on translation (or protein accumulation), miRNPs can have other effects on targeted mRNAs, including promoting deadenylation, which might result in degradation (increased turnover) (Nilsen, 2007). Recent reports have also indicated that miRNA, with or without perfect sequence complementarity, can cause an increase in mRNA degradation by endonucleolytic cleavage or deadenylation respectively (Figure 5.1) (Jackson & Standart, 2007), or changes in proteins associated with RISC can cause a shift from translational inhibition to translational enhancement (Vasudevan, Tong, & Steitz 2007; Orom, Nielsen, & Lund, 2008). Translational repression and/or deadenylation occurs followed by decapping and exonuclease mediated degradation if base-pairing is partially complementary, or in the case of perfect complementarity, and provided the miRNP contains specifically Ago2, may result in endonucleolytic cleavage of the mRNA at the site where the miRNA is annealed (Standart & Jackson, 2007). Those mRNAs that are repressed by miRNAs are further stored in cytoplasmic foci called P-bodies (Liu, Valencia-Sanchez et al., 2005; Rehwinkel et al., 2005; Liu, Rivas et al., 2005). miRNAs have been found to play an integral part of animal gene regulatory networks as one of the most abundant classes of gene regulators. Roughly

30% of all animal genes are predicted to be targeted by miRNAs. An algorithm that attempts to identify miRNA target sites without relying on cross-species conservation or miRNA sequences (Miranda et al., 2006) predicts even higher numbers of miRNA regulated genes.

5.4 miRNA in Germ Cell Differentiation and Oogenesis

Primordial germ cells (PGCs) become germline stem cells (GSCs), and their differentiating products undergo gametogenesis. Asymmetrical division and self-renewal of GSCs for oogenesis requires both intrinsic signaling mechanisms and extrinsic signals from neighboring cells. PGCs are found to have many factors required for the processing and functioning of mature miRNAs including Dicer and members of the Argonaute family (eIF2C) (Kaneda et al., 2009; Tang et al., 2007; Harris & Macdonald 2001; Hayashi et al., 2008). Emerging evidence suggests that miRNA-mediated translational regulation may control different stages of germline development and oogenesis in mammals. A critical function for Dicer has been demonstrated in regulating mouse oogenesis by targeted disruption of Dicer in growing oocytes (Murchison et al., 2007; Tang et al., 2007). Oocytes lacking Dicer were found to be unable to complete meiosis and arrested with multiple defects of meiotic spindle organization and chromosome congression. These studies revealed that miRNAs could be one of the factors that are involved in turnover of many maternal transcripts whose degradation is essential for successful meiotic maturation. Oocytes after germline-specific knockouts of Dicer were found to be arrested in meiosis I, and displayed defective chromosomal segregation as well as disorganized spindles, including multiple spindles (40%), misaligned chromosomes (80%), apparent monopolar spindle attachments, (affecting either single or many chromosomes; 70%), decondensed chromatin, and anaphase bridges (3%) (Murchison et al., 2007). It has been assumed that the defects could be either due to the loss of centromeric repeat-derived miRNAs, which could prevent the establishment of appropriate centromeric chromatin structure for assembly of the kinetochore, or the loss of miRNAs leading to deregulation of essential gene products. Furthermore, analysis of the transcripts found in developing oocytes lacking Dicer were found to be enriched for genes implicated in microtubule-related processes, including the predicted miRNA target gene whose activity is known to be directly regulated by the condensation state of chromosomes (Murchison et al., 2007).

In another study, ablation of Dicer in the PGCs shows the microRNA pathway to be essential for PGCs' proliferation. Interestingly, several conserved miRNA clusters (miR-17–92 cluster and miR-290–295 cluster) important for promoting cell cycle were found to be enriched in PGCs, whereas the expression of other miRNAs (miR-141, miR-200a, miR-200c and miR-323) were found to be progressively decreased as PGC development proceeds. In addition, preferential expression of let-7a, let-7d, let-7e, let-7f, let-7g, miR-125a, and miR-9 has been observed in male PGCs compared to that in female ones over the course of embryonic development (Hayashi et al., 2008). Recently, conditional deletion of Argonaute2 (Ago2) in the developing mouse oocyte also revealed abnormal spindles and misaligned chromosomes with the ability to develop mature oocytes similar to the phenotype of Dicer-deficient oocytes (Kaneda et al., 2009). This study suggests that Ago2 plays a key function in the mouse oocyte through global regulation of miRNA stability affecting gene expression in developing oocytes. Until recently, it was clear that the most common findings from studies on the effect of disrupting global miRNA biosynthesis in the oocyte by deletion or knocking out processing genes (Dicer1, Dgcr8, or Ago2) are defective spindles and chromosomal alignment. It is obvious that all the miRNAs or genes are not contributing to this outcome, but functional analysis of selected candidate miRNAs is still missing to elucidate the specific role and mechanism of miRNA-mediated assembly of meiotic spindles and the establishment of proper meiotic spindle–chromosome interactions as major cytological events involved in various aspects of meiotic processes, including positioning of the cell division plane during cytokinesis.

5.5 Expression and Regulation of miRNA in Oocyte Development

Mammalian oocyte development begins during fetal life when primordial germ cells proliferate by mitosis, and DNA synthesis proceeds and leads the oogonia to transform into primary oocytes. The oocytes progress into the first meiotic prophase consisting of several transient stages. On reaching the diplotene stage of meiosis, the oocyte becomes surrounded by a single layer of 4–8 pre-granulosa cells and an intact basal lamina to form the resting primordial follicle (Fair, 2003). Once a primordial follicle oocyte is activated to grow, it embarks on a complex passage involving numerous molecular and morphological changes to the oocyte and surrounding follicular cells. In general, oocyte growth and development are the result of various cellular and molecular changes occurring throughout ovarian follicular development and controlled by closely coordinated endocrine and paracrine factors contributed by both the oocyte and surrounding somatic cells. All of these processes are controlled by tightly regulated expression and interaction of a multitude of genes in different compartments of the ovary (oocyte, granulosa, and theca cells) to attain oocyte development (Bonnet, Dalbies-Tran, & Sirard 2008), and are likely regulated by miRNAs.

Microarray experiments show that Dicer1 is highly expressed and functionally important in oocytes during folliculogenesis, as well as in mature oocytes (Su et al., 2002; Choi et al., 2007; Murchison et al., 2007). Moreover, transcriptional analysis through microarray experiments has demonstrated that the majority of transcripts are misregulated in Dicer1-deficient oocytes. The role of Dicer1 in mouse ovarian follicle development was explored in another study by using Dicer1 conditional knockout (cKO) mouse ovarian tissue. Subsequently, follicular development and its related gene expression were assessed comparatively between wild-type (WT) and Dicer1 cKO mouse ovaries (Lei et al., 2010). This study revealed an important role of the Dicer1-regulated miRNA signaling pathway in mouse follicular development through the regulation of follicle cell proliferation, differentiation, and apoptosis. Recently we have localized the expression of Dicer1 and Ago2 in different stages of bovine whole mount ovarian follicles to understand stage-specific miRNA-mediated regulation in oocyte development (Figure 5.2). This study reveals their constant expression during progression of follicular development.

Several studies highlight the expression and regulation of miRNAs in oocytes. The first attempt to clone individual miRNAs was made in 2006, and identified a number of miRNAs as well as some other small noncoding RNAs (rasiRNAs, gsRNAs) in mouse oocytes (Watanabe et al., 2006). However, further identification of miRNAs in oocytes through direct cloning methodology is lacking; rather, greater initiative has been seen in microarray or RT-PCR based miRNAs detection through homologous or heterologous approaches. Compared to other species, human miRNA sequence register in miRbase is much advanced with only few miRNA entries for species like sheep. As miRNAs showed a high degree of conservation between various species, heterologous approaches i.e for example the use of human PCR array platforms for investigation of the expression of miRNA homologs in other species, with only few known miRNA sequences, can be an option to identify new miRNAs and investigate expression profile of the already known ones. For example, the differential expression of miRNAs has been identified during bovine oocyte maturation *in vitro* using the heterologous approach (Tesfaye et al., 2009).

Reported miRNAs that are abundantly expressed in the oocyte as well as other follicular cells are presented in Figure 5.3. All these miRNAs were profiled for their expression in different

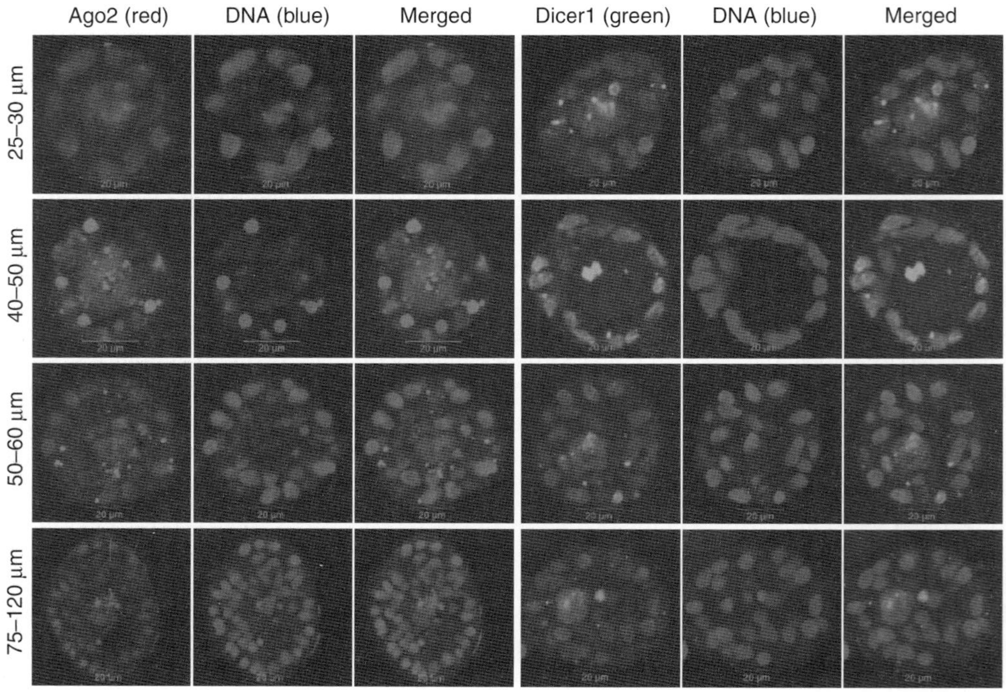

Figure 5.2 Expression of Dicer1 (green) and Ago2 (red) at different stages of bovine ovarian follicles categorized based on their size. Both Ago2 and Dicer are found to be expressed in both granulose and theca cells of the follicle at different stages of follicular development. (For color detail, see color plate section.)

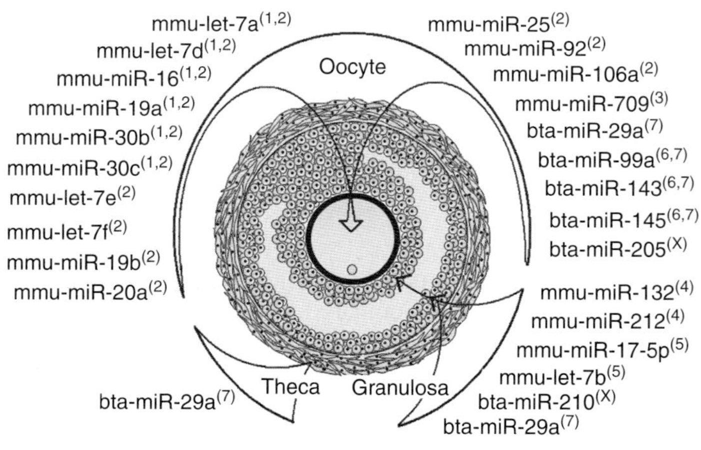

Figure 5.3 Highly expressed miRNAs in oocytes and other follicular cells in different species. 1-(Murchison et al. 2007), 2-(Tang et al. 2007), 3-(Choi et al. 2007), 4-(Fiedler et al. 2008), 5-(Otsuka et al. 2008), 6-(Tesfaye et al. 2009), 7-(Hossain et al. 2009), X-(from own unpublished data).

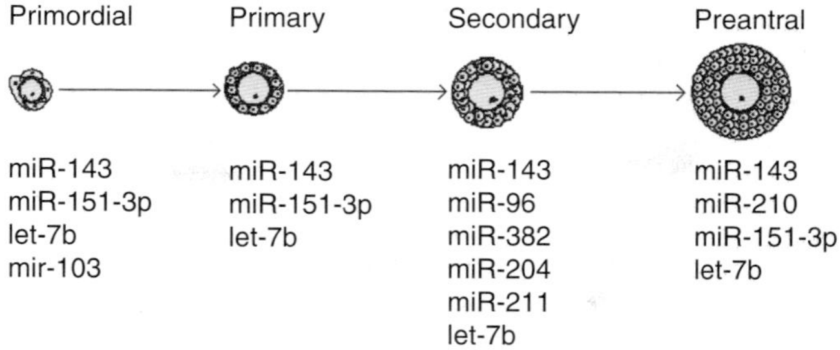

Figure 5.4 MiRNAs predominantly expressed throughout bovine follicular development stages (own unpublished data). Expression analysis of miRNAs in follicles of various developmental stages in bovine showed preferential expression of candidate miRNAs in stage-specific manner.

stages of development. We have identified several miRNAs that are preferentially expressed at specific follicular developmental stages in theca and granulosa cells (Figure 5.4). Constant and abundant expression of let-7b and miR-143 has been observed across different stages of preantral bovine follicle development. This suggests that miR-143 and let-7b might be associated with the development of the oocyte during the preantral stage. Where as, exclusive expression of miR-103 in the primordial stage; miR-96, miR-382, miR-204, and miR-211 in the secondary stage; and miR-210 in the tertiary stage indicates their stage-specific roles, and specifically the transition of one stage to another throughout oocyte growth as well as proliferation of granulosa cells. However, the mechanism and characterization of their targets for the proposed functions need to be further elucidated. Although all these efforts provide initial evidence for the role of miRNAs in oocyte development, in-depth functional analysis of individual miRNAs is still required.

5.6 miRNAs in Oocyte Maturation and Competence

The final development of an oocyte to be competent for fertilization usually occurs during the process of meiosis and is referred to as oocyte maturation. Before maturation, the oocyte contains a large germinal vesicle (GV) with decondensed, dispersed, and transcriptionally active chromosomes (Smith & Richter, 1985). With the initiation of maturation, transcription ceases, the chromosomes begin to condense, the GV breaks down, and nucleoli disperse (Masui & Clarke, 1979). As maturation progresses, the paired homologous chromosomes align in the middle of the forming meiotic spindle during metaphase I. Separation of the paired homologous chromosomes is followed by the first polar body formation. Afterward, the chromosomes in the oocyte are again arranged on a meiotic spindle at metaphase II. With the second meiotic division, chromatids separate and the second polar body is formed (Voronina & Wessel, 2003). Advances in molecular analysis revealed that miRNA could be involved in different processes of oocyte maturation. The importance of miRNA in the maturation of oocytes has initially been demonstrated in Dicer knockout or knockdown studies, and it has become clear that Dicer is required for meiotic spindle integrity and completion of meiosis I (Murchison et al., 2007). Depleting Dicer1 in oocytes resulted in meiotic failure and revealed that Dicer1-conditioned ooplasm of primary oocytes contains Dicer1-dependent factors that are crucial for chromosome segregation and for the progression of metaphase to anaphase during meiotic maturation (Mattiske et al., 2009). More recently, Dicer knockdown in mouse oocytes resulted in a significant reduction in oocyte maturation with an increase in abnormal spindles and chromosomal disorganization (Liu et al., 2010). In addition, transcripts related to spindle formation proteins (plk1 and AURKA) and spindle check points (Bub1, Bublb) were found to be reduced remarkably. Thus, Dicer appear to play an important role during oogenesis and are essential for meiotic completion through regulation of miRNA biogenesis. However, it is likely that several biological factors along with miRNAs are being affected by Dicer1 knockout or knockdown so as to result in defective meiotic maturation.

Recent investigation on the expression of miRNAs in immature and *in vitro* matured bovine oocytes using a heterologous approach revealed differential expression of 59 miRNAs (Tesfaye et al., 2009). Among them 31 and 28 miRNAs were found to be preferentially expressed in immature and matured oocytes respectively. The expression of seven microRNAs (miR-496, miR-297, miR-292-3P, miR-99a, miR-410, miR-145, and miR-515-5p) in matured and two microRNAs (miR-512-5p and miR-214) in immature oocytes showed higher abundance by at least a twofold difference (Tesfaye et al., 2009). miR-496 (5.2-fold change) and miR-512-5p (2.3-fold change) revealed the maximum upregulated fold changes in matured and immature oocytes respectively. Similarly, miR-145 has been identified to have the most altered expression during oocyte maturation, where it was more abundant in mature bovine oocytes. In addition, expression of miRNAs in fully grown and growing oocytes (selected by Brilliant Cresyl Blue (BCB) staining) by qRT-PCR analysis revealed higher expression of miRNAs in BCB (−) (growing oocytes) as compared to BCB (+) (fully grown oocytes). Further study on miRNA itself rather than on Dicer is needed to elucidate specific miRNA-mediated regulation of oocyte maturation.

5.7 miRNAs as Temporal Regulatory Cascades of Maternal mRNA Translation

Oocytes contain all the messenger RNAs necessary to establish a new life upon transition to the fertilized egg, which essentially involves many changes, including protein synthesis and protein and RNA degradation, as well as organelle remodeling (Stitzel & Seydoux, 2007). These changes take place concomitantly with the meiotic divisions that construct the haploid maternal genome. It is well-recognized that the regulation of maternal messages during meiotic maturation requires exquisite control of mRNA stability and translation. Maternal deposits of mRNAs during oogenesis are translated before the onset of zygotic transcription. miRNAs have also been found to be maternally deposited in flies and mammals (Tang et al., 2007; Liu et al., 2007). It is suggested that maternally deposited miRNAs are needed for zygotic function in the mouse (Tang et al., 2007). When oocytes were depleted of miRNA by ablating Dicer in the germline, the resultant zygotes failed to complete the first cell division. However, this defect could be due to either a lack of miRNA-mediated translational control or disorganization of the oocyte caused by lack of Dicer during oogenesis. Many miRNAs are present in mature Dicer + oocytes (Tang et al., 2007). In addition, Dicer is central to a regulatory network that controls oocyte gene expression, as well as essential turnover of a considerable subset of maternal transcripts that are normally degraded during oocyte maturation to promote genomic integrity (Murchison et al., 2007). The degree of contribution of miRNAs to translational repression in oocytes is not yet well understood. A recent analysis of thousands of proteins in the oocyte of Drosophila revealed only 4% of them to be increased in abundance in Dicer mutants (Nakahara et al., 2005). Another study discovered a cluster of miRNAs (miR-309) that destabilizes a large number of putative target maternal mRNAs in Drosophila (Bushati et al., 2008). In the absence of this miRNA cluster about 410 maternal mRNAs were upregulated. A similar study suggested the same phenomena, where miRNAs are found to promote the deadenylation of target mRNAs, thereby establishing the link with the RNA degradation machinery (Giraldez et al.,

2006). This topic has attracted much attention, and more discoveries on miRNA-mediated regulatory cascades of maternal mRNA translation are expected in the years to come.

5.8 miRNAs in Oocyte Development in Relation to Endocrine Control

The growth and differentiation of the oocyte during follicular development are regulated by the activity of granulosa cells, whose functions are in turn regulated by various hormones and other secreted factors. Several mechanisms for triggering oocyte maturation have been proposed, including hormones as well-documented stimulators. Steroid hormones induce oocyte maturation in a number of vertebrate species. In mammals, hormones affecting oocyte development exert their action on somatic cells of the follicle, which then relay the signal to the oocyte. The pituitary gland releases the gonadotropins FSH and LH. FSH causes the follicle cells to proliferate and differentiate, and LH initiates the oocyte's progression through meiosis and ovulation (Voronina & Wessel, 2003).

Several studies reveal an interesting relationship between ovarian steroids and miRNAs. LH/hCG regulates the expression of selected miRNAs, which affect post-transcriptional gene regulation in mouse within ovarian granulosa cells (Fiedler et al., 2008). Estrogen was found to suppress the level of a set of miRNAs in mice and human cultured cells through estrogen receptor α (ERα) by associating with the Drosha complex and preventing the conversion of pri-miRNAs into pre-miRNAs (Yamagata et al., 2009). Several miRNAs play an important role in ovarian steroidogenesis (Sirotkin et al., 2009). Genome-wide screening of miRNAs revealed the involvement of miRNAs in controlling release of the ovarian steroid hormones progesterone, androgen, and estrogen in human ovarian cells (Sirotkin et al., 2009). They evaluated the effect of transfection of cultured primary ovarian granulosa cells with gene constructs encoding the majority of identified human pre-miRNAs on release of progesterone, testosterone, and estradiol. Accordingly, although 36 of 80 tested miRNA constructs inhibited, 10 miRNAs promoted progesterone release in these granulosa cells. Subsequent transfection of cells with antisense constructs to two selected miRNAs (mir-15a and mir-188) revealed induction of progesterone output due to a lack of blockage of progesterone release. Although 57 tested miRNAs were found to inhibit testosterone release, only one miRNA (mir-107) enhanced testosterone output. Fifty-one miRNAs suppressed estradiol release, whereas none of the 80 miRNAs tested were found to stimulate it (Sirotkin et al., 2009). However, the complex regulatory mechanisms of the interrelationship between miRNAs and the steroids are still unclear. The involvement of miRNAs in ovarian endocrine regulation to mediate oocyte growth and maturation could be very interesting in future studies.

5.9 miRNA Regulation of Epigenetic Mechanisms in the Oocyte

The term "epigenetics" refers to all heritable changes in gene expression that are not associated with concomitant alterations in the DNA sequence. Reversible DNA methylation and histone modifications are known to have profound effects on controlling gene expression. Correct DNA methylation patterns are paramount for the generation of functional gametes. Progression of meiosis in mammalian oocytes entails a complex and dynamic epigenetic mechanism (De La Fuente et al., 2010). Evidence suggests that chromatin remodeling in the germline is essential to modulate chromosome structure (Matsui & Hayashi, 2007; Surani et al., 2007). Differentiation of chromatin structure and function during postnatal oocyte growth is also critical for the acquisition of meiotic and developmental competence (De La Fuente, 2006). Remarkably, different noncoding RNAs including miRNAs have recently become known as one of the key regulators of chromatin and other epigenetic processes (De La Fuente et al., 2010). Conditional loss-of-function for Dicer in mammalian cells demonstrates a critical role in the epigenetic regulation of heterochromatin function through the induction of specific histone modifications (Fukagawa et al., 2004; Kanellopoulou et al., 2005). Dicer has been found to induce histone methylation in somatic and embryonic stem cells (reviewed in De La Fuente et al., 2010). This post-translational modification may directly or indirectly affect the patterns of DNA methylation by affecting the binding of DNA methyltransferase enzymes (Goldberg, Allis, & Bernstein 2007). The relation of miRNA and epigenetics is presently being elucidated. Much less is known about the specific miRNAs and their targets that regulate epigenetic machinery, or epigenetic regulation of specific miRNAs that are required for normal physiology and development of oocytes, but this area of research is rapidly moving forward.

5.10 Strategic Approaches and Challenges to Study the Role of miRNAs in Oocytes

Identification of the entire set of miRNAs in oocytes through cloning and advanced sequencing is the first approach, which is of paramount importance for further functional studies. Extensive expression profiling of miRNAs concerning or targeting specific processes involved in oocyte growth, development, maturation, and competence could be the easiest way to functional screening of miRNAs. Later on, this information could be applied to the *in vivo* or *in vitro* condition. In that case, specific candidate miRNAs could be studied in detail. The experimental approach will move away from the concept of knocking out hundreds of miRNAs by disrupting Dicer, to knocking out single candidate miRNAs or a family of miRNAs to study a specific phenotypic effect. However, it is not simple to carry out such experiments *in vivo* in all species, particularly in large animals where time and several technical problems are involved. In addition, knowledge about the function of specific miRNAs from mouse knockout models cannot be systematically applied to human and ruminants. *In vitro* models for whole follicle culture could be utilized, and would be an excellent approach for elucidating such miRNA-mediated regulation that might overcome the technical difficulties inherent in *in vivo* experiments in large animals. In that case targeted knockdown of specific miRNA (loss of function) by transfection (or by other suitable methods) of miRNA inhibitor molecules against specific miRNA, or forced overexpression of specific miRNA (gain of function) by using miRNA mimics/precursor molecules could be very useful. Thereafter, it might be possible to draw a fine description of the role of specific miRNA in the molecular mechanisms of the dynamic processes occurring during oocyte growth, development, and maturation. In such an *in vitro* model, it will be essential to support and consider the complex structure and composition of the follicle, where different follicular cells are functionally related and constantly changing and differentiating. More important, experiments should be designed in consideration of oocyte growth and maturation time as well as ordinary global changes of transcripts, including either synthesis of hundreds of new products or disappearance of many transcripts and proteins during oocyte growth and maturation after germinal vesicle breakdown. Moreover, the large number of target genes for a single miRNA and multiple miRNAs targeting the expression of one gene will be one of the major challenges in the assessment of the role of specific miRNAs and establishment of precise miRNA-target networks. However, we are confident that such a multiplicity of complementary approaches and tools will most likely allow in the very near future a better understanding of the role of miRNAs, and will significantly contribute to the characterization of the miRNA-mediated cellular and molecular basis of oocyte physiology and development.

5.11 Concluding Remarks

As an important regulator of genes, miRNAs have been shown to be involved in oocyte physiology, as has been demonstrated in other physiological processes and disease conditions. Although the miRNA-mediated mechanisms controlling oocyte physiology are not fully understood, a growing body of evidence from Dicer and Ago2 studies, as well as identification or expression profiling of miRNAs, suggests the importance of miRNAs in different processes of oocyte growth, development, and maturation. To date, much of the work on miRNAs is focused on expression profiling rather than their regulation and functional characterization. However, this area of research is quickly moving forward. We anticipate rapid discovery of the specifics of microRNA function through modern approaches and tools in the very near future. This will be an interesting avenue to assist in our understanding of oocyte physiology, and answer numerous questions regarding the complex biological processes involved.

References

Bonnet, A., Dalbies-Tran, R., & Sirard, M. A. (2008). Opportunities and challenges in applying genomics to the study of oogenesis and folliculogenesis in farm animals. *Reproduction, 135*(2), 119–128.

Bracht, J., Hunter, S., Eachus, R., Weeks, P., & Pasquinelli, A. E. (2004). Trans-splicing and polyadenylation of let-7 microRNA primary transcripts. *RNA, 10*(10), 1586–1594.

Bushati, N., Stark, A., Brennecke, J., & Cohen, S. M. (2008). Temporal reciprocity of miRNAs and their targets during the maternal-to-zygotic transition in Drosophila. *Current Biology, 18*(7), 501–506.

Cai, X., Hagedorn, C. H., & Cullen, B. R. (2004). Human microRNAs are processed from capped, polyadenylated transcripts that can also function as mRNAs. *RNA, 10*(12), 1957–1566.

Cheloufi, S., Dos Santos, C. O., Chong, M. M., & Hannon, G. J. (2010). A dicer-independent miRNA biogenesis pathway that requires Ago catalysis. *Nature, 465*(7298), 584–589.

Chendrimada, T. P., Gregory, R. I., Kumaraswamy, E., Norman, J., Cooch, N. et al. (2005). TRBP recruits the Dicer complex to Ago2 for microRNA processing and gene silencing. *Nature, 436*(7051), 740–744.

Choi, Y., Qin, Y., Berger, M. F., Ballow, D. J., Bulyk, M. L., & Rajkovic, A. (2007). Microarray analyses of newborn mouse ovaries lacking Nobox. *Biology of Reproduction, 77*(2), 312–319.

Cifuentes, D., Xue, H., Taylor, D. W., Patnode, H., Mishima, Y. et al. (2010). A novel miRNA processing pathway independent of Dicer requires Argonaute2 catalytic activity. *Science, 328*(5986), 1694–1698.

De La Fuente, R. (2006). Chromatin modifications in the germinal vesicle (GV) of mammalian oocytes. *Developmental Biology, 292*(1), 1–12.

De La Fuente, R., Baumann, C., Yang, F., & Viveiros, M. M. (2010). Chromatin remodelling in mammalian oocytes. In M. H. Verlhac & A. Villeneuve (eds.), *Oogenesis: The Universal Process*. West Sussex: John Wiley & Sons Ltd.

Denli, A. M., Tops, B. B., Plasterk, R. H., Ketting, R. F., & Hannon, G. J. (2004). Processing of primary microRNAs by the Microprocessor complex. *Nature, 432*(7014), 231–235.

Diederichs, S., & Haber, D. A. (2007). Dual role for argonautes in microRNA processing and posttranscriptional regulation of microRNA expression. *Cell, 131*(6), 1097–1108.

Fair, T. (2003). Follicular oocyte growth and acquisition of developmental competence. *Anim Reproductive Science, 78*(3–4), 203–216.

Fiedler, S. D., Carletti, M. Z., Hong, X., & Christenson, L. K. (2008). Hormonal regulation of MicroRNA expression in periovulatory mouse mural granulosa cells. *Biology of Reproduction, 79*(6), 1030–1037.

Forstemann, K., Tomari, Y., Du, T., Vagin, V. V., Denli, A. M. et al. (2005). Normal microRNA maturation and germ-line stem cell maintenance requires Loquacious, a double-stranded RNA-binding domain protein. *PLoS Biology, 3*(7), e236.

Fukagawa, T., Nogami, M., Yoshikawa, M., Ikeno, M., Okazaki, T. et al. (2004). Dicer is essential for formation of the heterochromatin structure in vertebrate cells. *Nature Cell Biology, 6*(8), 784–791.

Giraldez, A. J., Mishima, Y., Rihel, J., Grocock, R. J., Van Dongen, S. et al. (2006). Zebrafish MiR-430 promotes deadenylation and clearance of maternal mRNAs. *Science, 312*(5770), 75–79.

Goldberg, A. D., Allis, C. D., & Bernstein, E. (2007). Epigenetics: A landscape takes shape. *Cell, 128*(4), 635–638.

Gregory, R. I., Chendrimada, T. P., Cooch, N., & Shiekhattar, R. (2005). Human RISC couples microRNA biogenesis and posttranscriptional gene silencing. *Cell, 123*(4), 631–640.

Gregory, R. I., Yan, K. P., Amuthan, G., Chendrimada, T., Doratotaj, B. et al. (2004). The Microprocessor complex mediates the genesis of microRNAs. *Nature, 432*(7014), 235–240.

Han, J., Lee, Y., Yeom, K. H., Kim, Y. K., Jin, H., & Kim, V. N. (2004). The Drosha-DGCR8 complex in primary microRNA processing. *Genes Development, 18*(24), 3016–3027.

Han, J., Lee, Y., Yeom, K. H., Nam, J. W., Heo, I. et al. (2006). Molecular basis for the recognition of primary microRNAs by the Drosha-DGCR8 complex. *Cell, 125*(5), 887–901.

Harris, A. N., & Macdonald, P. M. (2001). Aubergine encodes a Drosophila polar granule component required for pole cell formation and related to eIF2C. *Development, 128*(14), 2823–2832.

Hayashi, K., Chuva de Sousa Lopes, S. M., Kaneda, M., Tang, F. Hajkova, P. et al. (2008). MicroRNA biogenesis is required for mouse primordial germ cell development and spermatogenesis. *PLoS One, 3*(3), e1738.

Hossain, M. M., Ghanem, N., Hoelker, M., Rings, F., Phatsara, C. et al. 2009. Identification and characterization of miRNAs expressed in the bovine ovary. *BMC Genomics, 10*(1), 443.

Hutvagner, G., McLachlan, J., Pasquinelli, A. E., Balint, E., Tuschl, T., & Zamore, P. D. (2001). A cellular function for the RNA-interference enzyme Dicer in the maturation of the let-7 small temporal RNA. *Science, 293*(5531), 834–838.

Hutvagner, G., & Zamore, P. D. (2002). A microRNA in a multiple-turnover RNAi enzyme complex. *Science, 297*(5589), 2056–2060.

Jackson, R. J., & Standart, N. (2007). How Do MicroRNAs Regulate Gene Expression? *Science STKE 2007, 367*(367), re1.

Jiang, F., Ye, X., Liu, X., Fincher, L., McKearin, D., & Liu, Q. (2005). Dicer-1 and R3D1-L catalyze microRNA maturation in Drosophila. *Genes Development, 19*(14), 1674–1679.

Kaneda, M., Tang, F., O'Carroll, D., Lao, K., & Surani, M. A. (2009). Essential role for Argonaute2 protein in mouse oogenesis. *Epigenetics Chromatin, 2*(1), 9.

Kanellopoulou, C., Muljo, S. A. Kung, A. L., Ganesan, S. Drapkin, R. et al. (2005). Dicer-deficient mouse embryonic stem cells are defective in differentiation and centromeric silencing. *Genes Development, 19*(4), 489–501.

Ketting, R. F., Fischer, S. E., Bernstein, E., Sijen, T., Hannon, G. J., & Plasterk, R. H. (2001). Dicer functions in RNA interference and in synthesis of small RNA involved in developmental timing in C. elegans. *Genes Development, 15*(20), 2654–2659.

Landthaler, M., Yalcin, A., & Tuschl, T. (2004). The human DiGeorge syndrome critical region gene 8 and Its D. melanogaster homolog are required for miRNA biogenesis. *Current Biology, 14*(23), 2162–2167.

Lee, Y., Ahn, C., Han, J., Choi, H., Kim, J. et al. (2003). The nuclear RNase III Drosha initiates microRNA processing. *Nature, 425*(6956), 415–419.

Lee, Y., Hur, I., Park, S. Y., Kim, Y. K., Suh, M. R., & Kim, V. N. (2006). The role of PACT in the RNA silencing pathway. *EMBO Journal, 25*(3), 522–532.

Lee, Y., Jeon, K., Lee, J. T., Kim, S., & Kim, V. N. (2002). MicroRNA maturation: Stepwise processing and subcellular localization. *EMBO Journal, 21*(17), 4663–4670.

Lee, Y., Kim, M., Han, J., Yeom, K. H., Lee, S. et al. (2004). MicroRNA genes are transcribed by RNA polymerase II. *EMBO Journal, 23*(20), 4051–4060.

Lei, L., Jin, S., Gonzalez, G., Behringer, R. R., & Woodruff, T. K. (2010). The regulatory role of Dicer in folliculogenesis in mice. *Molecular and Cellular Endocrinology, 315*(1–2), 63–73.

Leuschner, P. J., & Martinez, J. (2007). In vitro analysis of microRNA processing using recombinant Dicer and cytoplasmic extracts of HeLa cells. *Methods, 43*(2), 105–109.

Liu, H. C., Tang, Y., He, Z., & Rosenwaks, Z. (2010). Dicer is a key player in oocyte maturation. *Journal of Assistive Reproduction and Genetics, 27*(9–10), 571–580.

Liu, J., Rivas, F. V., Wohlschlegel, J., Yates, J. R., Parker, R., & Hannon, G. J. (2005). A role for the P-body component GW182 in microRNA function. *Nature Cell Biology, 7*(12), 1261–1266.

Liu, J., Valencia-Sanchez, M. A., Hannon, G. J., & Parker, R. (2005). MicroRNA-dependent localization of targeted mRNAs to mammalian P-bodies. *Nature Cell Biology, 7*(7), 719–723.

Liu, X., Park, J. K., Jiang, F., Liu, Y., McKearin, D., & Liu, Q. (2007). Dicer-1, but not Loquacious, is critical for assembly of miRNA-induced silencing complexes. *RNA, 13*(12), 2324–2329.

Maniataki, E., & Mourelatos, Z. (2005). A human, ATP-independent, RISC assembly machine fueled by pre-miRNA. *Genes Development, 19*(24), 2979–2990.

Martinez, J., & Tuschl, T. (2004). RISC is a 5′ phosphomonoester-producing RNA endonuclease. *Genes Development, 18*(9), 975–980.

Masui, Y., & Clarke, H. J. (1979). Oocyte maturation. *International Review of Cytology, 57*, 185–282.

Matranga, C., Tomari, Y., Shin, C., Bartel, D. P., & Zamore, P. D. (2005). Passenger-strand cleavage facilitates assembly of siRNA into Ago2-containing RNAi enzyme complexes. *Cell, 123*(4), 607–620.

Matsui, Y., & Hayashi, K. (2007). Epigenetic regulation for the induction of meiosis. *Cellular and Molecular Life Science, 64*(3), 257–262.

Mattiske, D. M., Han, L., & Mann, J. R. (2009). Meiotic maturation failure induced by DICER1 deficiency is derived from primary oocyte ooplasm. *Reproduction, 137*(4), 625–632.

Miranda, K. C., Huynh, T., Tay, Y., Ang, Y. S., Tam, W. L. et al. (2006). A pattern-based method for the identification of MicroRNA binding sites and their corresponding heteroduplexes. *Cell, 126*(6), 1203–1217.

Murchison, E. P., Stein, P., Xuan, Z., Pan, H., Zhang, M. Q. et al. (2007). Critical roles for Dicer in the female germline. *Genes Development, 21*(6), 682–693.

Nakahara, K., Kim, K., Sciulli, C., Dowd, S. R., Minden, J. S., & Carthew, R. W. 2005. Targets of microRNA regulation in the Drosophila oocyte proteome. *Proceedings of the National Academy of Science USA, 102*(34), 12023–12028.

Nilsen, T. W. (2007). Mechanisms of microRNA-mediated gene regulation in animal cells. *Trends in Genetics, 23*(5), 243–249.

O'Carroll, D., Mecklenbrauker, I., Das, P. P., Santana, A., Koenig, U. et al. (2007). A Slicer-independent role for Argonaute 2 in hematopoiesis and the microRNA pathway. *Genes Development, 21*(16), 1999–2004.

Orom, U. A., Nielsen, F. C., & Lund, A. H. (2008). MicroRNA-10a binds the 5′UTR of ribosomal protein mRNAs and enhances their translation. *Molecules and Cells, 30*(4), 460–471.

Otsuka, M., Zheng, M., Hayashi, M., Lee, J. D., Yoshino, O. et al. (2008). Impaired microRNA processing causes corpus luteum insufficiency and infertility in mice. *Journal of Clinical Investigations, 118*(5), 1944–1954.

Pillai, R. S., Bhattacharyya, S. N., Artus, C. G., Zoller, T., Cougot, N. et al. (2005). Inhibition of translational initiation by Let-7 MicroRNA in human cells. *Science, 309*(5740), 1573–1576.

Rehwinkel, J., Behm-Ansmant, I., Gatfield, D., & Izaurralde, E. (2005). A crucial role for GW182 and the DCP1:DCP2 decapping complex in miRNA-mediated gene silencing. *RNA, 11*(11), 1640–1647.

Saito, K., Ishizuka, A., Siomi, H., & Siomi, M. C. (2005). Processing of pre-microRNAs by the Dicer-1-Loquacious complex in Drosophila cells. *PLoS Biology, 3*(7), e235.

Schwarz, D. S., Hutvagner, G., Du, T., Xu, Z., Aronin, N., & and Zamore, P. D. (2003). Asymmetry in the assembly of the RNAi enzyme complex. *Cell, 115*(2), 199–208.

Sirotkin, A. V., Ovcharenko, D., Grossmann, R., Laukova, M., & Mlyncek, M. (2009). Identification of microRNAs controlling human ovarian cell steroidogenesis via a genome-scale screen. *Journal of Cell Physiology, 219*(2), 415–420.

Smith, L. D., & Richter, J. D. (1985). Synthesis, accumulation, and utilization of maternal macromolecules during oogenesis and oocyte maturation. In C. B. Metz & A. Monroy (eds.), *Biology of fertilization* (pp. 141–188). Orlando, FL: Academic Press.

Standart, N., & Jackson, R. J. (2007). MicroRNAs repress translation of m7Gppp-capped target mRNAs in vitro by inhibiting initiation and promoting deadenylation. *Genes Development, 21*(16), 1975–1982.

Stitzel, M. L., & Seydoux, G. (2007). Regulation of the oocyte-to-zygote transition. *Science, 316*(5823), 407–408.

Su, A. I., Cooke, M. P., Ching, K. A., Hakak, Y., Walker, J. R. et al. (2002). Large-scale analysis of the human and mouse transcriptomes. *Proceedings of the National Academy of Science USA, 99*(7), 4465–4470.

Surani, M. A., Hayashi, K., & Hajkova, P. (2007). Genetic and epigenetic regulators of pluripotency. *Cell, 128*(4), 747–762.

Tang, F., Kaneda, M., O'Carroll, D., Hajkova, P., Barton, S. C. et al. (2007). Maternal microRNAs are essential for mouse zygotic development. *Genes Development, 21*(6), 644–648.

Tesfaye, D., Worku, D., Rings, F., Phatsara, C., Tholen, E. et al. (2009). Identification and expression profiling of microRNAs during bovine oocyte maturation using heterologous approach. *Molecular Reproduction and Development, 76*(7), 665–677.

Vasudevan, S., Tong, Y., & Steitz, J. A. (2007). Switching from repression to activation: MicroRNAs can up-regulate translation. *Science, 318*(5858), 1931–1934.

Voronina, E., & Wessel, G. M. (2003). The regulation of oocyte maturation. *Current Topics in Developmental Biology, 58*, 53–110.

Watanabe, T., Takeda, A., Tsukiyama, T., Mise, K., Okuno, T. et al. (2006). Identification and characterization of two novel classes of small RNAs in the mouse germline: Retrotransposon-derived siRNAs in oocytes and germline small RNAs in testes. *Genes Development, 20*(13), 1732–1743.

Yamagata, K., Fujiyama, S., Ito, S., Ueda, T., Murata, T. et al. (2009). Maturation of microRNA is hormonally regulated by a nuclear receptor. *Molecules and Cells, 36*(2), 340–347.

Yi, R., Qin, Y., Macara, I. G., & Cullen, B. R. (2003). Exportin-5 mediates the nuclear export of pre-microRNAs and short hairpin RNAs. *Genes Development, 17*(24), 3011–3016.

6 Bovine Oocyte Gene Expression: Identification of Functional Regulators of Early Embryogenesis[1]

Swamy K. Tripurani, Jianbo Yao, and George W. Smith

6.1 Introduction

Oocyte developmental competence is defined as the capacity of the oocyte to resume meiosis, cleave after fertilization, help promote embryonic development and implantation, and bring a pregnancy to term in good health (Picton et al., 1998, Krisher, 2004, Sirard et al., 2006). Oocytes gradually and sequentially acquire developmental competence during the course of folliculogenesis by synthesizing and accumulating transcripts and proteins that are of importance for successful folliculogenesis, germ cell maturation, fertilization, and early embryogenesis and are the functional mediators of oocyte competence (Eppig, 2001, Eppig et al., 2002, Hussein et al., 2006). The development of transgenic mouse models and availability of vast arrays of DNA sequence information in the form of complete genomes, expressed sequence tags libraries and suppressive subtractive hybridization (SSH), and microarray data has been critical in identifying and revealing how oocyte-specific transcripts are essential to the above oocyte developmental milestones and how they could be functionally associated with poor oocyte competence (Andreu-Vieyra et al., 2006). Our laboratories have placed considerable emphasis in recent years on understanding the functional role of oocyte-derived gene products in promoting early embryogenesis and determination of oocyte competence in cattle. We have utilized a combination of functional studies of candidate genes of interest selected based on published reports in the mouse, and functional studies of genes identified using genomics tools in the bovine model system hypothesized to be important for oocyte competence and/or early embryogenesis. Given the relative lack of data in other mammalian species, this chapter summarizes current knowledge obtained from studies in the mouse of the functional role of select candidate oocyte-specific genes (transcriptional and post-transcriptional regulators) potentially relevant to oocyte competence in cattle. Available information and gaps in knowledge in the bovine model system (Table 6.1) where much of the published data are descriptive, are emphasized. In the second part of this chapter, major emphasis is placed on recent insights gained (using functional genomics approaches) into novel oocyte-expressed genes that confer oocyte competence and/or are critical for early embryogenesis in cattle.

Oocyte Physiology and Development in Domestic Animals, First Edition. Edited by Rebecca L. Krisher.

Table 6.1 Oocyte-specific genes addressed in this chapter potentially required for early embryonic development in cattle.

Gene Name/Symbol	Year Discovered	Protein	Domain/Motif	Species First Identified	Key Function	Published Work in Bovine
Factor in the germline alpha (*Figla*)	1997	Transcription factor	Helix-loop-helix	Mouse	Folliculogenesis	Tripurani et al. (2010)
Newborn ovary homeobox gene (*Nobox*)	2002	Transcription factor	Homeobox domain	Mouse	Folliculogenesis and early embryogenesis	Tripurani et al. (2010)
Spermatogenesis and Oogenesis specific basic helix–loop–helix 1 and 2 (*Sohlh1*and *Sohlh2*)	2006	Transcription factor	Helix-loop-helix	Mouse	Early Folliculogenesis	None
Oocyte-specific homeobox gene family (*Obox*)	2002	Transcription factor	Homeobox domain	Mouse	Unknown	None
Deleted in Azoospermia Like gene (*Dazl*)	1996	RNA binding protein	RNA binding domain	Mouse	Folliculogenesis and oocyte maturation	Liu et al. (2007); Zhang et al. (2008)
Y box protein 2 (*Ybx2*)	1990	RNA binding protein	S1-like cold-shock domain	Xenopus	Folliculogenesis and early embryogenesis	Vigneault et al. (2004); Lingenfelter et al. (2007)
Cytoplasmic polyadenylation element-binding protein1 (*Cpeb1*)	1994	RNA binding protein	RNA binding domain	Xenopus	Folliculogenesis, oocyte maturation and early embryogenesis	Uzbekova et al. (2008)
Maternal antigen that embryo require (*MATER*)	2000	Antigen	NACHT NTPase; Leu-rich repeat	Mouse	Early embryonic development	Pennetier et al. (2004, 2006)
Factor located in oocytes permitting embryonic development (*FLOPED*)	2008	RNA-binding protein	A typical KH	Mouse	Early embryonic development	None
Zygote arrest 1 (*Zar1*)	2003	Transcriptional regulator	A typical PHD domain	Mouse	Early embryonic development	Brevini et al. (2004); Uzbekova et al. (2006)
Nucleoplamin2 (*Npm2*)	1980	Nuclear chaperone	Nucleoplasmin	Xenopus	Early embryonic development	Lingenfelter et al. (2008)
Developmental pluripotency associated 3 (*Dppa3*)	2002	Basic protein	SAP-like domain	Mouse	Early embryonic development	Thélie et al. (2007)
Octamer binding transcription factor 4 (*Oct4*)	1990	Transcription factor	POU specific and Homeobox	Mouse	Early embryonic development and pluripotency	Gandolfi et al. (1997); van Eijk (1999); Kurosaka (2005)
JY1	2007	Secreted Protein	No significant matches	Bovine	Early embryonic development	Bettegowda et al. (2007)
Importin alpha 8 (*Kpna7*)	2009	Nuclear transport receptor	Importin-β binding domain and armadillo (ARM) motifs	Bovine	Early embryonic development	Tejomurthula et al. (2009)

6.2 Potential Contribution of Oocyte-Specific Transcriptional and Post-Transcriptional Regulators to Bovine Oocyte Competence: Available Evidence and Gaps in Knowledge

The ovarian follicle is the reproductive unit of the ovary; it consists of the oocyte and its associated somatic cells (granulosa, theca, and cumulus cells). Folliculogenesis involves a series of highly regulated sequential steps in which a growing follicle either develops to the ovulatory stage or dies by atresia. Progression through follicular development is a prerequisite for acquisition of oocyte competence, as more than 99% of bovine follicles become arrested in development at various stages of folliculogenesis, die via atresia, and never have the opportunity to release an oocyte to be fertilized. The major steps in folliculogenesis include the primordial/primary transition, the primary/secondary transition, selection, and atresia. The delicate interplay of developing oocytes and somatic cells is controlled and orchestrated through several endocrine factors (follicle-stimulating hormone (FSH) and luteinizing hormone (LH) (Danforth, 1995; Moley & Schreiber, 1995)], autocrine and paracrine regulators such as TGFβ family members (Elvin et al., 2000), the insulin-like growth factor system (Schams et al., 1999), inhibins/activins (Knight & Glister, 2001) and gap junctional communication (connexins) (Kidder & Mhawi, 2002). With the advent of genomics and gene knockout studies, several oocyte-specific transcription factors, RNA-binding proteins, and growth factors were identified and proven to play essential role in mammalian folliculogenesis (Epifano & Dean, 2002, Andreu-Vieyra et al., 2006, Choi & Rajkovic, 2006b, Pangas, 2007). The list of oocyte-specific genes required for normal development of germ cells and surrounding somatic cells during folliculogenesis continues to expand, and currently includes *Figla* (factor in the germline alpha) (Soyal et al., 2000), *Nobox* (newborn ovary homeobox) (Rajkovic *et al.*, 2004), *Sohlh1* and *Sohlh2* (spermatogenesis and oogenesis-specific basic helix–loop–helix 1 and 2) (Ballow et al., 2006; Pangas et al., 2006), *Obox* (oocyte-specific homeobox) (Rajkovic et al., 2002), *Dazla* (deleted in azoospermia-like autosomal) (Ruggiu et al., 1997), *Ybx2* (Y box protein 2) (Gu et al., 1998), *Cpeb1* (cytoplasmic polyadenylation element binding protein 1) (Racki & Richter, 2006), *Gdf9* (growth differentiation factor 9) (Dong et al., 1996), and *Bmp15* (bone morphogenetic protein15) (Dube et al., 1998). These genes encode for a compilation of transcription factors, mRNA-binding proteins, and secreted proteins expressed in the oocyte (and in some cases in the surrounding somatic cells). Such genes play important regulatory roles in control of follicle formation and germ cell development, and regulation of granulosa cell proliferation and steroidogenesis, and in some instances also impact early embryonic development. The functional role of oocyte-specific growth factors is discussed in Chapter 4. To date, the functional role of the above oocyte/germ cell–specific transcription factors and post-transcriptional regulators (RNA-binding proteins, translational regulators) during the initial stages of early embryogenesis has largely been overlooked, often due to defects in folliculogenesis in gene targeting models in mice. However, this potential functional role may be of significance particularly to species such as cattle with a longer interval from fertilization to transfer of control of development to products of the embryonic genome (maternal-to-embryonic transition). The known functional role of such transcriptional and post-transcriptional regulators in the mouse and potential significance to early embryonic development in the bovine species is outlined below.

6.2.1 Factor in the Germline Alpha (Figla)

Factor in the germline alpha (*Figla*) encodes a germ cell–specific basic-helix-loop-helix (bHLH) transcription factor (Liang et al., 1997). *Figla* is a major regulator of, and was initially identified through, studies of transcriptional regulation of the zona pellucida protein genes (*Zp2, Zp3*) (Liang et al., 1997). In the mouse, *Figla* transcript is expressed as early as embryonic day (E) 13.5 in the female gonad and persists throughout folliculogenesis (Soyal et al., 2000). Female *Figla* knockout mice are infertile (Soyal et al., 2000). Germ cell migration and proliferation and development of embryonic gonads appear normal in female *Figla* knockout mice (Soyal et al., 2000). However, the formation of primordial follicles is blocked and oocytes are completely lost by day 7 after birth (Soyal *et al.,* 2000). As expected *Zp1, Zp2,* and *Zp3* transcripts are not detectable in the ovaries of the *Figla* knockout mice (Soyal et al., 2000), but important genes such as *Gdf9*, *Bmp15*, *Kit, Kitl, Cx43* (connexin 43), and *Fgf8* (fibroblast growth factor 8) are present in both knockout and wild-type newborn ovaries. These observations indicate that *Figla* plays a key regulatory role in at least two independent oocyte-specific pathways: those that initiate folliculogenesis and those that regulate expression of the zona pellucida protein genes (Soyal et al., 2000). A comparison of normal and *Figla* knockout mice ovaries by microarray and serial analysis of gene expression (SAGE) identified several oocyte-specific genes, including maternal effect genes, that are directly or indirectly regulated by *Figla* (Joshi et al., 2007). Recently, it was shown that expression of *Figla* is critical not only for activation of oocyte- associated genes but also for repression of sperm-associated genes during postnatal oogenesis (Hu et al., 2010).

In cattle, our work indicates *Figla* mRNA is restricted to gonadal tissues and is present in fetal ovaries harvested as early as 90 days of gestation when primordial follicles start to form (Tripurani et al., 2010). Expression of *Figla* mRNA and protein is abundant in germinal vesicle and metaphase II stage oocytes, as well as in embryos from pronuclear to 8-cell stages, but barely detectable in embryos collected at the morula and blastocyst stages. These results suggest that this transcription factor (*Figla*) might also function as a maternal effect gene important for bovine early embryonic development and promote gene expression at embryonic genome activation (Tripurani et al., 2010). However, future studies will be required to test this hypothesis.

Evidence also indicates that *Figla* may undergo post-transcriptional regulation by miRNA during bovine early embryogenesis. A microRNA-212 (miR-212) binding site is present in the 3′untranslated region (UTR) of the bovine *Figla* transcript. Alignment of the 3′-UTR of *Figla* mRNAs from bovine, human, and mouse shows complete conservation of the "seed" region indicating that miR-212 might be a post-transcriptional regulator of *Figla* and that the microRNA:mRNA interaction is evolutionary conserved. Expression analysis indicates that bovine miR-212 is expressed in oocytes and tends to increase in 4-cell and 8-cell stage embryos followed by a decline at the morula and blastocyst stages, indicating that miR-212 is presumably of maternal origin and potentially involved in maternal transcript degradation during the maternal-to-embryonic transition (Tripurani et al., 2010). Transfection experiments and luciferase reporter assays revealed that miR-212 specifically binds to the predicted microRNA recognition element in the 3′UTR region of the *Figla* transcript and inhibits its expression (Tripurani et al., 2010). Furthermore, injection of a miR-212 mimic into bovine zygotes down-regulates *Figla* mRNA and protein in early embryos (Tripurani et al., unpublished). Collectively, our results indicate miR-212 is a potential post-transcriptional regulator of *Figla* during the maternal-to-embryonic transition in bovine embryos (Tripurani et al., 2010). Results also suggest that coordinated regulation of *Figla* may be critical for early embryonic development in cattle, but further research is needed to verify the proposed functional role.

6.2.2 Newborn Ovary Homeobox Encoding Gene (Nobox)

Nobox was identified by *in silico* subtraction of expressed sequence tags (ESTs) derived from the newborn ovaries of mice (Suzumori et al., 2002). *Nobox* mRNA is preferentially expressed in germ cells, is detectable in embryonic ovaries as early as E15.5, and is present in oocytes throughout all stages of folliculogenesis (Suzumori et al., 2002, Rajkovic et al., 2004). Female mice lacking *Nobox* have apparently normal embryonic ovarian development and germ cell migration, and form primordial follicles perinatally (Rajkovic et al., 2004). However, growth of primordial follicles and development from the primordial to primary follicle stage is severely compromised (Rajkovic et al., 2004). *Nobox* deletion also accelerates the loss of oocytes, such that by 14 days postpartum only a few remain in ovaries of mutant mice (Rajkovic et al., 2004). Gene expression analysis in newborn ovaries (before pronounced germ cell depletion) revealed an abundance of mRNA for numerous genes preferentially expressed in oocytes (such as *Mos*, *Oct4*, *Rfpl4*, *Fgf8*, *Zar1*, *Dnmt1o*, *Gdf9*, *Bmp15*, and *H1oo*) is reduced in ovaries of *Nobox* knockout mice, whereas transcripts for genes important in germ cell migration (*Kitl* and *Kit*), apoptosis (*Bcl2*, *Bcl2l2*, *Casp2*, and *Bax*), and meiosis (*Mlh1* and *Msh5*) display similar expression levels in ovaries of wild-type and *Nobox* knockout mice (Rajkovic et al., 2004). Furthermore, *Nobox* has been shown to bind to putative *Nobox*-binding elements with high affinity and to regulate transcriptional activity of mouse *Gdf9* and *Oct4* genes (Choi & Rajkovic, 2006a). These findings suggest that *Nobox* is an important transcriptional regulator of folliculogenesis that either directly or indirectly regulates a subset of genes preferentially expressed in oocytes, some of which have been shown to play essential roles in oogenesis and early embryogenesis.

In the bovine ovary and early embryos, *Nobox* is also expressed in a temporal and cell-specific fashion (Tripurani et al., 2011). *Nobox* mRNA is expressed specifically in bovine ovarian tissue, is present in fetal ovaries harvested as early as day 100 of gestation (when primary follicles start to form), and is highly abundant in the fetal ovaries of late gestation. *Nobox* protein is localized specifically to bovine oocytes and is present throughout all stages of folliculogenesis. *Nobox* mRNA and protein are abundant in germinal vesicle and metaphase II stage oocytes as well as in pronuclear, 2-cell and 4-cell stage embryos, but the immunostaining intensity for the Nobox protein drops by the 8-cell stage and is barely detectable at the morula and blastocyst stages (Tripurani et al., 2011). Culture of embryos in the presence of the transcriptional inhibitor alpha amanitin revealed that *Nobox* mRNA present in early embryos is of maternal origin.

In addition to its presumed role in regulation of folliculogenesis in cattle, our results (Tripurani et al., 2011) support an important functional role for oocyte-derived Nobox in early embryos (Figure 6.1).

It is of no doubt that conventional gene-targeting technology has improved our understanding of the function of various genes in oocyte development. However, the functional roles of some of the key oocyte-specific genes in early embryogenesis goes undetected in mice models with an ovarian phenotype due to disruptions in follicular development and to early embryonic genome activation (around the 1-cell stage) compared to monoovulatory species, such as cattle and primates (around the 8-cell stage). Cytoplasmic microinjection of double-stranded RNA into 1-cell embryos can result in specific and effective depletion of the targeted gene product during later stages of preimplantation embryo development and can reveal important information on gene function in early embryogenesis that may not be detectable using conventional gene-targeting strategies.

Ablation of *Nobox* in early embryos using siRNA established that *Nobox* is required for embryonic development to the blastocyst stage (Figure 6.1A). Depletion of *Nobox* in early embryos caused significant down-regulation of genes associated with transcriptional regulation, signal transduction, and cell cycle regulation during embryonic genome activation, suggesting that *Nobox* either directly or indirectly upregulates expression from the embryonic genome of key genes linked to early development. Further, *Nobox* depletion in early embryos reduced expression of pluripotency genes (*Pou5f1/Oct4* and *Nanog*) and the number of inner cell mass cells in embryos that reached the

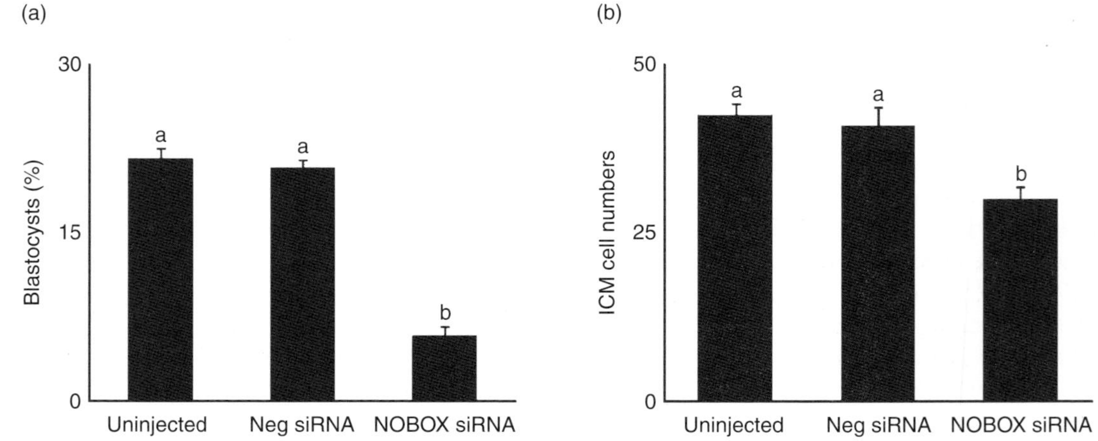

Figure 6.1 Effect of RNAi mediated depletion of NOBOX on bovine early embryonic development and blastocyst cell allocation. **(a)** Proportion of embryos developing to blastocyst stage (determined on d 7). **(b)** Number of inner cell mass (ICM) cells in Nobox siRNAs and control embryos that reached blastocyst stage on d 7 post insemination. Uninjected embryos and embryos injected with a nonspecific siRNA (Neg siRNA) were used as controls. Data are expressed as mean ± SEM from four replicates (n = 25–30 zygotes per treatment per replicate). Values with different letters across treatments indicate significant differences ($P < 0.05$) (Tripurani *et al.* 2011).

blastocyst stage (Figure 6.1B), indicating a functional role for *Nobox* in regulating pluripotency genes and cell allocation in bovine blastocysts (Tripurani et al., 2011). Our studies (Tripurani et al., unpublished) also indicate that *Nobox* abundance in early embryos is regulated by a specific miRNA. Collectively, these findings from mouse models and the bovine model system suggest that *Nobox* may function as a critical transcriptional regulator required for maintenance of gene expression both during folliculogenesis and during early embryonic development. Our studies of *Nobox* and *Figla,* described here, collectively suggest that the abundance of oocyte-derived transcription factors in early embryos is highly regulated and of functional significance to embryonic genome activation and normal early embryonic development in cattle.

6.2.3 Spermatogenesis and Oogenesis Specific Basic Helix–loop–helix 1/2 and LIM Homeodomain Transcription Factor 8 (Sohlh1, Sohlh2 and Lhx8)

Sohlh1 and *Sohlh2* encode germ cell–specific basic helix-loop-helix transcription factors. These genes were discovered using an *in silico* subtraction strategy to identify genes that are preferentially expressed during early folliculogenesis in mice (Ballow et al., 2006; Pangas et al., 2006). *Sohlh1* and *Sohlh2* transcripts are expressed in E15.5 and E13.5 embryonic ovaries respectively (Ballow et al., 2006; Pangas et al., 2006). *Sohlh1* and *Sohlh2* proteins are detectable in germ cell cysts, primordial follicles, and primary follicles, but undetectable by the secondary follicle stage, suggesting that regulation of *Sohlh1* and *Sohlh2* may be important for early folliculogenesis (Ballow et al., 2006; Pangas et al., 2006).

Female mice lacking *Sohlh1* appear to have normal germ cell migration and embryonic gonadogenesis, but form imperfect primordial follicles that do not progress to the primary follicle stage (Pangas et al., 2006). Ovaries of *Sohlh1* knockout mice contain significantly lower amounts of *Figla* transcripts and transcripts for the *Figla* target genes *Zp1* and *Zp2* (Pangas *et al.*, 2006).

Nobox transcripts are also reduced approximately fourfold in *Sohlh1* knockout ovaries. In addition, oocyte-specific genes that were down-regulated in ovaries of *Nobox* knockout mice were also down-regulated in *Sohlh1* knockout ovaries, confirming that the *Nobox* pathway is compromised in *Sohlh1* mutant mice and that *Nobox* functions downstream of *Sohlh1* (Pangas et al., 2006).

Microarray analysis of ovaries of *Sohlh1* knockout mice also revealed that the LIM homeodomain transcription factor, *Lhx8*, is down-regulated in the absence of *Sohlh1* (Pangas et al., 2006). *Lhx8* is preferentially expressed in testes and ovaries and localizes to oocytes of germ cell cysts and primordial, primary, and antral follicles (Pangas *et al.*, 2006). *Lhx8* mRNA expression is detectable as early as E13.5 and mimics embryonic ovary expression of *Sohlh1* (Pangas et al., 2006). Ovaries of *Lhx8* knockout mice fail to maintain primordial follicles, and the transition from primordial to growing follicles does not occur (Choi et al., 2008a). Oocyte- specific genes, such as *Nobox*, *Gdf9*, *Oct4*, and *Zp3*, also show abnormal expression in the ovaries of *Lhx8* knockout mice (Choi et al., 2008a). Furthermore, chromatin immunoprecipitation and luciferase reporter assay revealed that *Sohlh1* binds to the conserved E box element found in the proximal promoter region of *Lhx8*, *Nobox, Zp1*, and *Zp3* (Pangas et al., 2006). Collectively, results indicate that *Sohlh1* and *Lhx8* are master oocyte transcription factors functionally linked to folliculogenesis, and that *Sohlh1* acts upstream of *Lhx8*, *Figla*, and *Nobox*.

Sohlh2 deficiency in female mice accelerates postnatal oocyte loss in the ovary and causes infertility (Choi et al., 2008b). *Sohlh2* knockout mice form primordial follicles, and despite limited oocyte growth, do not differentiate surrounding granulosa cells into cuboidal and multilayered structures (Choi et al., 2008b). In addition, *Sohlh2* deficiency affects the expression of numerous genes (*Nobox, Figla, Gdf9, Pou5f1, Zp1, Zp3, Oosp1, Nlrp14, H1foo, and Stra8)* in oocytes including *Sohlh1* (Choi et al., 2008b), indicating that *Sohlh1* and *Sohlh2* share a similar pathway and potentially regulate each other's expression directly or indirectly. It is still unclear whether *Sohlh1* and *Sohlh2* can function as heterodimers to affect oocyte differentiation and survival. These findings show that *Sohlh1* and *Sohlh2* are critical for early follicle formation and development and oocyte survival, and that *Sohlh1* and *Sohlh2* play distinct yet important roles (Choi et al., 2008b).

To our knowledge, there have been no published studies on *Sohlh1, Sohlh2*, and *Lhx8* expression in bovine oocytes and early embryos. It would be interesting to know whether these transcription factors are also linked to normal early embryonic development in cattle and help mediate gene expression changes coincident with embryonic genome activation.

6.2.4 Oocyte-specific Homeobox Gene Family (Obox)

Obox (oocyte-specific homeobox) genes have been referred to as the first homeobox gene family preferentially expressed in mouse adult germ cells (Rajkovic et al., 2002, Cheng et al., 2007). *Obox1* and *Obox2* were identified in adult mouse germ cells using *in silico* subtraction strategies (Rajkovic et al., 2002). *Obox1* and *Obox2* transcripts encode homeodomain proteins and share 97% identity with each other (Rajkovic et al., 2002). Nucleotide sequence analysis of BACs encoding *Obox1* and *Obox2* and BLAST searches against the publicly available mouse genome database identified *Obox3*, *Obox4*, *Obox5*, and *Obox6*, which share significant homology to *Obox1* and *Obox2*. Northern blot and RT-PCR analysis revealed that five of six *Obox* genes are preferentially expressed in gonads (ovary) (Rajkovic et al., 2002). *In situ* hybridization analysis of *Obox1* and *Obox6* revealed their transcripts are exclusively expressed in oocytes throughout folliculogenesis (Rajkovic et al., 2002). Mice lacking *Obox6* undergo normal embryonic development and are fertile, indicating a functional *Obox6* gene is not obligatory for fertility (Cheng et al., 2007). However the roles of other *Obox* gene family members are currently unknown in mice, and to our knowledge, no information is available on expression of *Obox* genes in bovine oocytes and early embryos. Lack of phenotype in *Obox6* knockout mice suggests potential functional redundancy among *Obox* gene family members, but studies are needed to understand their expression and potential contribution to early embryogenesis in cattle.

6.2.5 Deleted in Azoospermia-Like Gene (Dazl)

The Deleted in Azoospermia-Like (*Dazl*) gene is a member of the Deleted in Azoospermia (*Daz*) family, expressed exclusively in germ cells. Protein products for *Daz* gene family members contain a highly conserved RNA binding motif and a unique Daz repeat (Yen, 2004). In humans and mice, the *Dazl* gene is expressed in both the testis and the ovary *(Cooke et al., 1996; Ruggiu et al., 1997; Brekhman et al., 2000)*. *Dazl* knockout mice ovaries contain steroidogenically active cells capable of producing estradiol and inhibin, despite a total loss of oocytes during fetal life (McNeilly et al., 2000) and absence of follicular structures. The *Dazl* protein in mice functions in regulating protein translation in germ cells by interacting with PABP-binding protein (Collier et al., 2005; Padmanabhan & Richter, 2006) and also in maintaining pluripotency and genetic and epigenetic modifications during germ cell development (Haston et al., 2009).

There is limited information on *Dazl* expression and function in oocytes of other species relevant to a potential functional role in cattle. In cattle, the *Dazl* gene is expressed exclusively in testis and ovary and thought to play an important role in spermatogenesis (Liu et al., 2007; Zhang et al., 2008). In pigs, *Dazl* mRNA and protein are localized to the oocyte throughout folliculogenesis and oocyte maturation (Liu et al., 2009). Furthermore, addition of factors such as glial cell line–derived neurotrophic factor (GDNF), epidermal growth factor (EGF), and follicle-stimulating hormone (FSH) that are known to enhance oocyte maturation and developmental competence, increased the expression of *Dazl* in oocytes derived from small and large antral follicles during *in vitro* maturation. Such results indicate that these factors play a potential role in promoting *Dazl* expression, which may subsequently affect translational regulation of key proteins associated with oocyte maturation and subsequent embryonic development (Liu et al., 2009). Interestingly, in human blastocysts, the presence of *Dazl* transcripts was detectable only in blastocysts of good quality (Cauffman et al., 2005), indicating that *Dazl* serves as biomarker for blastocyst quality and could potentially influence embryo survival and implantation. Together, results from other species indicate *Dazl* functions in germ cell differentiation, folliculogenesis, and oocyte maturation. Given the importance of translational regulatory mechanisms in the maternal-to-embryonic transition (Bettegowda & Smith, 2007), investigation of the link between *Dazl* expression and bovine oocyte competence holds merit.

6.2.6 Y Box Protein 2 (Msy2)

Y box proteins are multifunctional proteins that are implicated in translational regulation via their ability to stabilize and/or prevent translation of specific mRNAs. In addition, some members are also involved in the subcellular localization or transport of mRNAs by interacting with cytoskeletal proteins *(Matsumoto et al., 1996; Sommerville & Ladomery, 1996; Ruzanov et al., 1999)*. The Y box protein 2 (synonym, *Msy2*) is a germ cell–specific DNA/RNA-binding protein and a member of a cold shock domain protein superfamily that is conserved from bacteria to humans (Wolffe et al.,

1992). *Msy2* contains the cold shock domain that is highly conserved among all Y box proteins and four basic/aromatic islands that are closely related to the other known germline Y box proteins from Xenopus (F*rgy*2), (Tafuri & Wolffe, 1990) and goldfish (G*fyp*2) (Katsu et al., 1997). In the ovary, *Msy2* is present exclusively in diplotene stage and mature oocytes (Gu et al., 1998). *Msy2* constitutes about 2% of the total protein in the fully grown oocyte, but after fertilization, it is totally degraded by the late 2-cell stage in mice. Female mice lacking *Msy2* are infertile due to increased oocyte loss, anovulation, and multiple follicular defects (Yang et al., 2005). Furthermore transgenic RNAi-mediated reduction of *Msy2* results in abnormal intracellular Ca^{2+} oscillations, chromatin morphology, meiotic spindle formation, and protein synthesis during maturation (Yu et al., 2004). Collectively, evidence in mice supports a requirement of Msy2 for oocyte survival, follicular development, and fertility, presumably by stabilization and translational regulation of mRNAs critical to oocyte development and survival.

In cattle, the abundance of Msy2 transcripts was significantly lower in oocytes from persistent than from growing follicles, suggesting that Msy2 is important for oocyte competence (Lingenfelter et al., 2007). A similar *Msy2* expression pattern as observed in the mouse is characteristic of bovine embryos (Vigneault et al., 2004) where expression is low to undetectable after embryonic genome activation (8-cell stage). Observed temporal expression patterns suggest *Msy2* may also play a functional role during initial stages of early embryonic development (Yu et al., 2001, Yu et al., 2002, Vigneault et al., 2004), particularly in cattle. However, direct evidence of such a requirement for *Msy2* during early embryogenesis is lacking.

6.2.7 Cytoplasmic Polyadenylation Element Binding Protein 1 (Cpeb1)

Cpeb is an RNA recognition motif (RRM) and zinc-finger-containing sequence-specific RNA-binding protein that is found in a wide range of vertebrates and invertebrates (Richter, 2007). *Cpeb* proteins are often referred to as *Cpeb1*. *Cpeb1* regulates mRNA translation and, through this activity, influences gametogenesis and early development (Mendez & Richter, 2001; Tay et al., 2003, Racki & Richter, 2006). Adult female *Cpeb1* knockout mice contain vestigial ovaries that are devoid of oocytes, but ovaries from mid-gestation embryos contain oocytes that are arrested at the pachytene stage due to the absence of protein components (*SCP1* and *SCP2*), which are targets of *Cpeb1* and critical for formation of synaptonemal complexes (Tay & Richter, 2001). Further, to assess the function of *Cpeb1* later during oocyte development, transgenic mice were generated in which short hairpin RNA (shRNA) against *Cpeb1* was placed under the control of the zona pellucida 3 (*Zp3*) promoter, which is transcribed after the pachytene stage (Racki & Richter, 2006). Oocytes from the ZP3-*Cpeb1* shRNA transgenic mice do not develop normally. Such oocytes undergo parthenogenetic cell division in the ovary, exhibit abnormal polar bodies, and are detached from the cumulus granulosa cell layer and display spindle and nuclear anomalies. Follicular development is also impacted as follicles containing such oocytes exhibit apoptotic granulosa cells (Racki & Richter, 2006). In addition, *Cpeb1* has been shown to bind RNA transcripts for several key oocyte-expressed genes (Racki & Richter, 2006), and *Cpeb1* knockdown oocytes displayed reduced expression of *Gdf9*, which is an important determinant of follicular development.

Some information is available on *Cpeb1* expression in bovine oocytes. *Cpeb1* is localized in oocyte cytoplasm and is hyperphosphorylated during the prophase/metaphase I transition. Most *Cpeb1* is degraded in metaphase II bovine oocytes, and *Cpeb1* degradation is prevented in the presence of meiotic inhibitors such as roscovitine (Uzbekova et al., 2008). Like data for *Msy2* and *Dazl* knockout animals, evidence from Cpeb1 knockout/knockdown studies illustrates the importance of translational regulatory mechanisms and specific molecular mediators to oocyte development and acquisition of competence. To our knowledge, Cpeb1 expression in early bovine embryos has not been examined. However, a role for maternally derived Cpeb1 in translational regulation during the maternal-to-embryonic transition post-fertilization in cattle seems unlikely given the massive degradation of Cpeb1 that accompanies progression of bovine oocytes to metaphase II.

6.3 Maternal Oocyte-Derived Factors Required Specifically for Early Embryogenesis

In numerous species, early embryonic development is characterized by important developmental transitions that occur following fertilization (Schultz et al., 1999), including the replacement of maternal with zygotic RNAs, compaction, the first lineage differentiation into inner cell mass, and trophectoderm and implantation.

First chronologically, and hence most important to the focus of this review is the maternal-to-embryonic transition (Figure 6.2), defined as the time period during embryonic development

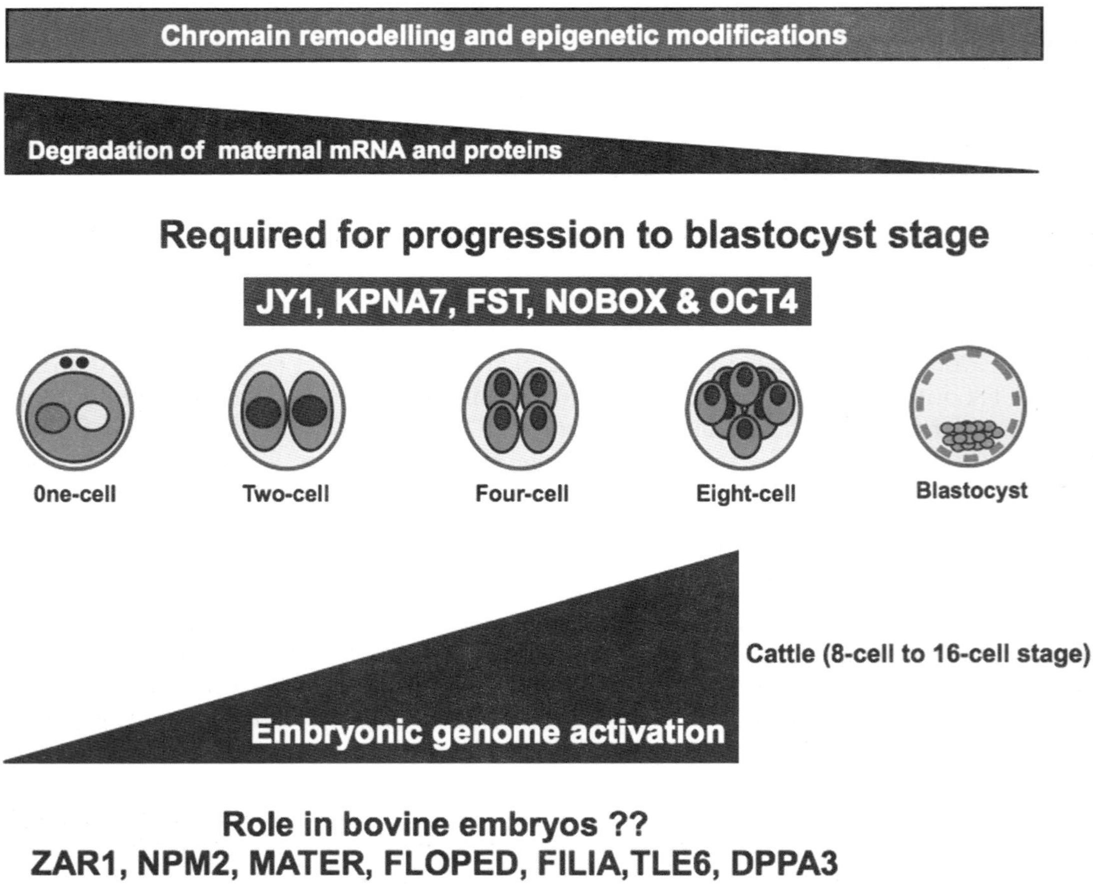

Figure 6.2 Oocyte-derived regulators and regulatory events during the bovine maternal-to-embryonic transition and subsequent stages of early embryogenesis. (For color detail, see color plate section.)

beginning at fertilization until control of early embryogenesis changes from regulation by oocyte-derived factors to regulation by products of the embryonic genome (Bettegowda et al., 2008a). In mice, embryonic genome activation occurs during the 2-cell stage, but it begins during the 4-cell stage in humans, rats, and pigs, and during the 8-cell to 16-cell stage in cattle and sheep (Telford et al., 1990). In bovine embryos, transcription is initiated as early as the 1-cell stage and referred to as minor genome activation (Memili & First, 1999), but the significance of minor genome activation to development is not clear. Progression to the 8-cell stage is not dependent on new transcription as development can occur when embryos are cultured in the presence of inhibitors of transcription. Several studies using techniques such as microarray, gene knockout, and RNAi have proved that products of numerous maternal effect genes transcribed and stored during oogenesis mediate several prominent functions during maternal-to-embryonic transition (Wianny & Zernicka-Goetz, 2000; Paradis et al., 2005; Cui et al., 2007; Sun et al., 2008; Schultz, 2002; Li et al., 2010). The first function is the removal of maternal RNA and protein. The second role is to promote dramatic reprogramming of both male and female genomes from a repressed chromatin state to one favoring transcription, and the third function of maternal factors is robust activation of the embryonic genome. Such factors also help mediate initial cleavage divisions post-fertilization.

In the mouse, the above-described developmental transitions are accompanied by orchestrated expression of transcription factors (*Hsf1*, *Bnc1*, *Ctcf*, *Oct4*, and *Sox2*), chromatin remodeling factors (*Ube2a*, *Npm2*, *Trim24*, *Smarca4*, and *Brwd1*), DNA methylation machinery (*Dntm1*, *Dppa3*, and *Zfp57*), genes involved in degradation of maternal factors (*Dicer1*, *Ago2*, and *Atg5*), and genes involved in preimplantation development (*Zar1*, *Mater* (*Nlrp5*), *Floped*, *Padi6*, and *Filia*). However, only those that are uniquely expressed in oocytes and functionally linked to early embryonic development (*Mater*, *Floped*, *Zar1*, *Npm2*, *Dppa3*, and *Oct4*) will be discussed below.

6.3.1 Maternal Antigen that Embryos Require (Mater)

Maternal antigen that embryos require (*Mater*) was identified as an antigen associated with autoimmune oophoritis in the mouse and was thought to play a role in autoimmune premature ovarian failure (Tong & Nelson, 1999). Subsequent studies demonstrated an important functional role for *Mater* in early embryonic development. *Mater* is a single-copy gene exclusively expressed by germ cells during oogenesis; it is present in early cleavage stage embryos, but barely detectable at the blastocyst stage in several species (Tong et al., 2002; Pennetier et al., 2004; Tong et al., 2004; Pennetier et al., 2006; Ma et al., 2009). *Mater* knockout mice have normal folliculogenesis and ovulate oocytes that can be fertilized (Tong et al., 2000). Although the resulting zygotes can progress through the first cleavage division, subsequent development is arrested at the 2-cell stage leading to a sterile phenotype in homozygous null females (Tong et al., 2000). De novo RNA transcription is decreased in the 1- and 2-cell embryos lacking *Mater*. Nevertheless, 2-cell embryos lacking *Mater* are able to synthesize the transcription requiring complex, indicating that *Mater* is not critical for initiation of embryonic genome activation (Tong et al., 2000).

Li et al. (2008) recently identified a subcortical maternal complex (SCMC) that assembles during oocyte growth and is essential for zygotes to progress beyond the first embryonic cell division. The SCMC is excluded from regions of cell–cell contact in the cleavage stage embryo and segregates to the outer cells of the morula and blastocyst. At least four maternally encoded proteins contribute to this complex: *Floped*, *Mater*, and *Tle6* interact with each other, and *Filia* binds independently to *Mater* (Li et al., 2008). *Filia* was originally identified as a *Mater* binding partner preferentially expressed in growing oocytes (Ohsugi et al., 2008). Both *Filia* and *Mater* colocalize to the cytocortex of ovulated oocytes, where the stability of *Filia* is dependent on the presence of *Mater* (Ohsugi et al., 2008). Additionally, depletion of maternal stores of *Filia* impairs preimplantation embryo development, with a high incidence of aneuploidy that results from abnormal spindle assembly, chromosome misalignment, and spindle assembly checkpoint (SAC) inactivation (Zheng & Dean, 2009).

Available evidence indicates *Mater* may also function as an oocyte-specific maternal effect gene in cattle. *Mater* mRNA and protein are present in bovine oocytes from the primary follicle stage onward (Pennetier et al., 2006). *Mater* mRNA is decreased during meiotic maturation and through initial cleavage divisions, and is low to undetectable in morula and blastocyst stage bovine embryos, whereas Mater protein persists through the blastocyst stage (Pennetier et al., 2006). To our knowledge, a potential functional role of *Mater* and its binding partner *Filia* during bovine oocyte and early embryonic development has not been investigated. However, no relationship between *Mater* transcript abundance and oocyte quality was observed in two established models (prepubertal calf and brilliant cresyl blue staining) used to identify oocytes of reduced developmental competence in cattle (Mota et al., 2010; Romar et al., 2011). Investigation of the potential existence of a SCMC and a functional role for *Mater* and *Filia* in early bovine embryos merits further investigation.

6.3.2 Factor Located in Oocytes Permitting Embryonic Development (Floped)

Floped was identified by screening SAGE libraries derived from ovaries of normal and *Figla* null mice. *Floped* is exclusively expressed in the ovary, and within the ovary, expression is restricted to growing oocytes (Li et al., 2008). During early embryonic development *Floped* transcripts are present in 1-cell stage embryos, but barely detectable from 2-cell to blastocyst stage. Female *Floped* null mice are infertile, but have normal ovarian histology, folliculogenesis, and oogenesis (Li et al., 2008). Oocytes from *Floped* mutant mice can be fertilized and normal-appearing 1-cell zygotes recovered. However, progression from the 1- to 2-cell stage is delayed, and blastomeres within embryos from *Floped* mutant mice often appear unequal in size with attenuated contact regions (Li et al., 2008).

As stated above, *Mater*, *Tle6*, and *Filia* were identified as potential binding partners of *Floped* and components of the SCMC (Li et al., 2008). *Tle6*, a mammalian homolog of the Groucho gene from Drosophila, belongs to the Groucho/Tle superfamily of transcriptional corepressors known to play a critical role in a range of developmental processes (Bajoghli, 2007). *Tle6* is predominantly expressed in ovaries similar to *Floped* and *Mater* (Li et al., 2008). There is no SCMC, and subcortical localization of *Mater*, *Tle6*, and *Filia* is absent in the oocytes from *Floped* null mutant animals. Subcortical localization of *Tle6*, *Filia*, and *Floped* is also absent in oocytes from *Mater*-null mutant mice indicating that existence of the SCMC depends upon the presence of *Floped* and *Mater* (Li et al., 2008). The similar expression pattern of the four genes, the physical interactions of the four cognate proteins, and their colocalization in conjunction with the sterile phenotype of *Floped* and *Mater* mutant female mice have established that the existence of the SCMC is potentially required for progression through early cleavage stages of preimplantation development. Comparative studies of the functional role of the above genes, and the existence and function of the SCMC in early embryogenesis in cattle, both merit further investigation to determine potential species specificity in the function of maternal effect genes during early embryogenesis.

6.3.3 *Zygote Arrest 1* (Zar1)

Zar1 was identified in mice by subtractive hybridization and cDNA library screening (Wu et al., 2003a). *Zar1* mRNA and protein are preferentially expressed in oocytes throughout folliculogenesis and persist in 1-cell embryos, but are markedly diminished by the 2-cell stage and absent in embryos from the 4-cell through the blastocyst stages (Wu et al., 2003a). Orthologs of *Zar1* have been identified and characterized for human, rat, frog, zebrafish, bovine, ovine, and chicken (Wu et al., 2003b; Brevini et al., 2004; Bebbere et al., 2008; Michailidis et al., 2010). *Zar1* knockout mice have normal ovarian development and folliculogenesis, and ovulated eggs lacking *Zar1* can be fertilized (Wu et al., 2003a). However, in contrast to the *Mater* null embryos, *Zar1* null embryos block predominantly at the 1-cell stage (Wu et al., 2003a). Additional analysis of the arrest in development showed that the maternal and paternal genomes remain separate in discrete pronuclei in *Zar1* null zygotes; thus the two haploid genomes do not unite, resulting in incomplete fertilization (Wu et al., 2003a). Despite containing a PHD (plant homeodomain) motif, the cytoplasmic localization of *Zar1* does not support a role in gene regulation, and its absence in embryos does not affect synthesis of the transcription-requiring complex associated with embryonic genome activation. However the precise mechanisms and potential interacting proteins associated with *Zar1* modulation of fertilization and early embryogenesis have yet to be determined (Wu et al., 2003a).

RT-PCR studies in cattle indicate expression of ZAR1 is gonad specific and that Zar1 is present in early embryos (Uzbekova et al., 2006). To our knowledge, evidence supporting a functional requirement of Zar1 for normal early embryogenesis in cattle has not been reported. However, abundance of Zar1 mRNA is lower in oocytes harvested from prepubertal animals (model of poor oocyte quality) compared to oocytes harvested from adults (Romar et al., 2011). The potential functional contribution of Zar1 to progression through early embryonic development in cattle merits further investigation.

6.3.4 *Nucleoplasmin2* (Npm2)

The nucleoplasmin (NPM) family of nuclear chaperones has three members: *Npm1*, *Npm2*, and *Npm3* (Frehlick et al., 2007). These proteins have a conserved N terminal protease-resistant core region domain, a classical bipartite nuclear localization signal, and a C terminal tail domain, containing up to two additional amino acid tracts of variable length between the family members (Frehlick et al., 2007). *Npm1* and *Npm3* are ubiquitously expressed in various tissues (Frehlick et al., 2007). Npm2 is an oocyte-specific nuclear protein that binds to histones and mediates the assembly of nucleosomes from DNA and histone proteins (Laskey et al., 1978). It also binds sperm nuclear binding proteins (SNBPs) in order to facilitate decondensation and remodeling of paternal chromatin following fertilization (Philpott et al., 1991; Philpott & Leno, 1992). In the mouse, *Npm2* transcripts are present exclusively in oocytes, as well as early embryos, but barely detectable in embryos at the blastocyst stage (Burns et al., 2003). However, *Npm2* protein is present at the blastocyst stage in mice (Vitale et al., 2007).

Female *Npm2* knockout mice are subfertile or infertile (Burns et al., 2003). Oogenesis and folliculogenesis are normal, except the DNA in the *Npm2* null oocytes is amorphous and diffuse with no condensation around the nucleolus, but this defect does not interfere with the progression of oocytes to metaphase II and ovulation (Burns et al., 2003). However, *Npm2*-deficient embryos display arrested development. Although sperm DNA decondensation proceeds without *Npm2*, abnormalities are evident in oocyte and early embryonic nuclei (Burns et al., 2003). These defects include an absence of coalesced nucleolar structures and loss of heterochromatin and deacetylated histone H3 that normally circumscribe nucleoli in oocytes and early embryos respectively (Burns et al., 2003). The mechanisms whereby failure of heterochromatin organization eventually leads to the mitotic failure in *Npm2*-deficient embryos are not known.

In cattle, *Npm2* expression is restricted to ovaries and is abundant in germinal vesicle and metaphase II stage oocytes. *Npm2* transcript abundance decreases in early-cleavage stage embryos, and is barely detectable in morula and blastocysts (Lingenfelter et al., 2007b). Although the functional requirement of *Npm2* for early embryogenesis in the bovine to our knowledge has not been investigated, expression of bovine *Npm2* mRNA is significantly lower in oocytes from persistent dominant follicles (model for poor oocyte quality) compared to oocytes from growing dominant follicles, suggesting that *Npm2* transcript abundance is associated with oocyte competence (Lingenfelter et al., 2007a). Furthermore, a conserved microRNA (miRNA-181a) binding site was identified in the 3′UTR of the bovine *Npm2* transcript. Transfection experiments showed that expression of *Npm2* protein is reduced dramatically in HeLa cells expressing bovine miR-181a compared to the control cells without miR-181a, indicating that translation of *Npm2* is repressed by miR-181a (Lingenfelter et al., 2007b). Elucidation of the role of maternal *Npm2* in bovine early embryogenesis and as a functional determinant of oocyte competence merits further investigation

6.3.5 Developmental Pluripotency Associated 3 (Dppa3)

Dppa3 (also known as *Stella* and *Pgc7*) was identified while gene expression patterns in mouse primordial germ cells and embryonic stem cells using a modified SAGE protocol was being analyzed (Sato et al., 2002). It is specifically expressed in primordial germ cells, oocytes, preimplantation embryos, and pluripotent stem cells (Sato et al., 2002). During early embryonic development, soon after the formation of the zygote, *Dppa3* accumulates in the pronuclei, although it is also detected in the cytoplasm (Payer et al., 2003). Both cytoplasmic and nuclear staining continue during cleavage stages until the blastocyst stage, at which time *Dppa3* is down-regulated until its reappearance in primordial germ cells (Sato et al., 2002; Payer et al., 2003). *Dppa3* protein has a conserved SAP-like domain known to be involved in chromosomal organization (Aravind & Koonin, 2000) and a splicing factor motif-like structure. Female mice lacking *Dppa3* display severely reduced fertility. Embryos without maternally derived *Dppa3* are compromised in preimplantation development and rarely reach the blastocyst stage (Payer et al., 2003). Taken together, these observations implicate *Dppa3* as an important regulator of early embryogenesis in the mouse.

In cattle, two variants of *Dppa3* have been identified. Variant 1 is present in testis, ovary and oocyte, whereas variant 2 is present only in oocytes (Thélie et al., 2007). Expression analysis during preimplantation stages indicates that bovine *Dppa3* present in early embryos is likely of maternal origin (Thélie et al., 2007), but the functional role of Dpp3a in early embryogenesis in cattle has not been determined.

6.3.6 Octamer Binding Transcription Factor 4 (Oct4)

Oct4 is a member of the POU (Pit, Oct, Unc) domain transcription factor family and can activate or repress the expression of target genes through binding to cis acting elements that contain an octameric DNA sequence motif (Ovitt & Scholer, 1998). *Oct4* is well studied for its role in maintenance of embryonic stem cell self-renewal and pluripotency (Boyer et al., 2006; Lengner et al., 2008;

Pei 2009). *Oct4* shows a dynamic expression pattern in mouse oocytes and embryos. It is expressed during oogenesis and folliculogenesis (Ovitt & Scholer, 1998). Maternal *Oct4* RNA and protein are present in fertilized oocytes until the 2-cell stage, and zygotic *Oct4* gene expression starts at the 4- to 8-cell stage (Ovitt & Scholer, 1998). *Oct4*-deficient mouse embryos develop to the blastocyst stage, but the inner cell mass cells in such embryos are not pluripotent leading to peri-implantation lethality (Nichols et al., 1998). Furthermore, in the absence of a true inner cell mass, trophoblast proliferation is not maintained in *Oct4*-deficient embryos (Nichols et al., 1998). *Oct4*-directed expression of *Fgf4* provides a paracrine signal that couples expansion of the extra embryonic trophoblast lineage with development of the embryonic primordium.

Collectively, these results indicate embryonic *Oct4* expression plays a pivotal role in promoting formation of a pluripotent inner cell mass layer in the mammalian embryo and expansion of the trophectoderm layer (Nichols et al., 1998). However, microinjection of *Oct4* antisense morpholino oligonucleotides into 1-cell mouse embryos revealed a key role for maternal *Oct4* in early events during embryogenesis. Results of such studies suggest that maternal *Oct4* is required for early embryonic development and that it plays a critical role in embryonic genome activation by regulating genes that encode transcriptional and post-transcriptional regulators as early as the 2-cell stage (Foygel et al., 2008). Therefore, the transcription factor *Oct4* is a master regulator of multiple aspects of early embryonic development in the mouse.

Temporal expression of bovine *Oct4* during oocyte and early embryonic development is similar to its human and murine orthologues. Bovine *Oct4* transcript is present at low levels in the bovine oocyte; it increases soon after zygotic genome activation, followed by a sharp increase subsequent to compaction (van Eijk et al., 1999; Kurosaka et al., 2004). Furthermore, selective degradation of *Oct4* in bovine embryos by injection of double-stranded RNA (RNAi) resulted in a significant reduction in numbers of inner cell mass cells in bovine blastocysts (Nganvongpanit et al., 2006). Our results suggest that oocyte-derived *Nobox* is an important regulator of zygotic expression of *Oct4* following embryonic genome activation (Tripurani et al., 2011). Collectively, results support an important functional role for *Oct4* in bovine blastocyst development, but specific *Oct4* target genes critical to early embryogenesis and the functional role of *Oct4* of maternal versus embryonic origin remain to be elucidated in cattle.

6.4 Functional Genomics Studies of Bovine Oocyte Competence and Early Embryogenesis: Identification of Novel Mediators

The evidence described above documents the clear contribution of gene targeting studies in mice to enhanced understanding of the functional contribution of oocyte-derived transcriptional and post-transcriptional regulators to early embryogenesis, and many areas where complementary comparative studies in the bovine model system are warranted. However, inherent species-specific differences in the duration and number of cell cycles required for embryonic genome activation and completion of the maternal-to-embryonic transition in mice versus cattle may suggest the potential for species specificity in the regulatory mechanisms and genes mediating this transition (Bettegowda et al., 2008a). Thus, comparative genomics approaches coupled to functional studies in the bovine model system are needed to address dissimilarities in transcriptome composition between model organisms and to provide information on existence of genes or gene families that may play important regulatory roles in early embryogenesis and may contribute to oocyte competence. In recent years, our laboratories have utilized expressed sequence tag sequencing and microarray approaches as tools to identify potential novel mediators of oocyte competence and the maternal-to-embryonic transition in cattle, coupled with pharmacological and gene knockdown strategies to determine the functional contribution of such genes to early embryogenesis (Figure 6.2). The remainder of this chapter will illustrate insight into oocyte competence and maternal regulation of early embryogenesis gleaned from such studies. This focus is not however meant to minimize the contributions of many excellent functional studies from other laboratories using similar approaches that have, for example, provided new functional understanding of the role of specific maternally derived transcription factors (Tesfaye et al., 2010) and chromatin remodeling enzymes (Canovas et al., 2012) in bovine early embryogenesis.

6.4.1 JY-1

To gain a better understanding of the bovine oocyte transcriptome, a cDNA library was generated from a pool of 200 immature, germinal vesicle and mature, metaphase II stage oocytes. Although only a limited number of ESTs were sequenced (Yao et al., 2004) from this library, important novel information on oocyte gene expression was obtained. The initial 230 ESTs represented 102 unique sequences. Forty-six of such sequences displayed significant similarity to sequences for known genes present in the Genbank database. Several ESTs represented housekeeping genes (e.g., ribosomal protein L15 [RPL15]). Some represented genes with previously documented expression both in oocytes and other tissues (e.g., CDC 28 protein kinase regulatory subunit 1B [CKS1B]). However, most of the ESTs encoded either for genes whose expression in mammalian oocytes was previously unknown or for genes of unknown function (Yao et al., 2004). Among the oocyte ESTs encoding for genes of unknown function, one sequence (represented by 14 fully sequenced clones of two different sizes) was selected for further investigation because it was completely novel and showed no significant homology to sequences of any known genes or ESTs in Genbank (Yao et al., 2004). The name *JY-1* was assigned to the putative novel gene encoding for this transcript. This novelty of this sequence was considered significant because at the time there were approximately 4.9 million human, 3.7 million murine, and 228,000 bovine EST sequences available in Genbank.

Our published studies (Bettegowda et al., 2007) established that the *JY-1* gene encodes for a species-specific secreted protein belonging to a novel protein family and supports an important functional role for oocyte-derived *JY-1* in promoting early embryogenesis. *JY-1* mRNA and protein are expressed in an ovary-specific fashion, present throughout follicular development in primordial through antral follicles, and expression within ovarian tissues is restricted exclusively to the oocyte. Within early embryos, abundance of *JY-1* transcripts is maximal at the germinal vesicle stage and declines thereafter to nearly undetectable levels in 16-cell embryos.

Results of embryo culture experiments in the presence of the transcription inhibitor α-amanitin indicate the *JY-1* gene is not transcribed during the first and second embryonic cell cycles; thus *JY-1* mRNA detected in early bovine embryos is oocyte-derived (Bettegowda et al., 2007).

To test the functional requirement of *JY-1* for bovine early embryogenesis, siRNA-mediated gene silencing was performed. Microinjection of siRNA targeting *JY-1* into zygote stage embryos results in reduced *JY-1* mRNA and protein in resulting embryos and a dramatic reduction in proportion of embryos developing to the 8–16 cell and blastocyst stages (Figure 6.3) relative to uninjected, sham-injected, and negative-control siRNA-injected embryos (Bettegowda et al., 2007). Furthermore, addition of recombinant JY-1 protein during the initial 72 h of embryo culture rescues development of *JY-1* siRNA-injected embryos to the blastocyst stage (Lee et al., unpublished). Results indicate that the novel oocyte-specific protein *JY-1* is obligatory for bovine early embryonic development.

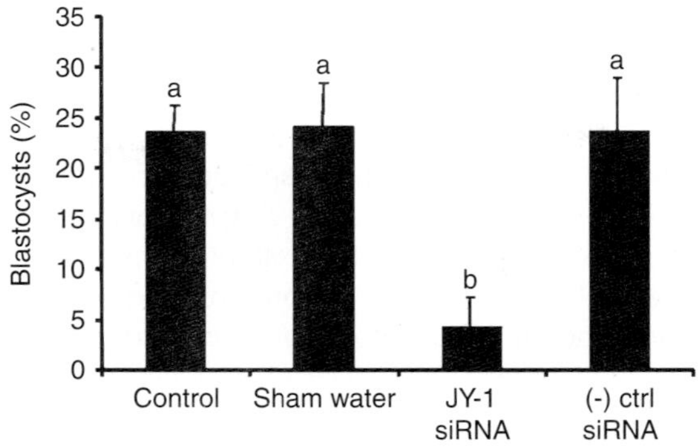

Figure 6.3 Effect of siRNA-mediated JY-1 knockdown on development of *in vitro* fertilized bovine embryos to the blastocyst stage. Presumptive zygotes were microinjected with sham water, JY-1 siRNA cocktail, negative (−) control (Ctrl) siRNA or served as uninjected controls. Average rates of blastocyst development were calculated and data are shown as mean ± SEM. Treatments without a common superscript are significantly different, $P < 0.05$ (Bettegowda *et al.*, 2007).

However, it is not yet known whether *JY-1* levels are limiting in established models of poor oocyte competence in cattle or the relationship between oocyte *JY-1* levels and infertility in a production (farm) setting.

Given the fact that *JY-1* sequences identified in the bovine oocyte cDNA library were totally novel, the presence of *JY-1* orthologues in other species was investigated (Bettegowda et al., 2007) using available genome sequence resources. *JY-1*–like sequences are present at chromosomal locations in other vertebrate species (e.g., mice, rats, humans) that are syntenic to the *JY-1* locus on bovine chromosome 29. However, putative *JY-1* loci in other species lack exons 1 and 2 and do not encode for a functional protein (Bettegowda et al., 2007). Thus, results to date indicate pronounced species specificity in evolution and functional requirement of the novel oocyte-specific *JY-1* gene for early embryonic development in cattle versus the other species noted above.

6.4.2 Importin Alpha 8 (Kpna7)

Further analysis of initial EST sequences obtained from the above-described bovine oocyte cDNA library resulted in identification of a novel transcript with similarity to genes of the importin α family. *Kpna7* (importin α8) is a new member of the importin α family (Tejomurtula et al., 2009). Members of this family are well studied for their role in nuclear transport in several species (Goldfarb et al., 2004). *Kpna7* mRNA is specifically expressed in bovine ovaries; it is abundant in germinal vesicle and metaphase II oocytes as well as in early cleavage stage bovine embryos collected before embryonic genome activation, but is barely detectable in morula and blastocyst stage embryos (Tejomurtula et al., 2009). RNAi-mediated knockdown of *Kpna7* in early embryos resulted in a decreased proportion of embryos reaching the blastocyst stage, indicating a functional requirement of *Kpna7* for early embryonic development (Tejomurtula et al., 2009). Furthermore, GST pull-down assays revealed that *Kpna7* has a stronger binding affinity for the nuclear protein *Npm2* relative to that of other importin α protein family members, suggesting that *Kpna7* may have an important role in the transport of key oocyte-specific nuclear proteins (e.g., chromatin remodeling and transcription factors) during early development (Tejomurtula et al., 2009). Elucidation of *Kpna7* binding partners critical to early embryogenesis and their specific functional role merits further investigation.

6.4.3 Follistatin

We have utilized functional genomics approaches to identify differences in RNA transcript profiles of both the oocyte and adjacent cumulus cells associated with poor oocyte competence (Patel et al., 2007; Bettegowda et al., 2008b) using the prepubertal calf model of poor oocyte quality (Revel et al., 1995; Damiani et al., 1996). A total of 193 genes encoding for transcripts displaying higher mRNA abundance in oocytes collected from adult animals and 223 genes encoding for transcripts displaying greater mRNA abundance in compromised oocytes from prepubertal calves were detected. Such results formed the foundation for a series of subsequent studies designed to elucidate the diagnostic and functional significance of putative markers identified. Of particular interest from the differentially expressed genes revealed from oocyte microarray studies were genes in the regulation of hormone secretion ontology category, which were overrepresented in the good quality oocytes harvested from adult (control) versus prepubertal animals. Messenger RNA for one such gene from this ontology category (follistatin; FST) was also greater in 2-cell bovine embryos that cleaved early and developed to the blastocyst stage at a rate fourfold greater than their late-cleaving counterparts, which displayed reduced follistatin mRNA. Given embryos were collected prior to completion of the maternal-to-embryonic transition and initiation of robust transcription from the embryonic genome (Bettegowda et al, 2008a); such differences likely reflect inherent differences in maternal (oocyte-derived) follistatin mRNA content post-fertilization and suggest a potential functional role for oocyte-derived follistatin in early embryogenesis.

Based on the above results, we hypothesized that maternal (oocyte-derived) follistatin abundance is a key determinant of early embryonic development *in vitro* and conducted studies to test this hypothesis (Lee et al., 2009). Exogenous follistatin addition to bovine embryo culture (during the first 72 h; until ~ time of embryonic genome activation) enhanced the proportion of embryos that cleaved early (within 30 h post-insemination; indicator of embryo developmental capacity) and the proportion of embryos developing to the blastocyst stage in a dose-dependent fashion. Furthermore, follistatin supplementation increased blastocyst total cell numbers specifically through an increase in trophectoderm cells, as no effect on the number of ICM cells was observed. An increase in mRNA for the trophectoderm specific transcription factor *CDX2* was also observed in response to follistatin treatment. Collectively, results indicate that exogenous follistatin treatment of early bovine embryos can enhance multiple indices of embryo developmental capacity (Lee et al., 2009). Follistatin is best known for its ability to bind the growth factor activin at a high affinity and inhibit its activity (Nakamura *et al.*, 1990). However, multiple lines of evidence indicate that the stimulatory effects of follistatin on embryo developmental capacity are nonclassical and not mediated by inhibition of activity of endogenous activin (Lee et al., 2009).

Loss-of-function experiments were also done (Lee et al., 2009) to determine the requirement for endogenous follistatin in bovine early embryonic development. Microinjection of follistatin siRNA into bovine zygotes reduced follistatin mRNA abundance in 4-cell stage embryos by $> 80\%$ relative to uninjected, sham-injected, and negative control siRNA-injected embryos and resulted in lower follistatin protein in 8-cell embryos. Furthermore, follistatin siRNA microinjection caused a $> 50\%$ reduction in blastocyst development, which can be rescued with exogenous follistatin treatment (Figure 6.4).

Follistatin treatment of siRNA-injected embryos also restored blastocyst *CDX2* mRNA abundance to control levels. Collectively, results (Lee *et al.*, 2009) strongly support a functional role for follistatin in control of time to first cleavage, blastocyst development, and blastocyst cell allocation in bovine embryos and suggest that follistatin may be an important functional determinant of bovine oocyte competence. However, the molecular mechanisms whereby follistatin can enhance early embryonic development remain to be elucidated.

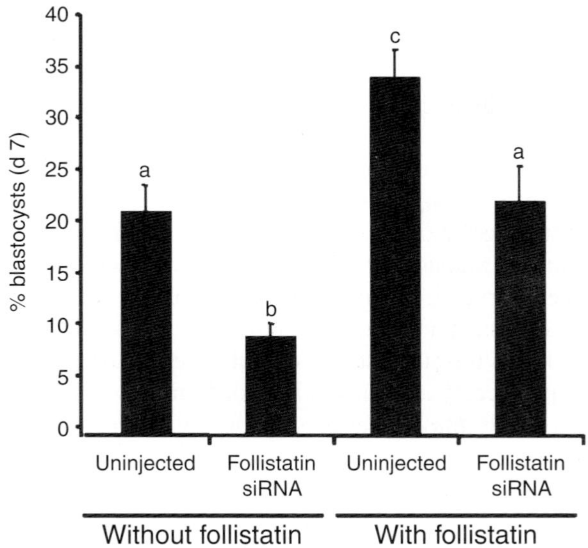

Figure 6.4 Effect of siRNA-mediated follistatin ablation and (or) replacement with exogenous follistatin (10 ng/ml) on proportion of embryos developing to blastocyst stage (determined on day 7). Values with different superscripts across treatments indicate significant differences, $P < 0.05$ (Lee *et al.*, 2009).

6.4.4 Cumulus Cell Cathepsins and Oocyte Competence

We have utilized a similar microarray approach to identify cumulus markers associated with poor-quality oocytes in the same prepubertal model system (Bettegowda et al., 2008b). Microarray experiments revealed 110 genes encoding for transcripts displaying greater mRNA abundance in cumulus cells surrounding germinal vesicle–stage oocytes collected from adult animals and 45 genes encoding for transcripts displaying greater mRNA abundance in cumulus cells surrounding compromised prepubertal oocytes. Genes in the cysteine-type endopeptidase activity category (cathepsins B, K, S, and Z) were overrepresented in cumulus cell samples harvested from oocytes of prepubertal animals and thus, based on microarray studies, tentatively associated with poor oocyte competence. Real-time PCR analysis confirmed that greater amounts of mRNA for cathepsins B, K, S, and Z are present in cumulus cells surrounding poor-quality oocytes harvested from ovaries of prepubertal animals (Bettegowda et al., 2008b). We hypothesized that cumulus cell cathepsin B, K, S, and Z expression may also be relevant to oocyte quality in adult animals. To test this hypothesis, parthenogenesis was used as an experimental tool to measure the quality of an oocyte as assessed by its potential to develop into a blastocyst following activation and the relationship between oocyte quality and cathepsin mRNA abundance determined retrospectively. Parthenogenesis was used because presence of cumulus cells during IVF is required for subsequent bovine embryonic development (Zhang et al., 1995; Luciano et al., 2005). As hypothesized, the relative abundance of mRNAs for cathepsins B, S, and Z is ~1.5-6–fold higher in cumulus cells collected from oocytes with low developmental competence versus those collected from oocytes of high developmental competence (Bettegowda et al., 2008b) providing further evidence that cumulus cell cathepsin expression is potentially predictive of an oocyte's embryo development potential.

We also conducted functional studies of the influence of cumulus cell cathepsin activity on oocyte competence. Given the observed association of cathepsin B, S, and Z transcript abundance with oocyte competence, we investigated the effects of treatment with an irreversible, cell-permeable, and highly selective cysteine proteinase inhibitor (E-64: inhibits cathepsin B) on oocyte meiotic maturation and early embryonic development. E-64 treatment during *in vitro* maturation does not affect progression to metaphase II, but results in a 40 to 50% increase in development to the blastocyst stage (d 7) after parthenogenetic activation or *in vitro* fertilization (Bettegowda et al., 2008b).

Given the negative association of cumulus cell cathepsin expression with oocyte competence, stimulatory effects of cathepsin inhibitor (E-64) treatment during meiotic maturation on oocyte competence, and the reported proapoptotic role of cathepsins (Broker et al., 2005; Stoka et al., 2005), we hypothesized that stimulatory effects of E-64 on oocyte competence may be mediated via promoting cumulus cell survival. Meiotic maturation of bovine oocytes in the presence of a cathepsin inhibitor (E-64) results in a pronounced reduction in numbers of apoptotic cumulus cells (Bettegowda et al., 2008b). Results suggest that cumulus cell cathepsin expression is functionally linked to poor oocyte competence by a negative effect on cumulus cell survival, and that a threshold number of viable cumulus cells may be necessary during *in vitro* maturation to maximize the acquisition of oocyte developmental competence and success of subsequent early embryogenesis. The mechanisms whereby cumulus cells directly impact developmental competence during meiotic maturation in cattle are directly relevant to a greater understanding of oocyte gene expression and developmental capacity following fertilization.

6.5 Conclusions

From a practical standpoint, oocyte competence is a key limiting factor in the efficiency of *in vitro* embryo production in cattle (Lonergan, 2007). Thus, gaining a greater understanding of the oocyte-expressed genes critical to various milestones in acquisition of developmental competence in general, and specifically to progression through early embryogenesis, is warranted. Discoveries made possible through functional genomics and gene targeting technologies in mice have greatly increased the understanding of genes/gene products obligatory to maternal control of early embryogenesis and completion of the maternal-to-embryonic transition in the mouse. However, comparative data establishing a functional role of such genes, particularly the transcriptional and post-transcriptional regulators described in the previous sections, is lacking in most instances for the bovine model system. This is well illustrated by studies of the role of *Nobox* in bovine early embryonic development described and the new functional insight gained.

It is conceivable that disruptions in the maternal-to-embryonic transition attributed to deficiencies in abundance or activity of key regulatory molecules contribute at least in part to the less-than-desirable efficiency of *in vitro* embryo production, even for those embryos that proceed in development beyond the 8-cell stage and embryonic genome activation. Such perturbations may also be relevant to the high rates of embryonic loss *in vivo* experienced in bovine species, particularly dairy cattle (Sreenan & Diskin, 1983; Inskeep & Dailey, 2005). Information obtained from mouse models described is a relevant foundation for comparative studies. However, because of inherent species-specific differences in the duration and number of cell cycles required for embryonic genome activation and completion of the maternal-to-embryonic transition in cattle, it is also possible that the regulatory mechanisms and maternal effect genes involved may not be identical to those described for the mouse. For example, available evidence indicates the functional role of the *JY-1* gene in early embryogenesis is species specific as a functional *JY-1* gene is not present in the murine genome. Furthermore, oocyte-derived follistatin does not make a functional contribution to control of early embryogenesis in mice as the follistatin gene is not expressed in oocytes, and follistatin mutant embryos display normal progression through early embryonic development (Matzuk et al., 1995). Thus, a systematic search for additional regulatory molecules mediating this key window in early embryonic development in cattle is warranted along with studies to test their functional contribution to early embryogenesis. Such scientific advancements will greatly increase knowledge of maternal control of early embryonic development in cattle with potential practical application to *in vitro* embryo production and potentially reproductive efficiency in cattle in general.

NOTE

1. Published studies from author's laboratories were supported in part by National Research Initiative Competitive Grant #2008-35203-19094 from the USDA National Institute of Food and Agriculture (GWS) and Agriculture and Food Research Initiative Competitive Grant #2009-65203-05706 from the USDA National Institute of Food and Agriculture (JY).

References

Andreu-Vieyra, C., Lin, Y.-N., & Matzuk, M. M. (2006) Mining the oocyte transcriptome. *Trends in Endocrinology and Metabolism, 17*, 136–143.

Aravind, L., & Koonin, E. V. (2000) SAP – a putative DNA-binding motif involved in chromosomal organization. *Trends in Biochemical Science, 25*, 112–114.

Bajoghli, B. (2007). Evolution of the Groucho/Tle gene family: Gene organization and duplication events. *Developments in Genes and Evolution, 217*, 613–618.

Ballow, D. J., Xin, Y., Choi, Y., Pangas, S. A., & Rajkovic, A. (2006). Sohlh2 is a germ cell-specific bHLH transcription factor. *Gene Expressions and Patterns, 6*, 1014–1018.

Bebbere, D., Bogliolo, L., Ariu, F., Fois, S., Leoni, G. G. et al. (2008) Expression pattern of zygote arrest 1 (ZAR1), maternal antigen that embryo requires (MATER), growth differentiation factor 9 (GDF9) and bone morphogenetic protein 15 (BMP15) genes in ovine oocytes and in vitro-produced preimplantation embryos. *Reproduction and Fertility Development, 20*, 908–915.

Bettegowda, A., Lee, K.-B., & Smith, G. W. (2008a). Cytoplasmic and nuclear determinants of the maternal-to-embryonic transition. *Reproduction and Fertility Development, 20*, 45–53.

Bettegowda, A., Patel, O., Lee, K., Park, K., Salem, M. et al. (2008b). Identification of novel bovine cumulus cell molecular markers predictive of oocyte competence: Functional and diagnostic implications. *Biology of Reproduction, 79*, 301–309.

Bettegowda, A., & Smith, G. W. (2007). Mechanisms of maternal mRNA regulation: Implications for mammalian early embryonic development. *Frontiers of Bioscience, 12*, 3713–3726.

Bettegowda, A., Yao, J., Sen, A., Li, Q., Lee, K.-B. et al. (2007). JY-1, an oocyte-specific gene, regulates granulosa cell function and early embryonic development in cattle. *Proceedings of the National Academy of Science USA, 104*, 17602–17607.

Boyer, L. A., Mathur, D., & Jaenisch, R. (2006). Molecular control of pluripotency. *Current Opinion in Genetic Development, 16*, 455–462.

Brekhman, V., Itskovitz-Eldor, J., Yodko, E., Deutsch, M., & Seligman, J. (2000). The DAZL1 gene is expressed in human male and female embryonic gonads before meiosis. *Molecular Human Reproduction, 6*, 465–468.

Brevini, T., Cillo, F., Colleoni, S., Lazzari, G., Galli, C., & Gandolfi, F. (2004). Expression pattern of the maternal factor zygote arrest 1 (Zar1) in bovine tissues, oocytes, and embryos. *Molecular Reproduction and Development, 69*, 375–380.

Broker, L. E., Kruyt, F. A., & Giaccone, G. (2005) Cell death independent of caspases: A review. *Clinical Cancer Research, 11*, 3155–3162.

Burns, K., Viveiros, M., Ren, Y., Wang, P., DeMayo, F. et al. (2003). Roles of NPM2 in chromatin and nucleolar organization in oocytes and embryos. *Science, 300*, 633–636.

Canovas, S., Cibelli, J. B., & Ross, P. J. (2012). Jumonji domain-containing protein 3 regulates histone 3 lysine 27 methylation during bovine preimplantation development. *Proceedings of the National Academy of Science USA, 109*, 2400–2405.

Cauffman, G., Van de Velde, H., Liebaers, I., & Van Steirteghem, A. (2005). DAZL expression in human oocytes, preimplantation embryos and embryonic stem cells. *Molecular Human Reproduction, 11*, 405–411.

Cheng, W.-C., Hsieh-Li, H. M., Yeh. Y.-J., & Li, H. (2007). Mice lacking the Obox6 homeobox gene undergo normal early embryonic development and are fertile. *Developmental Dynamics, 236*, 2636–2642.

Choi, Y., Ballow, D. J., Xin, Y., & Rajkovic, A. (2008a). Lim homeobox gene, lhx8, is essential for mouse oocyte differentiation and survival. *Biology of Reproduction, 79*, 442–449.

Choi, Y., & Rajkovic, A. (2006a). Characterization of NOBOX DNA binding specificity and its regulation of Gdf9 and Pou5f1 promoters. *Journal of Biology and Chemistry, 281*, 35747–35756.

Choi, Y., & Rajkovic, A. (2006b). Genetics of early mammalian folliculogenesis. *Cellular and Molecular Life Science, 63*, 579–590.

Choi, Y., Yuan, D., & Rajkovic, A. (2008b). Germ cell-specific transcriptional regulator sohlh2 is essential for early mouse folliculogenesis and oocyte-specific gene expression. *Biology of Reproduction, 79*, 1176–1182.

Collier, B., Gorgoni, B., Loveridge, C., Cooke, H. J., & Gray, N. K. (2005). The DAZL family proteins are PABP-binding proteins that regulate translation in germ cells. *EMBO Journal, 24*, 2656–2666.

Cooke, H. J., Lee, M., Kerr, S., & Ruggiu, M. (1996). A murine homologue of the human DAZ gene is autosomal and expressed only in male and female gonads. *Human Molecular Genetics, 5*, 513–516.

Cui, X. S., Li, X. Y., Yin, X. J., Kong, I. K., Kang, J. J., & Kim, N. H. (2007). Maternal gene transcription in mouse oocytes: genes implicated in oocyte maturation and fertilization. *Journal of Reproductive Development, 53*, 405–418.

Damiani, P., Fissore, R. A., Cibelli, J. B., Long, C. R., Balise, J. J., et al. (1996). Evaluation of developmental competence, nuclear and ooplasmic maturation of calf oocytes. *Molecular Reproduction and Development, 45*, 521–534.

Danforth, D. R. (1995). Endocrine and paracrine control of oocyte development. *American Journal of Obstetrics & Gynecology, 172*, 747–752.

Dong, J., Albertini, D. F., Nishimori, K., Kumar, T. R., Lu, N., & Matzuk, M. M. (1996). Growth differentiation factor-9 is required during early ovarian folliculogenesis. *Nature, 383*, 531–535.

Dube, J. L., Wang, P., Elvin, J., Lyons, K. M., Celeste, A. J., & Matzuk, M. M. (1998). The bone morphogenetic protein 15 gene is X-linked and expressed in oocytes. *Molecular Endocrinology, 12*, 1809–1817.

Elvin, J. A., Yan, C., & Matzuk, M. M. (2000). Oocyte-expressed TGF-beta superfamily members in female fertility. *Molecular and Cellular Endocrinology, 159*, 1–5.

Epifano, O. & Dean, J. (2002). Genetic control of early folliculogenesis in mice. *Trends in Endocrinology & Metabolism, 13*, 169–173.

Eppig, J. (2001). Oocyte control of ovarian follicular development and function in mammals. *Reproduction, 122*, 829–838.

Eppig, J. J., Wigglesworth, K., & Pendola, F. L. (2002). The mammalian oocyte orchestrates the rate of ovarian follicular development. *Proceedings of the National Academy of Sciences USA, 99*, 2890–2894.

Foygel, K., Choi, B., Jun, S., Leong, D. E., Lee, A., et al. (2008). A novel and critical role for Oct4 as a regulator of the maternal-embryonic transition. *PLOS ONE, 3*, e4109.

Frehlick, L. J., Eirín-López, J. M., & Ausió, J. (2007). New insights into the nucleophosmin/nucleoplasmin family of nuclear chaperones. *Bioessays, 29*, 49–59.

Goldfarb, D. S., Corbett, A. H., Mason, D. A., Harreman, M. T., & Adam, S. A. (2004). Importin alpha: a multipurpose nuclear-transport receptor. *Trends in Cellular Biology, 14*, 505–514.

Gu, W., Tekur, S., Reinbold, R., Eppig, J. J., Choi, Y. C., et al. 1998 Mammalian male and female germ cells express a germ cell-specific Y-Box protein, MSY2. *Biology of Reproduction, 59*, 1266–1274.

Haston, K. M., Tung, J. Y., & Reijo Pera, R. A. (2009). Dazl functions in maintenance of pluripotency and genetic and epigenetic programs of differentiation in mouse primordial germ cells in vivo and in vitro. *PLOS ONE, 4*, e5654.

Hu, W., Gauthier, L., Baibakov, B., Jimenez-Movilla, M., & Dean, J. (2010). FIGLA, a basic helix-loop-helix transcription factor, balances sexually dimorphic gene expression in postnatal oocytes. *Molecular and Cellular Biology, 30*, 3661–3671.

Hussein, T. S., Thompson, J. G., & Gilchrist, R. B. (2006). Oocyte-secreted factors enhance oocyte developmental competence. *Developmental Biology, 296*, 514–521.

Inskeep, E. K. & Dailey, R. A. (2005). Embryonic death in cattle. *Veterinary Clinics of North America: Food Animal Practice, 21*, 437–461.

Joshi, S., Davies, H., Sims, L. P., Levy, S. E., & Dean, J. (2007). Ovarian gene expression in the absence of FIGLA, an oocyte-specific transcription factor. *BMC Developmental Biology, 7*, 67.

Katsu, Y., Yamashita, M., & Nagahama, Y. (1997). Isolation and characterization of goldfish Y box protein, a germ-cell-specific RNA-binding protein. *European Journal of Biochemistry, 249*, 854–861.

Kidder, G. M. & Mhawi, A. A. (2002). Gap junctions and ovarian folliculogenesis. *Reproduction, 123*, 613–620.

Knight, P. G. & Glister, C. (2001). Potential local regulatory functions of inhibins, activins and follistatin in the ovary. *Reproduction, 121*, 503–512.

Krisher, R. L. (2004). The effect of oocyte quality on development. *Journal of Animal Science, 82*, E-Suppl E14–23.

Kurosaka, S., Eckardt, S., & McLaughlin, K. (2004). Pluripotent lineage definition in bovine embryos by Oct4 transcript localization. *Biology of Reproduction, 71*, 1578–1582.

Laskey, R. A., Honda, B. M., Mills, A. D., & Finch, J. T. (1978). Nucleosomes are assembled by an acidic protein which binds histones and transfers them to DNA. *Nature, 275*, 416–420.

Lee, K. B., Bettegowda, A., Wee, G., Ireland, J. J., & Smith, G. W. (2009). Molecular determinants of oocyte competence: potential functional role for maternal (oocyte-derived) follistatin in promoting bovine early embryogenesis. *Endocrinology, 150*, 2463–2471.

Lengner, C. J., Welstead, G. G., & Jaenisch, R. (2008). The pluripotency regulator Oct4: a role in somatic stem cells? *Cell Cycle, 7*, 725–728.

Li, L., Baibakov, B., & Dean, J. (2008). A subcortical maternal complex essential for preimplantation mouse embryogenesis. *Developmental Cell, 15*, 416–425.

Li, L., Zheng, P., & Dean, J. (2010). Maternal control of early mouse development. *Development, 137*, 859–870.

Liang, L., Soyal, S. M., & Dean, J. (1997). FIGalpha, a germ cell specific transcription factor involved in the coordinate expression of the zona pellucida genes. *Development, 124*, 4939–4947.

Lingenfelter, B. M., Dailey, R. A., Inskeep, E. K., Vernon, M. W., Poole, D. H., et al. (2007a). Changes of maternal transcripts in oocytes from persistent follicles in cattle. *Molecular Reproduction and Development, 74*, 265–272.

Lingenfelter, B. M. & Yao, J. (2007b). Bos taurus microRNA-181a oromotes translational silencing of nucleoplasmin 2. *Biology of Reproduction, 78*(60).

Liu, J., Linher, K., & Li, J. (2009). Porcine DAZL messenger RNA: its expression and regulation during oocyte maturation. *Molecular and Cellular Endocrinology, 311*, 101–108.

Liu, W. S., Wang, A., Uno, Y., Galitz, D., Beattie, C. W., & Ponce de León, F. A. (2007). Genomic structure and transcript variants of the bovine DAZL gene. *Cytogenetic and Genome Research, 116*, 65–71.

Lonergan, P. (2007). State-of-the-art embryo technologies in cattle. *Society for Reproduction and Fertility Supplement, 64*, 315–325.

Luciano, A. M., Lodde, V., Beretta, M. S., Colleoni, S., Lauria, A., & Modina, S. (2005). Developmental capability of denuded bovine oocyte in a co-culture system with intact cumulus-oocyte complexes: role of cumulus cells, cyclic adenosine 3',5'-monophosphate, and glutathione. *Molecular Reproduction and Development, 71*, 389–397.

Ma, J., Milan, D., & Rocha, D. (2009). Chromosomal assignment of the porcine NALP5 gene, a candidate gene for female reproductive traits. *Animal Reproduction Science, 112*, 397–401.

Matsumoto, K., Meric, F., & Wolffe, A. P. (1996). Translational repression dependent on the interaction of the Xenopus Y-box protein FRGY2 with mRNA. Role of the cold shock domain, tail domain, and selective RNA sequence recognition. *The Journal of Biological Chemistry, 271*, 22706–22712.

Matzuk, M. M., Lu, N., Vogel, H., Sellheyer, K., Roop, D. R., & Bradley, A. (1995). Multiple defects and perinatal death in mice deficient in follistatin. *Nature, 374*, 360–363.

McNeilly, J. R., Saunders, P. T., Taggart, M., Cranfield, M., Cooke, H. J., & McNeilly, A. S. (2000). Loss of oocytes in Dazl knockout mice results in maintained ovarian steroidogenic function but altered gonadotropin secretion in adult animals. *Endocrinology, 141*, 4284–4294.

Memili, E. & First, N. L. (1999). Control of gene expression at the onset of bovine embryonic development. *Biology of Reproduction, 61*, 1198–1207.

Mendez, R. & Richter, J. D. (2001). Translational control by CPEB: a means to the end. *Nature Reviews Molecular Cell Biology, 2*, 521–529.

Michailidis, G., Argiriou, A., & Avdi, M. (2010). Expression of chicken zygote arrest 1 (Zar1) and Zar1-like genes during sexual maturation and embryogenesis. *Veterinary Research Communications, 34*, 173–184.

Moley, K. H. & Schreiber, J. R. (1995). Ovarian follicular growth, ovulation and atresia. Endocrine, paracrine and autocrine regulation. *Advances in Experimental Medicine and Biology, 377*, 103–119.

Mota, G. B., Batista, R. I., Serapiao, R. V., Boite, M. C., Viana, J. H., et al. (2010). Developmental competence and expression of the MATER and ZAR1 genes in immature bovine oocytes selected by brilliant cresyl blue. *Zygote, 18*, 209–216.

Nakamura, T., Takio, K., Eto, Y., Shibai, H., Titani, K., & Sugino, H. (1990). Activin-binding protein from rat ovary is follistatin. *Science, 247*, 836–838.

Nganvongpanit, K., Müller, H., Rings, F., Hoelker, M., Jennen, D., et al. (2006). Selective degradation of maternal and embryonic transcripts in in vitro produced bovine oocytes and embryos using sequence specific double-stranded RNA. *Reproduction, 131*, 861–874.

Nichols, J., Zevnik, B., Anastassiadis, K., Niwa, H., Klewe-Nebenius, D., et al. (1998). Formation of pluripotent stem cells in the mammalian embryo depends on the POU transcription factor Oct4. *Cell, 95*, 379–391.

Ohsugi, M., Zheng, P., Baibakov, B., Li, L., & Dean, J. (2008). Maternally derived FILIA-MATER complex localizes asymmetrically in cleavage-stage mouse embryos. *Development, 135*, 259–269.

Ovitt, C. & Scholer, H. (1998). The molecular biology of Oct-4 in the early mouse embryo. *Molecular Human Reproduction, 4*, 1021–1031.

Padmanabhan, K. & Richter, J. D. (2006). Regulated Pumilio-2 binding controls RINGO/Spy mRNA translation and CPEB activation. *Gene Development, 20*, 199–209.

Pangas, S. A. (2007). Growth factors in ovarian development. *Seminars in Reproductive Medicine, 25*, 225–234.

Pangas, S. A., Choi, Y., Ballow, D. J., Zhao, Y., Westphal, H., et al. (2006). Oogenesis requires germ cell-specific transcriptional regulators Sohlh1 and Lhx8. *Proceedings of the National Academy of Sciences USA, 103*, 8090–8095.

Paradis, F., Vigneault, C., Robert, C., & Sirard, M-A. (2005). RNA interference as a tool to study gene function in bovine oocytes. *Molecular Reproduction and Development, 70*, 111–121.

Patel, O. V., Bettegowda, A., Ireland, J. J., Coussens, P. M., Lonergan, P., & Smith, G. W. (2007). Functional genomics studies of oocyte competence: Evidence that reduced transcript abundance for follistatin is associated with poor developmental competence of bovine oocytes. *Reproduction, 133*, 95–106.

Payer, B., Saitou, M., Barton, S. C., Thresher, R., Dixon, J. P. C., et al. (2003). Stella is a maternal effect gene required for normal early development in mice. *Current Biology, 13*, 2110–2117.

Pei, D. (2009). Regulation of pluripotency and reprogramming by transcription factors. The *Journal of Biological Chemistry, 284*, 3365–3369.

Pennetier, S., Perreau, C., Uzbekova, S., Thelie, A., Delaleu, B., et al. (2006). MATER protein expression and intracellular localization throughout folliculogenesis and preimplantation embryo development in the bovine. *BMC Developmental Biology, 6*, 26.

Pennetier, S., Uzbekova, S., Perreau, C., Papillier, P., Mermillod, P., & Dalbiès-Tran, R. (2004). Spatio-temporal expression of the germ cell marker genes MATER, ZAR1, GDF9, BMP15, and VASA in adult bovine tissues oocytes, and preimplantation embryos. *Biology of Reproduction, 71*, 1359–1366.

Philpott, A. & Leno, G. H. (1992). Nucleoplasmin remodels sperm chromatin in Xenopus egg extracts. *Cell, 69*, 759–767.

Philpott, A., Leno, G. H., & Laskey, R. A. (1991). Sperm decondensation in Xenopus egg cytoplasm is mediated by nucleoplasmin. *Cell, 65*, 569–578.

Picton, H., Briggs, D., & Gosden, R. (1998). The molecular basis of oocyte growth and development. *Molecular and Cellular Endocrinology, 145*, 27–37.

Racki, W. J. & Richter, J. D. (2006). CPEB controls oocyte growth and follicle development in the mouse. *Development, 133*, 4527–4537.

Rajkovic, A., Pangas, S. A., Ballow, D., Suzumori, N., & Matzuk, M. M. (2004). NOBOX deficiency disrupts early folliculogenesis and oocyte-specific gene expression. *Science, 305*, 1157–1159.

Rajkovic, A., Yan, C., Yan, W., Klysik, M., & Matzuk, M. (2002). Obox, a family of homeobox genes preferentially expressed in germ cells. *Genomics, 79*, 711–717.

Revel, F., Mermillod, P., Peynot, N., Renard, J. P., & Heyman, Y. (1995). Low developmental capacity of in vitro matured and fertilized oocytes from calves compared with that of cows. *Journal of Reproduction and Fertility, 103*, 115–120.

Richter, J. D. (2007). CPEB: a life in translation. *Trends in Biochemical Science, 32*, 279–285.

Romar, R., De Santis, T., Papillier, P., Perreau, C., Thelie, A., et al. (2011). Expression of maternal transcripts during bovine oocyte in vitro maturation is affected by donor age. *Reproduction in Domestic Animals, 46*, e23–30.

Ruggiu, M., Speed, R., Taggart, M., McKay, S. J., Kilanowski, F., et al. 1997 The mouse Dazla gene encodes a cytoplasmic protein essential for gametogenesis. *Nature, 389*, 73–77.

Ruzanov, P. V., Evdokimova, V. M., Korneeva, N. L., Hershey, J. W., & Ovchinnikov, L. P. (1999). Interaction of the universal mRNA-binding protein, p50, with actin: a possible link between mRNA and microfilaments. *Journal of Cell Science, 112*, 3487–3496.

Sato, M., Kimura, T., Kurokawa, K., Fujita, Y., Abe, K., et al. (2002). Identification of PGC7, a new gene expressed specifically in preimplantation embryos and germ cells. *Mechanisms of Development, 113*, 91–94.

Schams, D., Berisha, B., Kosmann, M., Einspanier, R., & Amselgruber, W. M. (1999). Possible role of growth hormone, IGFs, and IGF-binding proteins in the regulation of ovarian function in large farm animals. *Domestic Animal Endocrinology, 17*, 279–285.

Schultz, R. M. (2002). The molecular foundations of the maternal to zygotic transition in the preimplantation embryo. *Human Reproduction Update, 8*, 323–331.

Schultz, R. M., Davis, W., Stein, P., & Svoboda, P. (1999). Reprogramming of gene expression during preimplantation development. *Journal of Experimental Zoology, 285*, 276–282.

Sirard, M. A., Richard, F., Blondin, P., & Robert, C. (2006). Contribution of the oocyte to embryo quality. *Theriogenology, 65*, 126–136.

Sommerville, J. & Ladomery, M. (1996). Transcription and masking of mRNA in germ cells: involvement of Y-box proteins. *Chromosoma, 104*, 469–478.

Soyal, S. M., Amleh, A., & Dean, J. (2000). FIGalpha, a germ cell-specific transcription factor required for ovarian follicle formation. *Development, 127*, 4645–4654.

Sreenan, J. M. & Diskin, M. G. (1983). Early embryonic mortality in the cow: its relationship with progesterone concentration. *Veterinary Record, 112*, 517–521.

Stoka, V., Turk, B., & Turk, V. (2005). Lysosomal cysteine proteases: structural features and their role in apoptosis. *IUBMB Life, 57*, 347–353.

Sun, Q. Y., Liu, K., & Kikuchi, K. (2008). Oocyte-specific knockout: a novel in vivo approach for studying gene functions during folliculogenesis, oocyte maturation, fertilization, and embryogenesis. *Biology of Reproduction, 79*, 1014–1020.

Suzumori, N., Yan, C., Matzuk, M. M., & Rajkovic, A. (2002). Nobox is a homeobox-encoding gene preferentially expressed in primordial and growing oocytes. *Mechanisms of Development, 111*, 137–141.

Tafuri, S. R. & Wolffe, A. P. (1990). Xenopus Y-box transcription factors: molecular cloning, functional analysis and developmental regulation. *Proceedings of the National Academy of Sciences USA, 87*, 9028–9032.

Tay, J., Hodgman, R., Sarkissian, M., & Richter, J. D. (2003). Regulated CPEB phosphorylation during meiotic progression suggests a mechanism for temporal control of maternal mRNA translation. *Genes & Development, 17*, 1457–1462.

Tay, J. & Richter, J. D. (2001). Germ cell differentiation and synaptonemal complex formation are disrupted in CPEB knockout mice. *Developmental Cell, 1*, 201–213.

Tejomurtula, J., Lee, K-B., Tripurani, S. K., Smith, G. W., & Yao, J. (2009). Role of importin alpha8, a new member of the importin alpha family of nuclear transport proteins, in early embryonic development in cattle. *Biology of Reproduction, 81*, 333–342.

Telford, N. A., Watson, A. J., & Schultz, G. A. (1990). Transition from maternal to embryonic control in early mammalian development: a comparison of several species. *Molecular Reproduction and Development, 26*, 90–100.

Tesfaye, D., Regassa, A., Rings, F., Ghanem, N., Phatsara, C., et al. (2010). Suppression of the transcription factor MSX1 gene delays bovine preimplantation embryo development in vitro. *Reproduction, 139*, 857–870.

Thélie, A., Papillier, P., Pennetier, S., Perreau, C., Traverso, J. M., et al. (2007). Differential regulation of abundance and deadenylation of maternal transcripts during bovine oocyte maturation in vitro and in vivo. *BMC Developmental Biology, 7*, 125.

Tong, Z. B., Bondy, C. A., Zhou, J., & Nelson, L. M. (2002). A human homologue of mouse Mater, a maternal effect gene essential for early embryonic development. *Human Reproduction, 17*, 903–911.

Tong, Z. B., Gold, L., De Pol, A., Vanevski, K., Dorward, H., et al. (2004). Developmental expression and subcellular localization of mouse MATER, an oocyte-specific protein essential for early development. *Endocrinology, 145*, 1427–1434.

Tong, Z., Gold, L., Pfeifer, K., Dorward, H., Lee, E., et al. (2000). Mater, a maternal effect gene required for early embryonic development in mice. *Nat Genetics, 26*, 267–268.

Tong, Z. B. & Nelson, L. M. (1999). A mouse gene encoding an oocyte antigen associated with autoimmune premature ovarian failure. *Endocrinology, 140*, 3720–3726.

Tripurani, S., Lee, K., Smith, G., & Yao, J. (2010). Cloning and expression of bovine factor in the germline alpha (FIGLA) in oocytes and early embryos: A potential target of microRNA212. *Reproduction. Fertility and Development, 23*, 109–109.

Tripurani, S. K., Lee, K. B., Wang, L., Wee, G., Smith, G. W., et al. 2011 A novel functional role for the oocyte-specific transcription factor newborn ovary homeobox (NOBOX) during early embryonic development in cattle. *Endocrinology,* (e-pub ahead of print).

Uzbekova, S., Arlot-Bonnemains, Y., Dupont, J., Dalbiès-Tran, R., Papillier, P., et al. (2008). Spatio-temporal expression patterns of aurora kinases a, B, and C and cytoplasmic polyadenylation-element-binding protein in bovine oocytes during meiotic maturation. *Biology of Reproduction, 78*, 218–233.

Uzbekova, S., Roy-Sabau, M., Dalbies-Tran, R., Perreau, C., Papillier, P., et al. (2006). Zygote arrest 1 gene in pig, cattle and human: evidence of different transcript variants in male and female germ cells. *Reproductive Biology and Endocrinology, 4*, 12.

van Eijk, M. J., van Rooijen, M. A., Modina, S., Scesi, L., Folkers, G., et al. (1999). Molecular cloning, genetic mapping, and developmental expression of bovine POU5F1. *Biology of Reproduction, 60*, 1093–1103.

Vigneault, C., McGraw, S., Massicotte, L., & Sirard, M. A. (2004). Transcription factor expression patterns in bovine in vitro-derived embryos prior to maternal-zygotic transition. *Biology of Reproduction, 70*, 1701–1709.

Vitale, A. M., Calvert, M. E., Mallavarapu, M., Yurttas, P., Perlin, J., et al. (2007). Proteomic profiling of murine oocyte maturation. *Molecular Reproduction and Development, 74*, 608–616.

Wianny, F. & Zernicka-Goetz, M. (2000). Specific interference with gene function by double-stranded RNA in early mouse development. *Nature Cell Biology, 2*, 70–75.

Wolffe, A. P., Tafuri, S., Ranjan, M., & Familari, M. (1992). The Y-box factors: a family of nucleic acid binding proteins conserved from Escherichia coli to man. *New Biology, 4*, 290–298.

Wu, X., Viveiros, M., Eppig, J., Bai, Y., Fitzpatrick, S., & Matzuk, M. (2003a). Zygote arrest 1 (Zar1) is a novel maternal-effect gene critical for the oocyte-to-embryo transition. *Nature Genetics, 33*, 187–191.

Wu, X., Wang, P., Brown, C. A., Zilinski, C. A., & Matzuk, M. M. (2003b). Zygote arrest 1 (Zar1) is an evolutionarily conserved gene expressed in vertebrate ovaries. *Biology of Reproduction, 69*, 861–867.

Yang, J., Medvedev, S., Yu, J., Tang, L. C., Agno, J. E., et al. (2005). Absence of the DNA-/RNA-binding protein MSY2 results in male and female infertility. *Proceedings of the National Academy of Sciences USA, 102*, 5755–5760.

Yao, J., Ren, X., Ireland, J., Coussens, P., Smith, T., & Smith, G. (2004). Generation of a bovine oocyte cDNA library and microarray: resources for identification of genes important for follicular development and early embryogenesis. *Physiological Genomics, 19*, 84–92.

Yen, P. H. (2004). Putative biological functions of the DAZ family. *International Journal of Andrology, 27*, 125–129.

Yu, J., Deng, M., Medvedev, S., Yang, J., Hecht, N. B., & Schultz, R. M. (2004). Transgenic RNAi-mediated reduction of MSY2 in mouse oocytes results in reduced fertility. *Developmental Biology, 268*, 195–206.

Yu, J., Hecht, N. B., & Schultz, R. M. (2001). Expression of MSY2 in mouse oocytes and preimplantation embryos. *Biology of Reproduction, 65*, 1260–1270.

Yu, J., Hecht, N. B., & Schultz, R. M. (2002). RNA-binding properties and translation repression in vitro by germ cell-specific MSY2 protein. *Biology of Reproduction, 67*, 1093–1098.

Zhang, L., Jiang, S., Wozniak, P. J., Yang, X., & Godke, R. A. (1995). Cumulus cell function during bovine oocyte maturation, fertilization, and embryo development in vitro. *Molecular Reproduction and Development, 40*, 338–344.

Zhang, Q., Li, Q., Li, J., Li, X., Liu, Z., et al. (2008). b-DAZL: A novel gene in bovine spermatogenesis. *Progress in Natural Science, 18*, 1209–1218.

Zheng, P. & Dean, J. (2009). Role of Filia, a maternal effect gene, in maintaining euploidy during cleavage-stage mouse embryogenesis. *Proceedings of the National Academy of Sciences USA, 106*, 7473–7478.

7 Epigenetic Modifications during Mammalian Oocyte Growth and Meiotic Progression

Claudia Baumann, Maria M. Viveiros, and Rabindranath De La Fuente

7.1 Introduction

Mammalian oocyte growth and differentiation encompass three intrinsic developmental programs that are essential for the production of a mature egg competent to undergo fertilization and embryonic development to term. (1) Nuclear maturation comprises the mechanisms involved in conferring the oocyte with competence to resume and complete the first meiotic division and arrest at metaphase-II in order to await fertilization. During this process, the growing oocyte accumulates cell cycle–related molecules and dormant maternal mRNA stores. (2) Cytoplasmic maturation involves a series of biochemical and metabolic changes that prepare the fully grown oocyte for fertilization, and pronuclear formation in order to sustain the first cleavage division before activation of the embryonic genome. (3) Epigenetic maturation on the other hand comprises a series of hierarchical and combinatorial chromatin-based modifications taking place during oocyte growth that affect the control of gene expression. These changes can occur on a genome-wide basis or may affect the chromatin environment at the single gene level ultimately affecting the patterns of gene expression without exerting any changes in DNA sequence. By definition, epigenetic modifications induce stable and heritable changes in gene expression. Importantly, it is increasingly recognized that transient chromosomal marks such as DNA repair mechanisms or changes in chromosome architecture during meiosis and/or mitosis also constitute important examples of epigenetic events (Jablonka, 2002; Mager & Bartolomei, 2005; Bird, 2007; Goldberg et al., 2007).

In mammalian cells, epigenetic modifications are established by several molecular mechanisms such as DNA methylation as well as the expression of noncoding or structural RNAs that might provide sequence specificity for methylation marks. Chromatin structure and function can also be regulated by ATP-dependent chromatin remodeling proteins (Fry & Peterson, 2001; Varga-Weisz, 2001; Davis & Brachmann, 2003), incorporation of histone variants at specific chromosomal domains such as deposition of CENP-A (centromere associated protein-A) at the centromeres of metaphase chromosomes or the incorporation of histone H3.3 or H2A.Z during developmental transitions (Sarma & Reinberg, 2005; Polo & Almouzni, 2006), and most prominently through the induction of histone post-translational modifications such as histone acetylation, methylation, and poly (ADP) ribosylation (Bannister et al., 2002; Kouzarides, 2007).

Two prominent examples of epigenetic modifications include the process of X chromosome inactivation, as well as the establishment of parental-specific genomic imprinting marks in the mammalian germline (Jenuwein & Allis, 2001; Reik et al., 2001; Heard, 2004; Lucchesi et al.,

Oocyte Physiology and Development in Domestic Animals, First Edition. Edited by Rebecca L. Krisher.

2005; Matsui & Hayashi, 2007; Sasaki & Matsui, 2008). Importantly, a growing body of evidence indicates that epigenetic modifications are essential for maintenance of genome and chromosome stability during meiosis (Peters et al., 2001; Celeste et al., 2002; Bourc'his & Bestor, 2004; Webster et al., 2005; De La Fuente et al., 2006). Epigenetic modifications affecting local or global chromatin structure may act synergistically, in order to induce a fast response in the form of regulation of gene expression following a wide range of differentiation or environmental stimuli (Hashimshony et al., 2003; Jaenisch & Bird, 2003; Delaval & Feil, 2004; Jeffery & Nakielny, 2004; Goldberg et al., 2007).

Chromatin modifications in the mammalian germline are required for primordial germ cell specification, epigenetic silencing of transposons and homologous chromosome synapsis during prophase-I of meiosis, thus revealing a remarkable functional diversity (Sasaki & Matsui, 2008; Kota & Feil, 2010). However, their role in the functional differentiation of chromatin structure during oocyte growth and meiotic maturation is less clear. This section will focus on the mechanisms involved in the epigenetic maturation of the mammalian oocyte genome. Our developmental context is the critical window of postnatal oocyte growth in which the oocyte and ovarian follicular environment establish a critical bidirectional communication to confer meiotic competence and during which oocyte-intrinsic mechanisms establish the maternal-specific imprinting marks to confer full developmental potential. A better understanding of the mechanisms involved in the establishment of epigenetic modifications and their function in the oocyte genome will contribute to elucidating the molecular basis of aneuploidy, and will have direct implications for assisted reproductive technologies and fertility control in human and veterinary medicine.

7.2 Establishment of Epigenetic Modifications during Postnatal Oocyte Growth

Mammalian oocyte growth and differentiation ensues from a limited pool of primordial follicles established shortly after birth. In the early postnatal ovary, oocytes are maintained in meiotic arrest at the diplotene stage until puberty. The time of meiotic arrest varies according to the species but in human oocytes this may range from two to four decades. Oocyte growth and differentiation takes place within the ovarian follicle and requires a complex bidirectional communication between germ cells and somatic cells of the ovary (Matzuk et al., 2002). The oocyte-intrinsic mechanisms that regulate growth and acquisition of meiotic and developmental potential are largely unknown. However, recent studies indicate that Kit ligand produced by surrounding granulosa cells stimulates the phosphatidyl inositol (PI3K) pathway in the oocyte as an essential mechanism to initiate oocyte and follicular growth in a coordinated manner (Liu et al., 2006a). In turn, several gene-knockout mouse models indicate that PTEN (phophatase and tensin homolog deleted on chromosome 10) is a negative regulator of the PIK3 pathway and functions in synergy with the tumor suppressor tuberosclerosis complex (Tsc1) as well as the target of rapamycin complex 1 (mTORC1) to maintain oocyte and follicular quiescence in the neonatal ovary (Liu et al., 2006a; Adhikari et al., 2010). Additional inhibitory pathways regulated by the cyclin-dependent kinase p27Kip1 as well as the transcription factor FOXO3 have also been found to function downstream of the PTEN/PIK3 pathway to maintain follicular quiescence (Liu et al., 2006a; Adhikari et al., 2010). Oocyte growth and differentiation ensues as a consequence of follicular activation, a stage that in the mouse ovary lasts 2 to 3 weeks, during which oocytes will undergo an approximately 300-fold increase in volume (Liu et al., 2006a). In order to sustain such dramatic growth, oocytes maintain high levels of RNA synthesis required to ensure the sufficient accumulation of dormant maternal mRNA stores, ribosomes, and other cytoplasmic organelles to acquire meiotic and developmental potential

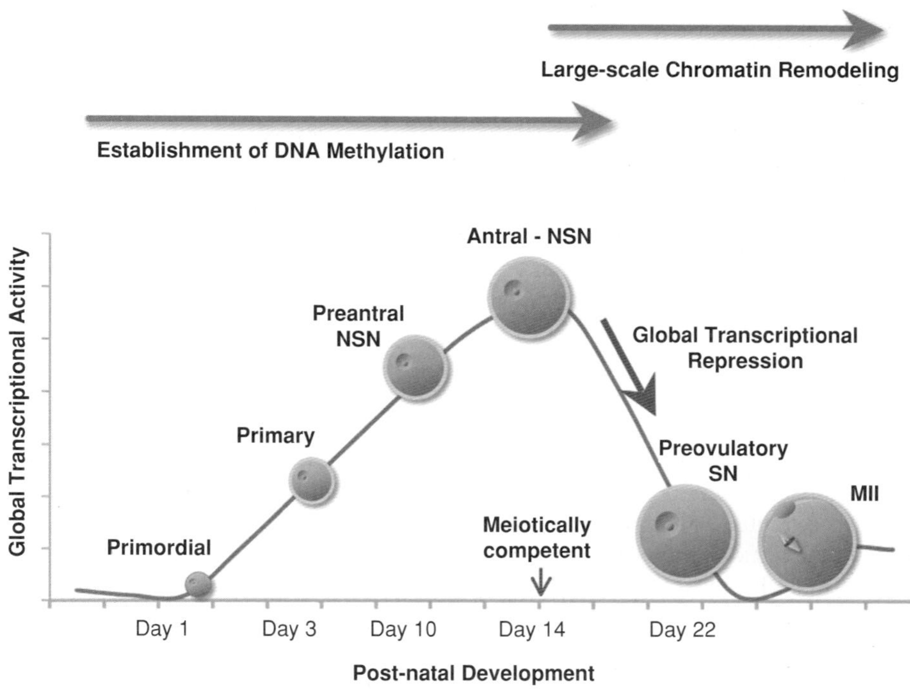

Figure 7.1 Temporal representation of global transcriptional activity in the growing and fully grown oocyte. Major chromatin remodeling in oocytes during the postnatal period confers the female gamete with full meiotic and developmental potential. Transcriptional and translational activity is under stringent control to allow for the timely synthesis and storage of dormant maternal mRNAs. In addition, the locus-specific methylation of CpG dinucleotides at single-copy genes contributes to the establishment of maternal imprints during this critical developmental window.

(Figure 7.1). A fraction of nascent transcripts is quickly translated to sustain the high levels of protein synthesis required for oocyte growth. In contrast, transcriptional activation of dormant maternal mRNAs in growing oocytes is coordinated with their translational repression by limiting the extent of polyadenylation at the 3′ untranslated region (UTR) (Richter, 2001). Polyadenylation and subsequent translation of oocyte-specific transcripts such as the synaptonemal complex proteins SYCP1 and SYCP3, growth differentiation factor 9 (GDF9), and the c-Mos proto-oncogene is regulated at different developmental transitions by the association of the cytoplasmic polyadenylation element binding protein (CPEB) with regulatory sequences at the 3′UTR region. Targeted deletion of CPEB in mice induced abnormal oocyte growth with progressive depletion of the ovarian reserve and infertility due to defective translation of multiple oocyte-specific factors (Racki & Richter, 2006). Thus, the timely synthesis and accumulation of maternal products such as cell cycle–related molecules is essential for the oocyte's acquisition of meiotic and developmental competence (Evsikov et al., 2006). This is also a critical window during which oocyte-specific epigenetic modifications take place, including the establishment of maternal methylation patterns for several imprinted genes (Fedoriw et al., 2004; Lucifero et al., 2004; Morgan et al., 2005). The primary mechanisms responsible for the establishment and maintenance of maternal imprints are not fully understood. However, the role of several factors including de novo DNA methyltransferases such as Dnmt3a and Dnmt3b as well as the DNA methyltransferase-like (Dnmt3L) (Bourc'his et al., 2001; Bourc"his & Bestor, 2006) are beginning to be elucidated (Ferguson-Smith & Surani, 2001;

Surani, 2001; Morgan et al., 2005; Surani et al., 2007). Initiation of oocyte growth is sufficient to trigger transcriptional activation of the maintenance methylase Dnmt1 as well as Dnmt3L, and loading of Dnmt1 to the nucleus seems to be an oocyte autonomous process mediated by stem cell factor-dependent stimulation of the phosphatidyl inositol 3 kinase (PI3K) pathway (Lees-Murdock et al., 2008).

7.3 Establishment and Maintenance of DNA Methylation during Oocyte Growth

In somatic cells, DNA methylation is maintained during cell division by the actions of the maintenance DNA methyltransferase (DNMT-1) as well as the lysine demethylase LSD1 also known as lysine demethylase 1B (KDM1B) (Bird & Wolffe, 1999; Chen & Li, 2006; Wang et al., 2009b). Most methylation marks are found at clusters of (CpG) dinucleotides called CpG islands where this chemical modification has been suggested to interfere with the binding of transcription factors, thus creating a transcriptionally repressive chromatin environment (Bird & Wolffe, 1999; Chen & Li, 2006). Global analysis of methylation patterns indicates that most cytosine methylation takes place at intergenic or nonregulatory regions as well as repetitive elements of the mammalian genome. Lack of DNA methylation results in a significant increase in the rates of spontaneous mutations as well as abnormal chromosome structure in somatic and meiotic cells, suggesting a prominent role for DNA methylation in epigenetic gene silencing and maintenance of genome stability (Chen et al., 1998; Bourc'his & Bestor, 2004; De La Fuente et al., 2006; Rollins et al., 2006; Weber & Schuebeler, 2007).

Establishment of de novo methylation in the oocyte genome takes place on a locus-by-locus basis during the critical window of postnatal oocyte growth and results in the establishment of a sex-specific mark or genomic imprint that is strictly required for normal preimplantation development (Barton, 1984; McGrath, 1984; Surani, 1984; Obata, 1998). Acquisition of methylation marks at different imprinted loci depends on oocyte growth. For example, analysis of eight imprinted genes following bisulfite modification revealed that establishment of methylation marks at different imprinted loci is an oocyte-size dependent process in which oocytes of 55–60 μM in diameter obtained from juvenile females had acquired the correct methylation pattern for Igf2r, Lit1, Zac1 and Snrpn, Peg1, and Grb10 respectively (Hiura *et al.*, 2006). Importantly, most currently identified parental-specific imprints are established in the growing oocyte as only a few loci have been described to acquire a paternal imprint in the male germline (Ferguson-Smith & Surani, 2001; Reik et al., 2001; Obata, 2002; Lucifero et al., 2004; Hiura et al., 2006). Differences in the patterns of DNA methylation between the paternal and maternal genomes are critical to regulate allele-specific gene expression and thus constitute the basis of genomic imprinting in mammals (Bourc'his et al., 2001; Bestor & Bourc'his, 2004; Kaneda et al., 2004). Notably, imprinted genes are required for proper placental differentiation, regulation of fetal growth, postnatal development, and maternal behavior in mice (Tilghman, 1999; Moore, 2001; Kelly & Trasler, 2004). Aberrant DNA methylation at specific Imprinting Control Regions (ICRs) is associated with several human syndromes such as the imprinting overgrowth disorder Beckwith-Wiedemann-Syndrome (BWS). BWS is, among other symptoms, characterized by increased birth weight and length, midline abdominal wall defects, and neonatal hypoglycemia (DeBaun & Tucker, 1998), and arises as consequence of demethylation of the *KvDMR1.* region within the KCNQ1OT gene on the maternal 11p15 allele. Several recent reports indicate, moreover, that standard assisted reproductive techniques, such as embryo transfer and *in vitro* culture of oocytes and embryos, are associated with DNA methylation changes at specific imprinted genes, such as H19, Snrpn, Igf2, Kcnq1ot1, Cdkn1c, Kcnq1, Mknr3, Ascl2,

Zim1, and Peg3 and pose a significantly increased risk for the development of imprinting disorders in the offspring (Doherty et al., 2000; Gicquel et al., 2003; Rivera et al., 2008; Boonen et al., 2008; DeBaun et al., 2003; Maher et al., 2003; van Montfoort et al., 2012).

DNA methylation patterns were initially thought to be established during oogenesis by the combined activity of de novo DNA methyltransferase enzymes Dnmt3a and Dnmt3b and their functional interaction with a DNA methyltransferase-like (Dnmt3L) enzyme, which bears structural similarities to Dnmt3a/b but lacks DNA methyltransferase activity (Bourc'his et al., 2001; Kaneda et al., 2004). However, studies subsequently demonstrated that Dnmt3b is dispensable for the establishment of maternal imprints during oogenesis and provided conclusive evidence that de novo methylation patterns at maternally imprinted loci are established by the functional interaction of Dnmt3a with Dnmt3L in the growing oocyte (Kaneda et al., 2009) where Dnmt3L is essential for directing the targeting of Dnmt3a2 to specific imprinted control regions (ICRs) (Nimura et al., 2006; Kaneda et al., 2009). The mechanism for specific targeting of this protein complex to ICRs of specific loci is not clear. However, recent studies have elegantly demonstrated that the patterns of histone methylation at ICRs might be critical for this process. Indeed, the first example of a functional relationship between histone methylation and DNA methylation during oogenesis was demonstrated after the analysis of methylation patterns in the oocytes of mice deficient for the histone demethylase (KDM1). Functional ablation of KDM1, an abundant nuclear protein in growing oocytes, results in higher levels of histone H3 di-methylation at lysine 4 (H3K4me2) and disrupts the establishment of maternal imprints in four out of seven loci in the mouse oocyte (Ciccone et al., 2009). These studies also provided interesting evidence indicating a critical role for a histone modification such as H3K4Me2 taking place at a global scale on the establishment of proper DNA methylation patterns for maternally imprinted genes. Additional factors such as the KRAB zinc finger protein Zfp57 may exhibit a direct interaction with DNA to establish the maternal imprinting signal at the Snrpn imprinted region (Li et al., 2008). Perhaps one of the most intriguing recent advancements toward unraveling the mechanisms responsible for the establishment of maternal imprinting is the potential role that transcription across imprinted control regions may play in the establishment and maintenance of imprinting marks during oogenesis (Chotalia et al., 2009). In these studies, transcripts originating from promoters within one of the DMRs of the GNAS locus as well as yet to be characterized histone modifications at regulatory sequences were found to be required for the establishment of maternal imprints (Chotalia et al., 2009). However, whether this mechanism is functional at ICRs of other maternally imprinted genes remains to be established.

Functional differentiation of global chromatin structure during postnatal oocyte growth is also critical for the acquisition of meiotic and developmental competence (Zuccotti et al., 1995; De La Fuente & Eppig, 2001; De La Fuente et al., 2004a; De La Fuente, 2006). However, little is known regarding the mechanisms regulating the dynamics of histone and chromatin modifications or the functional significance for the myriad combinations of potential acetylation or methylation changes at specific lysine residues during oogenesis. For example, previous studies indicate that global DNA methylation, as determined by 5-methyl cytosine staining can be initially detected in growing oocytes on day 10 of postnatal development (Kageyama et al., 2006; Kageyama et al., 2007). However, the relationship between global and locus-specific DNA methylation during oocyte growth remains to be determined. In addition, the levels of both methylation and acetylation at several lysine residues of histone H3 and H4 increase with oocyte growth together with the expression of several histone-modifying enzymes such as histone acetyltransferases and histone methyltransferases (Kageyama et al., 2006; Kageyama et al., 2007). An increasing number of histone and chromatin modifications have been described in the oocyte genome of several species, including human oocytes (Table 7.1). However, little is known concerning the functional interaction of these global epigenetic

Table 7.1 Chromatin modifications during meiotic progression in domestic animal oocytes according to the Brno nomenclature.

Species	Type of Histone Modification	GV	GVBD	MI	AT-I	MII	Reference
	H3S10ph	+	+++	+++	+++	+++	(Gu et al., 2008)
		–	+++	+++	+++	+++	(Jelinkova & Kubelka, 2006), (Bui et al., 2007)
	H3S28ph	+	+++	+++	+++	+++	(Jelinkova & Kubelka, 2006), (Bui et al., 2007)
		–	+++	+++	+++	+++	(Gu et al., 2008)
	H3K9ac	+++	+++	+++	+++	+++	Wang et al., 2006)
		+++	–	–	+	–	(Bui et al., 2007), (Jeong et al., 2007)
		+++	+++	–	n.d.	–	(Endo et al., 2005)
		+	–	–	–	n.d.	(Xue et al., 2010)
	H3K14ac	+++	+++	+++	+++	+++	(Wang et al., 2006)
		+++	+	–	+	–	(Bui et al., 2007)
		+++	+++	+	n.d.	+	(Endo et al., 2005), (Endo et al., 2008)
Pig	***H3K18ac***	+++	+	–	–	–	(Bui et al., 2007)
		–	+	++	+	n.d.	(Xue et al., 2010)
	H3K23ac	–	–	+	+	n.d.	(Xue et al., 2010)
	H3K9me2	+	++	++	++	n.d.	(Xue et al., 2010)
	H3K9me3	+++	+++	+++	+++	+++	(Endo et al., 2005), (Bui et al., 2007)
	H3R17me2	–	–	+	+	n.d.	(Xue et al., 2010)
	H4K5ac	+++	+/–	–	+++	–	(Wang et al., 2006)
		+++	+++	–	+	–	(Endo et al., 2005)
	H4K8ac	+++	+++	+++	+++	+++	(Wang et al., 2006)
		+++	+++	+	n.d.	+	(Endo et al., 2005), (Endo et al., 2008)
	H4K12ac	+++	+++	+++	+++	+++	(Wang et al., 2006)
		+++	+++	+	+++	+	(Endo et al., 2005)
		n.d.	n.d.	n.d.	n.d.	+	(Cui et al.)
	H4K16ac	+++	–	–	+++	–	(Wang et al., 2006)
	H3K9ac	–	n.d.	n.d.	n.d.	–	(Hou et al., 2008)
		+++	++	++	++	++	(Tang et al., 2007)
Sheep	***H4K5ac***	–	n.d.	+	+++	+ +	(Tang et al., 2007)
	H4K12ac	+++	n.d.	+	+++	+++	(Tang et al., 2007)
	H3K9me3	–	n.d.	n.d.	n.d.	+++	(Hou et al., 2008)
	H3S10ph	n.d.	n.d.	n.d.	n.d.	+	(Uzbekova et al., 2008)
	H3K9ac	+++	+++	–	n.d.	–	(Wee et al., 2010)
	H3K9me	+++	+++	+++	n.d.	+++	(Wee et al., 2010)
	H3K9me2	++	++	++	n.d.	++	(Racedo et al., 2009)
		+++	+++	+++	n.d.	+++	(Wee et al., 2010)
	H3K9me3	+++	+++	+++	n.d.	+++	(Wee et al., 2010)
Bovine	***H3K27me3***	+++	n.d.	n.d.	n.d.	++	(Ross et al., 2008)
	H4K5ac	+++	++	n.d.	n.d.	–	(Maalouf et al., 2008)
		+++	+++	–	n.d.	–	(Wee et al., 2010)
	H4K8ac	+++	+ +	n.d.	n.d.	+	(Maalouf et al., 2008)
	H4K12ac	+++	++	n.d.	n.d.	–	(Maalouf et al., 2008)
		+++	+++	+	n.d.	+	(Racedo et al., 2009)
	H4K16ac	+++	++	n.d.	n.d.	–	(Maalouf et al., 2008)
Horse	***H3K9me3***	n.d.	n.d.	n.d.	n.d.	+++	(Vanderwall et al., 2010)

+++: strong intensity of signals; ++: decreased intensity of signals; +: weak intensity of signals; +/–: detectable in some oocytes, but absent in other oocytes; –: not detectable; n.d.: not determined (ac – acetylation; me2 – dimethylation; me3 – trimethylation; ph – phosphorylation) modifications with the control of transcriptional activity and/or large-scale chromatin remodeling in the oocyte genome (Adenot et al., 1997; De La Fuente et al., 2004a; Kimura et al., 2004; Spinaci et al., 2004; Kageyama et al., 2007). Clearly this is a critical area requiring further investigation.

7.4 Large-Scale Chromatin Remodeling during Meiotic Division

7.4.1 Changes in Chromatin Structure and Function in Fully Grown Oocytes

Mammalian meiosis is an intricate cellular process consisting of two sequential cell divisions without an intervening S-phase in order to create haploid gametes from diploid progenitors. During the initial reductional division of meiosis-I in oocytes, one set of homologous chromosomes is extruded into the first polar body and the oocyte arrests at metaphase-II. Subsequently, in response to fertilization, sister chromatids become separated during the equatorial division of meiosis-II resulting in the formation of a second polar body (Hassold & Hunt, 2001; Eppig et al., 2004). Unique structural and functional properties are required to orchestrate the accurate segregation of homologous chromosomes during meiosis-I, and of sister chromatids during meiosis-II. For example, in addition to complex interchromosomal interactions, such as centromeric cohesion and formation of chiasmata, meiotic division also requires specialized spindle microtubule attachments with paired centromeres at homologous chromosomes during metaphase-I or with individualized centromeric domains of sister chromatids at metaphase-II (Page & Hawley, 2003; Petronczki et al., 2003). Therefore, mammalian oocytes undergo striking chromatin remodeling events in preparation for meiotic division with far-reaching implications for the individualization of chromosome bivalents and the maintenance of chromosome stability during meiosis. Importantly, a growing body of evidence suggests that these chromatin remodeling events, and hence the differentiation of chromatin structure and function, during oogenesis form an essential prerequisite for the acquisition of both meiotic and developmental competence (De La Fuente et al., 2004a; De La Fuente, 2006; Yang et al., 2009; Baumann et al., 2010).

With the onset of puberty, small groups of oocytes resume meiosis during each estrus cycle and are then characterized by striking and rapid morphological changes in global chromatin structure and function. For instance, chromatin in growing oocytes initially exhibits a highly decondensed chromatin distribution, the non-surrounded nucleolus (NSN) configuration (Mattson & Albertini, 1990; Debey et al., 1993). This stage is characterized by high levels of transcriptional activity in the oocyte genome required for the synthesis and storage of maternal transcripts essential for meiotic resumption and the first cleavage divisions. Subsequent oocyte growth and differentiation leads, however, to a dramatic nuclear reorganization characterized by progressive chromatin condensation, and the formation of a perinucleolar heterochromatin rim characteristic of the surrounded nucleolus (SN) configuration in fully grown preovulatory oocytes (Parfenov et al., 1989; Debey et al., 1993; Zuccotti et al., 1995; De La Fuente, 2006).

Concomitantly with these large-scale chromatin remodeling events, oocytes also undergo a remarkable transition from high levels of transcriptional activity found in NSN oocytes to global transcriptional repression at the SN stage (Figure 7.1). Interestingly, both processes, global chromatin remodeling and a shift toward transcriptional quiescence, have been shown to be of vital importance for the timely progression of meiotic maturation (Wickramasinghe et al., 1991; Debey et al., 1993; Schramm et al., 1993; De La Fuente, 2006) as well as for the oocyte's developmental potential. This notion is supported by studies reporting significantly higher rates of blastocyst formation in SN-GV oocytes following *in vitro* fertilization (Zuccotti et al., 1998).

The regulatory mechanisms set in place to control these complex processes remain poorly understood. Studies indicate the potential involvement of paracrine factors originating from surrounding granulosa cells (De La Fuente & Eppig, 2001; Liu & Aoki, 2002). In addition, analysis of animal models, such as transgenic mice lacking the nuclear chaperone nucleoplasmin 2 (Npm2), provided critical insight to determine whether there is a potential functional interaction between chromatin remodeling processes and global transcriptional repression in the mammalian oocyte. Accordingly, chromatin of oocytes lacking nucleoplasmin 2 remains in a decondensed, NSN-like status of growing oocytes, while a transition into the SN configuration is not observed following gonadotropin stimulation. In contrast, timely genome-wide transcriptional silencing remains unaffected in this model system as indicated by a lack of nascent transcripts in gonadotropin-stimulated Npm2 mutant oocytes (De La Fuente et al., 2004a). These findings provided the initial evidence that global transcriptional repression in GV stage oocytes can be experimentally separated from the chromatin remodeling events leading to the transition into the SN configuration (De La Fuente et al., 2004a). In support of this hypothesis, studies indicate that global transcriptional silencing is due to dissociation of the largest subunit of RNA polymerase II (RPB1) from the chromatin template following dephosphorylation of RPB1 in the GV of fully grown oocytes (Abe et al., 2010). Interestingly, oocytes lacking MLL2, a major histone H3K4 methyltransferase, fail to undergo global, RNA polymerase II-dependent, transcriptional repression despite major chromatin remodeling from the NSN to the SN chromatin configuration (Andreu-Vieyra et al., 2010). In addition, pharmacological stimulation of GV stage oocytes using the histone deacetylase inhibitor TSA results in prominent chromatin decondensation and severely affects the structure of the karyosphere (De La Fuente et al., 2004a). Nevertheless, TSA-induced chromatin decondensation is not sufficient to restore transcriptional activity as determined by transcription run-on assays (De La Fuente et al., 2004a; Abe et al., 2010). Collectively, these studies indicate that despite being temporally linked, large-scale chromatin remodeling and global transcriptional silencing in the oocyte genome are autonomous developmental processes under the control of distinct cellular pathways and that chromatin condensation into the SN configuration is not required for global transcriptional silencing.

Currently, most of the experimental evidence on chromatin remodeling events in female germ cells originates from studies with mouse oocytes. Nevertheless, similar large-scale changes in chromatin configuration have also been described in human (Parfenov et al., 1989), monkey (Lefevre et al., 1989; Schramm et al., 1993) and rat oocytes (Mandl, 1962). However, analysis of chromatin configuration in preovulatory oocytes in domestic animal species, such as pig {McGaughey, 1979 #1008), cattle (Liu et al., 2006b), sheep (Russo et al., 2007), dog (Lee et al., 2008a), horse (Hinrichs et al., 2005), goat (Sui et al., 2005) and rabbit (Wang et al., 2009a) showed a remarkable diversity in the type of chromatin configuration as well as the morphological appearance of the nucleolus.

Although chromatin transitions observed in rat oocytes show patterns equivalent to those in the mouse (Mandl, 1962; Zucker et al., 2000), the final stages of oogenesis in human (Combelles et al., 2009) and rabbit (Wang et al., 2009a) germ cells exhibit different chromatin patterns. In these species, oocyte chromatin converts successively from a non-surrounded nucleolus configuration in preantral follicles to a so-called netlike (type A (human)/NL (rabbit)), loosely condensed (B/LC), tightly condensed (C/TC), and finally a singly condensed (D/SC) configuration (Parfenov et al., 1989; Wang et al., 2009a). Types A through D in human as well as type NL, LC, TC, SC in rabbits correspond thereby to a surrounded nucleolus configuration, although transcriptional activity is known to cease only in type C and D (human) as well as type TC and SC (rabbit) oocytes (Parfenov *et al.*, 1989; Wang *et al.*, 2009a).

In monkeys, non-surrounded (NSN) oocytes in preantral follicles are termed GV1 and show dispersed and filamentous chromatin. Concurrently with follicular growth, chromatin condenses and transforms in this species into a configuration partially (GV2) or completely (GV3) surrounding the nucleolus (Lefevre et al., 1989; Schramm et al., 1993).

Small antral follicles in porcine ovaries contain oocytes with a decondensed chromatin configuration similar to the murine NSN stage, here termed GVO (Sun et al., 2004b). During follicular development chromatin condenses to form a nucleolar rim or, alternatively, a so-called horseshoe configuration in transcriptionally inactive GV1, GV2, or GV3 oocytes (Motlik & Fulka, 1976; Guthrie & Garrett, 2000; Sun et al., 2004b). By contrast, chromatin configuration in bovine oocytes is distinguished based on a diffuse (NSN), netlike (N), or clumped (C) chromatin appearance near the nuclear envelope as well as the nucleolus in non-surrounded GV stage oocytes (Liu et al., 2006b). Finally, transcriptionally quiescent bovine oocytes also exhibit a clumped chromatin configuration in close apposition to a small nucleolus, somewhat resembling the SN configuration observed in the mouse, and are transcriptionally quiescent (Liu et al., 2006b).

Interestingly, in some species such as the goat (Sui et al., 2005), the dog (Lee et al., 2008a) and the horse (Hinrichs & Williams, 1997), oogenesis culminates in a different chromatin structure. In the horse, the chromatin configuration corresponding to the NSN-equivalent stage in the mouse is termed "fibrillar" with strands of chromatin projecting throughout the GV, or "intermediate" with irregular masses of chromatin throughout half the volume of the GV. Chromatin in equine oocytes acquires, moreover, configurations of loosely (LCC) and tightly (TCC) condensed irregular or circular chromatin masses (Hinrichs et al., 2005). A further chromatin type, the "fluorescent nucleus" with homogeneous chromatin over the entire GV, the significance of which requires further investigation, is primarily found following prolonged time periods between tissue harvest and oocyte fixation (Hinrichs et al., 2005). Canine GV stage oocytes show chromatin configurations similar to the horse with a fibrillar/intermediate type termed GV1, as well as a loosely condensed type (GV2) and a tightly condensed type (GV4/5) (Lee et al., 2008a). Goat chromatin configurations are classified as GV1 containing a large nucleolus and diffuse chromatin, GV2 with netlike chromatin and a medium-sized nucleoli, and GV2c with clumped chromatin. Further types are distinguished based upon small nucleoli and netlike (GV3n) or clumped chromatin (GV3c) as well as clumped chromatin and the lack of a nucleolus (GV4). Goat GV stage oocytes represent an important exception to the general process of NSN to SN transitions in that a surrounded GV stage is not formed in this species. Nevertheless, in goat oocytes, transcription ceases from the GV2c stage onwards (Sui et al., 2005). Importantly, the systematic analysis of chromatin configuration as well as the position of the germinal vesicle in fully grown oocytes of several species might provide an important noninvasive marker of the oocyte's meiotic and developmental potential (Brunet & Maro, 2007; Inoue et al., 2008; Bellone et al., 2009). However, further studies are required so we may understand the significance of such diverse chromatin configurations in domestic species and their potential association with nucleolar structure and function. The regulatory pathways necessary to coordinate these global changes in nuclear architecture during the transition into the SN configuration in mammalian oocytes remain largely unexplored. In insects, a meiosis-specific organization of chromosomes (the karyosome) with notable functional similarities to the karyosphere of mammalian oocytes forms during prophase-I arrest (Ivanovska & Orr-Weaver, 2006). Karyosheres as well as karyosomes are both essential to maintain a specific meiotic chromosome arrangement in order to facilitate meiotic chromosome segregation (Gruzova & Parfenov, 1993). Interestingly, recent evidence obtained in this model organism suggests a critical role for a conserved histone H2A kinase, NHK-1, in nuclear chromatin remodeling that functions under control of the meiotic prophase-I checkpoint in Drosophila oocytes (Ivanovska et al., 2005). Lack of NHK-1 interferes with the formation of the karyosome and leads to complete sterility due to chromosomal defects and abnormal polar body formation. Moreover, failure to form a karyosome in nhk-1 mutant oocytes is known to result in absence of histone H2A phosphorylation (H2AT119ph), lack of histone H3 (H3K14ac), and histone H4 (H4K5ac) acetylation as well as failure to dissociate the synaptonemal complex and to load condensin complexes onto chromosomes (Ivanovska et al., 2005; Ivanovska & Orr-Weaver, 2006). These observations support the hypothesis that condensins play a critical role in the formation of the karyosphere in mammalian oocytes (Ivanovska & Orr-Weaver, 2006). The importance of this histone kinase in the establishment of the unique chromatin configuration in preparation for meiotic resumption in mammalian oocytes has been confirmed by functional deletion of the vaccinia-related kinase (VRK1), the mammalian homologue of Drosophila NHK-1 (Schober et al., 2011). In mice, lack of VRK1 function results in male sterility due to a proliferation defect in differentiating spermatogonia (Wiebe et al., 2010). Moreover, females deficient for VRK1 exhibit abnormal chromatin configuration in which only a small fraction of oocytes undergoes the transition into the SN configuration, in addition to a delay in meiotic progression, chromosome segregation defects, and fertilization failure (Schober et al., 2011).

7.4.2 Histone Modifications during the Resumption of Meiosis

Large-scale chromatin remodeling is defined as a series of genome-wide changes in nuclear structure recognized at the chromosomal level that occur in response to environmental or differentiation stimuli (Berger & Felsenfeld, 2001; Cremer & Cremer, 2001). Like many other cell types, chromatin remodeling in differentiating mammalian germ cells is regulated through the induction of histone post-translational modifications such as histone acetylation or histone methylation at different lysine residues (Bannister et al., 2002; Kouzarides, 2007), or through the incorporation of histone variants such as histone H3.3, CENP-A, and H2A.Z (Sarma & Reinberg, 2005; Polo & Almouzni, 2006), as well as the action of ATP-dependent chromatin remodeling proteins (Fry & Peterson, 2001; Varga-Weisz, 2001; Davis & Brachmann, 2003). However, a growing body of evidence from different model organisms points to the existence of control mechanisms specific to the germline that are essential to regulate global transcriptional activity and chromatin remodeling, and ultimately enable the unique structural and functional properties of meiotic chromosome arrangements necessary to form haploid gametes (Sassone-Corsi, 2002; Kimmins & Sassone-Corsi, 2005).

Covalent post-translational modifications on core histones have direct effects on higher chromatin structure and function by inducing alterations in the organization and molecular composition of the nucleosome (Langst, 2001; Tsukiyama, 2002; Luger, 2003). Histone modifications also have the potential to alter the degree of chromatin condensation and thereby, regulate the accessibility of chromatin-associated factors to regulatory sequences (Grunstein, 1997; Goldberg et al., 2007) generating, depending on the context of histone modifications, either a transcriptionally permissive or repressive chromatin environment (Cheung et al., 2000; Margueron et al., 2005). Multiple histone posttranslational modifications, such as methylation (Bannister et al., 2002), acetylation (Grunstein, 1997), phosphorylation (Peterson & Laniel, 2004), poly(ADP)-ribosylation (Faraone-Mennella, 2005), ubiquitylation (Zhang, 2003), and sumoylation (Gill, 2004) may occur simultaneously at different amino acid residues (i.e., lysine, serine, proline, arginine) of the core histones (H2A, H2B, H3, and H4). Therefore, covalent histone modifications form a larger chromatin context that confers structural and functional identity to individual nuclear domains in response to the cellular environment or the transcriptional status (Cleveland et al., 2003).

In turn, such nuclear domains become organized into functional compartments that are essential for the control of gene expression and proper chromosome segregation (Dundr & Misteli, 2001; Dillon & Festenstein, 2002) (Figure 7.2). A number of chromatin modifications associated with

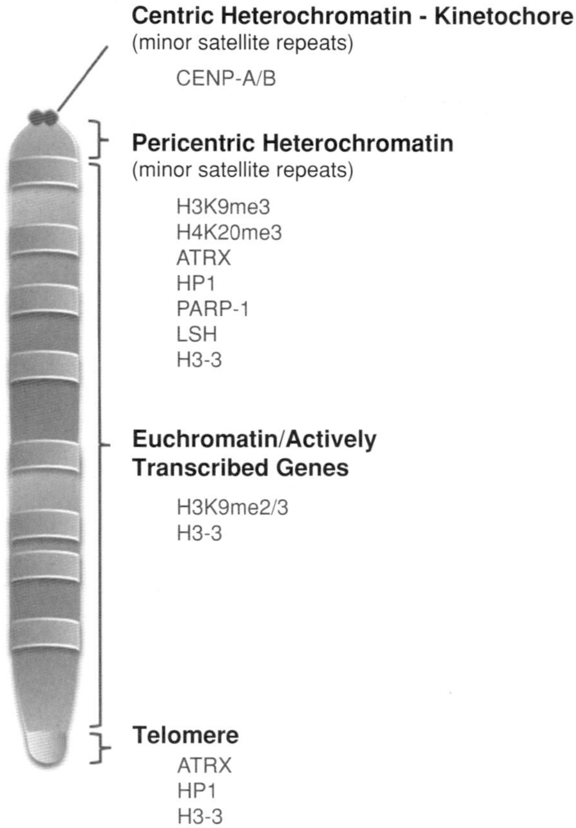

Figure 7.2 Schematic representation of chromatin organization in condensed chromosomes. Incorporation of histone post-translational modifications and chromatin-associated factors, such as H3K9me3 and ATRX, are essential for the proper chromatin organization and the formation of functional chromosomal subdomains, such as centric and pericentric heterochromatin, euchromatin, and telomeres. Characteristic chromatin components that are discussed in the text are listed for each individual subdomain.

meiotic chromosome condensation and segregation in oocytes of several mammalian species have been described to date (Table 7.1). Interestingly, histone methylation patterns appear to be highly conserved between domestic animal species, as exemplified by the levels of histone H3K9 di- and trimethylation during meiotic resumption in pigs, sheep, cows, and horses. However, other histone modifications, in particular histone acetylation, seem to have evolved species-specific differences in deacetylation dynamics in female gametes in the bovine and in pigs compared to sheep oocytes. Undoubtedly, future research will aid in elucidating a more complete picture of specific chromatin remodeling dynamics in individual domestic species. In addition, broadened insight into epigenetic chromatin modification patterns may assist in the development of optimized *in vitro* culture and embryo manipulation protocols and also serve as prognosis marker for gamete quality and pregnancy outcome.

Moreover, little is known concerning the enzymatic activities that are responsible for inducing these specific post-translational modifications during the different stages of meiotic maturation. Current evidence indicates the establishment of a subset of epigenetic marks, such as histone lysine methylation, during the oocyte growth period, which remain stably associated with specific chromatin domains of meiotic chromosomes throughout meiosis (Arney et al., 2002; Cowell et al., 2002; Hodges & Hunt, 2002; Fu et al., 2003; Kourmouli et al., 2004; Liu et al., 2004; Wang et al., 2006; Swain et al., 2007; Meglicki et al., 2008; Ooga et al., 2008). Importantly, epigenetic modifications such as histone acetylation exhibit highly dynamic localization patterns during meiotic resumption (Adenot et al., 1997; Kim et al., 2003; De La Fuente et al., 2004b; Sarmento et al., 2004; Huang et al., 2007).

Several histone methyltransferases have also been identified to date, and genetic loss of function analyses have provided critical insight into the implications of histone methylation during gametogenesis, meiosis, and preimplantation development. For instance, the lysine residues 4 and 9 of histone H3 (H3K4me, H3K9me) may be enzymatically modified to exist as mono-, di-, or tri-methylated forms (H3K4me, H3K4me2, H3K4me3, etc.). Experimental interference with mono- and di-methylation of H3K9 through targeted deletion of the methyltransferase G9A (KMT1C) leads to abnormal chromosome synapsis during prophase-I and widespread meiotic abnormalities (Tachibana et al., 2007). Loss of ESET, a methyltransferase targeting H3K9me to euchromatic domains, impairs peri-implantation development in mice (Dodge et al., 2004). Moreover, simultaneous inactivation of the methyltransferases SUV39H1/H2 hinders H3K9 methylation at pericentric heterochromatin, a crucial chromosomal subdomain for the regulation of centromere cohesion and the timely separation of individual chromatids during mitotic divisions in somatic cells (Guenatri et al., 2004), resulting in meiotic phenotypes in pachytene spermatocytes similar to those reported in G9A knockout mice (Peters et al., 2001). Pericentric heterochromatin is also critically important to support the dynamic and highly specialized centromeric interactions essential for homologue segregation during metaphase-I as well as the prevention of erroneous sister chromatid separation before the end of metaphase-II (Petronczki et al., 2003).

The observation that mammalian centromeres consist of repetitive DNA sequences at centric heterochromatin that are necessary, but not sufficient for centromere formation, suggested the existence of an epigenetic mechanism in the regulation of centromere function (Karpen & Allshire, 1997; Dillon & Festenstein, 2002). In fact, tri-methylation of H3K9 as a constitutive component of pericentric heterochromatin has been implicated in centromeric heterochromatin formation and the maintenance of a transcriptionally repressive chromatin environment. Moreover, H3K9me3, together with H4K20me3, act as docking sites for additional chromatin binding proteins such as heterochromatin protein 1 (HP1) and the chromatin remodeling protein ATRX (Rea et al., 2000; Bannister et al., 2001; Lachner et al., 2001; Schotta et al., 2004; Kourmouli et al., 2005). Centromere structure and function also entails the incorporation of histone variants, such as CENP-A, to induce the formation of a higher order chromatin structure through histone deacetylation and large-scale chromatin remodeling (Pluta et al., 1995; Karpen & Allshire, 1997; Murphy & Karpen, 1998; Wiens & Sorger, 1998; Henikoff et al., 2001) (Figure 7.2).

Post-translational modifications at lysine residues of histone H3 may also play a central role in epigenetic reprogramming events of nuclear domains other than pericentric heterochromatin. For example, the histone methyltransferase MLL2 is required for bulk H3K4 tri-methylation at interstitial chromosome segments in peri-ovulatory oocytes (Andreu-Vieyra et al., 2010). H3K4me3 is thereby commonly associated with promoter sequences of transcriptionally active genes. Consequently, loss of function of MLL2 has been shown to result in reduced levels of H3K4 tri-methylation at the GV stage, while other histone modifications such as H3K4me1 remained unaffected. Importantly, lack of MLL2 leads to impaired ovulation rates, increased proportions of chromosome misalignment during *in vitro* maturation, and female infertility. In this model system, reduction of H3K4 tri-methylation was also paralleled by an increased level of acetylation of histone H4 at lysine 12 (H4K12ac), indicating a lack of global transcriptional silencing in MLL2 deficient oocytes.

7.4.3 Global Histone Deacetylation during Meiotic Resumption is Required for Proper Chromosome Condensation and Segregation

The timely progression of developmental transitions in chromatin modifications during the resumption of meiosis is an essential epigenetic mechanism for the maintenance of chromosome stability in the female gamete. This process was initially reported after analysis of acetylation patterns at different lysine residues of histones H3 and H4 in mouse oocytes (Kim et al., 2003; De La Fuente et al., 2004b). Accordingly, although preovulatory oocytes with an intact germinal vesicle show high levels of histone H3 and H4 acetylation, a wave of global histone deacetylation occurs coincidently with germinal vesicle breakdown in mouse oocytes as well as in oocytes of several domestic species (Table 1). Among the histone modifications involved in this process, deacetylation of histone H4 at lysine 12 (H4K12ac) is particularly extensive and specific to meiotic chromosome condensation (Kim et al., 2003) as chromosomes of somatic cells remain acetylated at H4K12 during mitosis (Kruhlak et al., 2001). Global histone deacetylation during meiotic resumption also affects the lysine 5 residue of histone H4 (H4K5ac) (De La Fuente *et al.*, 2004b). Although the precise molecular mechanisms involved in this process are not fully understood, studies suggest that histone deacetylases (HDACs) play a critical role in global histone deacetylation during meiosis. For example, exposure of oocytes to the histone deacetylase inhibitor TSA prevents the onset of global deacetylation upon meiotic resumption and results in the formation of hyperacetylated, decondensed chromosomes (Kim et al., 2003; De La Fuente et al., 2004b). In addition, global histone deacetylation during meiosis has been demonstrated to play a critical role in the deposition of heterochromatin binding proteins such as ATRX to centromeric domains, as TSA-induced hyperacetylation of chromosomes is associated with lack of centromeric ATRX foci (De La Fuente et al., 2004b). Inhibition of histone deacetylation induces, moreover, the formation of elongated, decondensed chromosomes and highly abnormal meiotic figures including misaligned chromosomes/chromosome lagging in a large proportion of oocytes (De La Fuente et al., 2004a) leading to severe oocyte aneuploidy and early embryonic demise (Akiyama et al., 2006) (Figure 7.3). Although HDAC1, HDAC2, HDAC3, and HDAC4 are abundant nuclear proteins in GV stage mouse oocytes (Kageyama *et al.*, 2006; Ma & Schultz, 2008) further studies are required to determine the specific HDAC isoform that is responsible for the wave of global deacetylation of meiotic chromosomes during meiotic resumption.

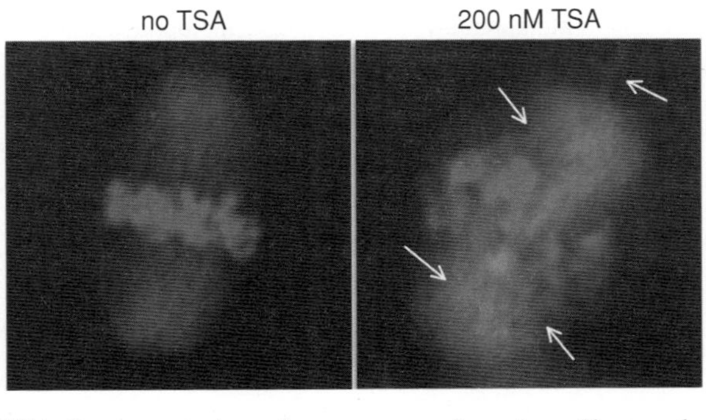

Figure 7.3 Treatment with TSA disrupts metaphase chromosome configurations. Pharmacological interference with histone deacetylation (200nM TSA) during the resumption of meiosis is associated with decondensed chromatin and severe chromosome misalignment (red, arrows) at the metaphase spindle (green), and ultimately leads to the formation of aneuploid gametes. (For color detail, see color plate section.)

7.4.4 Role of Chromatin Remodeling Factors in Chromosome Segregation

Centromeric heterochromatin in mice encompasses two distinct chromosomal subdomains with discrete higher-order chromatin structure and function (Guenatri et al., 2004; Sullivan & Karpen, 2004). Centric heterochromatin consists of several hundred kilobases of minor satellite repeat units and contributes, together with the incorporation of centromere-specific proteins, such as histone variant CENP-A, to the formation of a functional kinetochore (Karpen & Allshire, 2004; Maison & Almouzni, 2004; Sarma & Reinberg, 2005; Polo & Almouzni, 2006) (Figure 7.2). By contrast, pericentric heterochromatin is formed by repeats of major satellite sequences and contributes to the formation of large chromatin remodeling complexes through the recruitment of DNA-binding proteins such as heterochromatin protein 1 (HP1) and helicases of the switch/sucrose non-fermenting (SWI/SNF2) family such as ATRX and LSH (McDowell et al., 1999; Yan et al., 2003a; De La Fuente et al., 2004b; Maison & Almouzni, 2004). Importantly, regulation of homologous chromosome interactions, proper chromosome segregation and growth, and differentiation of the mammalian embryo depend critically on both functional centric as well as pericentric heterochromatin domains (Bernard et al., 2001; Peters et al., 2001; Bernard & Allshire, 2002; Houlard et al., 2006). Functional disruption of pericentric heterochromatin proteins involved in DNA methylation of major satellite sequences may, therefore, cause congenital conditions in humans such as ATRX and ICF (immunodeficiency-centromeric instability) syndromes. For example, mutations in the gene encoding the DNA methyltransferase DNMT3b result in aneuploidy, centromeric instability, and developmental anomalies (Xu, 1999; Robertson & Wolffe, 2000), whereas spontaneous mutations in the *Atrx* gene cause the X-linked alpha-thalassemia/mental retardation (ATRX) syndrome in males. In addition to facial dysmorphism these males may also exhibit gonadal dysgenesis.

ATRX belongs to a large protein family of ATP-dependent chromatin remodeling proteins and binds pericentric heterochromatin domains in human and mouse somatic cells. ATRX is essential for the establishment of proper DNA methylation patterns at repetitive sequences within the human genome (Gibbons et al., 1997; Picketts et al., 1998; McDowell et al., 1999; Gibbons et al., 2000). Original studies conducted in mouse oocytes indicate that ATRX binds to pericentric heterochromatin in metaphase-I and metaphase-II chromosomes and is involved in mediating chromosome alignment at the meiotic spindle (De La Fuente et al., 2004b) (Figure 7.4). Interestingly, pericentric heterochromatin integrity is essential to coordinate sister centromere cohesion during mitosis (Guenatri et al., 2004; Maison & Almouzni, 2004) and evidence suggests that chromatin remodeling

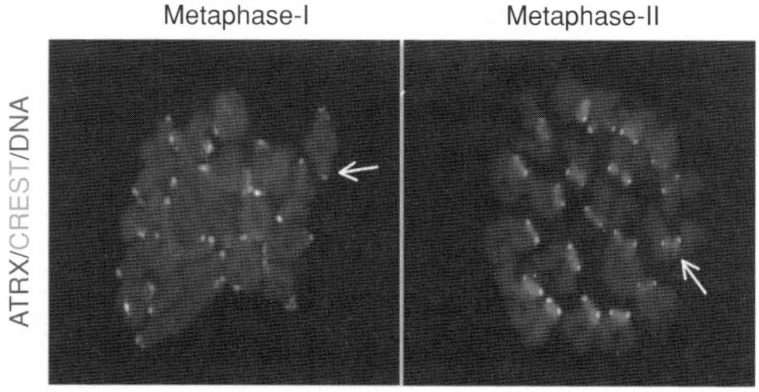

Figure 7.4 Chromatin remodeling proteins, such as ATRX, play an essential role in the functional differentiation of chromatin domains. ATRX (red) is a constitutive component of pericentric heterochromatin during metaphase-I as well as metaphase-II in mouse oocytes (arrows). Chromosomes are shown in blue, and the kinetochore marker CREST is shown in green. (For color detail, see color plate section.)

complexes such as SNF2h are essential to load the cohesin subunit Rad21 onto the centromeres in human mitotic cells (Hakimi et al., 2002). Moreover, loss of HP1 from pericentric heterochromatin in mouse somatic cells deficient for the SUV39 histone methyltransferase severely affects cohesion between sister chromatids (Peters et al., 2001; Maison et al., 2002). Collectively, these studies indicate that pericentric heterochromatin formation plays a major role in centromere cohesion in mitotic cells.

Whether similar functional implications of pericentric heterochromatin are set in place to coordinate the complex chromosome dynamics during meiosis is less clear. However, studies indicate that loss of maternal ATRX function in mouse oocytes contributes to chromosome misalignment at the metaphase-II spindle and pericentromeric DNA breaks leading to severe meiotic centromere instability, the formation of aneuploid embryos, and female subfertility (De La Fuente et al., 2004b; Baumann et al., 2010) (Figure 7.5). Interestingly, ATRX deficiency at pericentric heterochromatin domains results in abnormal individualization of chromosomes and is associated with reduced levels of histone H3 phosphorylation (H3S10ph), known to be an epigenetic hallmark of chromosome condensation. In addition, loss of ATRX function results in failure to recruit the transcriptional regulator DAXX to pericentric heterochromatin at the GV stage (Baumann et al., 2010). Taken together,

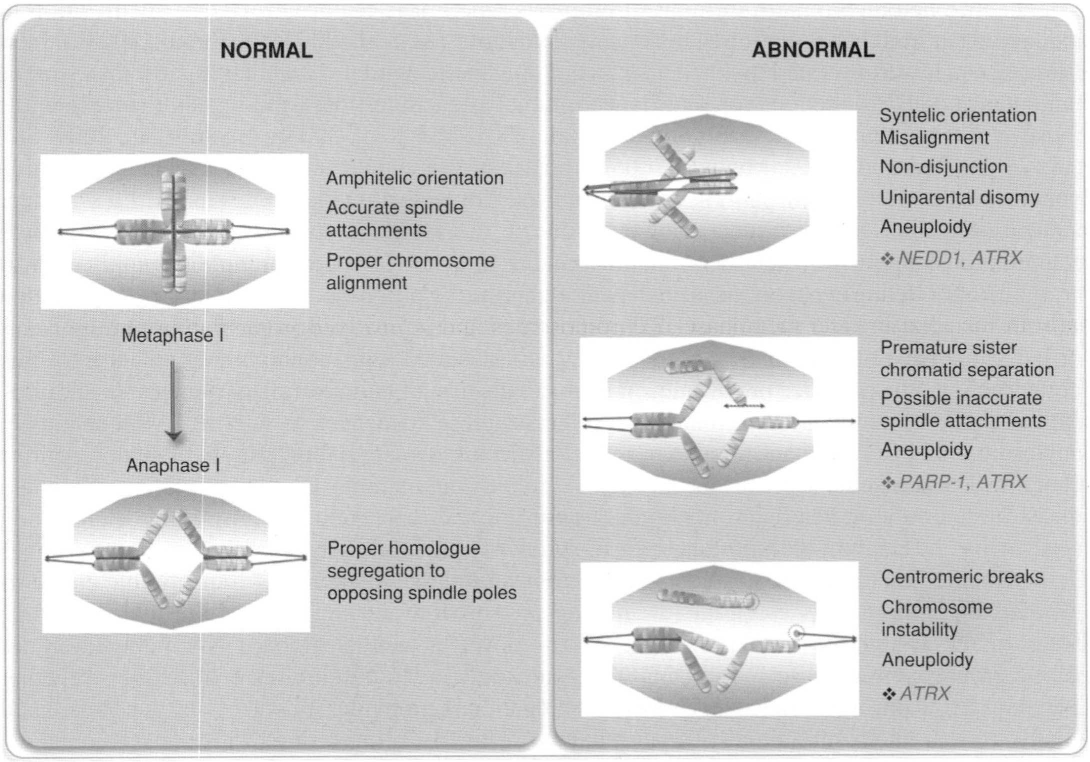

Figure 7.5 Schematic representation of normal (bi-oriented) as well as inappropriate chromosome-spindle microtubule interactions. Proper chromosome alignment at the metaphase spindle depends critically on the establishment of a proper epigenetic chromatin environment as well as on the formation of accurate and stable attachments between kinetochore domains and spindle microtubules. Loss-of-function analyses of key proteins (shown in red) lead to chromosome-microtubule attachment errors. (For color detail, see color plate section.)

these findings support a crucial role of ATRX in the formation and molecular composition of pericentric heterochromatin in the female germline, a process that is essential to mediate the complex chromatin remodeling events during meiosis in mammalian oocytes. Intriguingly, ATRX is a novel component of the epigenetic landscape at the inactive X chromosome in somatic cells and during imprinted X chromosome inactivation in trophoblast stem cells (Baumann & De La Fuente, 2009). These findings suggest that ATRX is an important component of both facultative and constitutive heterochromatin, and underscore a critical role for this chromatin remodeling factor in heterochromatin formation in mammalian cells. Notably, the chromosomal defects observed in ATRX-deficient oocytes provide direct evidence that epigenetic modifications in mammalian oocytes are essential for the maintenance of chromosome stability at metaphase-II and during the transition to the first mitosis. Elucidating the molecular mechanisms through which ATRX regulates chromosome condensation and phosphorylation of histone H3 (H3S10Ph) at pericentric heterochromatin and also throughout the entire length of meiotic chromatids will be an essential step toward our understanding of critical epigenetic factors contributing to the onset of aneuploidy in the female gamete.

The lymphoid-specific helicase (LSH), also known as helicase lymphoid-specific or Hells, is a member of the SWI/SNF2 family of chromatin-remodeling proteins and has recently been implicated in the control of DNA methylation of repetitive sequences in the mammalian genome (Jarvis et al., 1996; Geiman et al., 1998; Sun et al., 2004a; Muegge, 2005; Flaus et al., 2006). LSH protein localizes to pericentric heterochromatin in mouse fibroblasts where it is required for DNA methylation of tandem repeats and the regulation of histone methylation (Yan et al., 2003a; Yan et al., 2003b). Loss of function analyses revealed that LSH is a major epigenetic regulator and that it participates in the maintenance of DNA methylation and transcriptional silencing of parasitic repeat elements in the mouse genome (Dennis et al., 2001; Geiman et al., 2001; Yan et al., 2003b; Huang et al., 2004; Sun et al., 2004a; Fan et al., 2005; Muegge, 2005). Importantly, in female germ cells LSH exhibits highly dynamic nuclear localization patterns during prophase-I of meiosis in that it presents a diffuse nuclear localization at the leptotene stage. However, during the zygotene stage, LSH shows a transient accumulation at pericentric heterochromatin domains, indicating a specific functional implication in chromatin remodeling events at these nuclear domains during prophase-I of meiosis. Loss of function analyses revealed an essential role of LSH protein in mediating homologous chromosome synapsis. Further evidence suggests that LSH deficiency is associated with nonhomologous chromosome interactions and the presence of chromosome univalents in a high proportion of oocytes. In addition, analysis of DNA methylation at major and minor satellite sequences as well as IAP elements in oocytes obtained from mutant females revealed a striking demethylation pattern at such repetitive elements. These results suggest that through its role in maintaining DNA methylation at tandem repeats of centromeric heterochromatin and transcriptional repression of transposable elements in the germ line, LSH might be important to ensure proper synapsis between homologous chromosomes (De La Fuente, 2006). Previous studies also demonstrated that silencing of transposable elements mediated by DNMT3L is essential for meiotic progression and viability of the male germline (Bourc'his & Bestor, 2004; Webster et al., 2005). However, neither global DNA methylation nor meiotic progression was affected in DNMT3L knockout oocytes (Bourc'his et al., 2001). Whether LSH plays a role during meiotic chromosome segregation in preovulatory oocytes is, due to neonatal lethality of Lsh knockout animals, as yet unknown. Notably, allografting of testis tissue obtained from lsh null male fetuses into immunodeficient adult males indicates an essential role for Lsh in homologous chromosome synapsis during spermatogenesis (Zeng et al., 2011).

Poly(ADP) ribosylation, the covalent attachment of ADP-ribose molecules onto histone residues and other proteins, is an epigenetic mechanism that regulates a variety of cellular processes, such as transcription, DNA repair and recombination, chromatin remodeling, genome stability, and the formation of a proper mitotic spindle in somatic cells (D'Amours et al., 1999; Ame et al., 2004; Kim et al., 2004; Schreiber et al., 2006). During oogenesis, PARP-1 (Poly(ADP ribose) polymerase-1), which accounts for the primary enzymatic activity, has been shown to associate with nuclear speckles in fully grown GV-stage oocytes and to relocate dynamically to the spindle poles as well as to pericentric heterochromatin upon meiotic resumption (Yang et al., 2009). Importantly, recent evidence supports a role for poly(ADP) ribosylation in the control of centromere structure and function during oogenesis and meiotic divisions, which is underscored by a spectrum of phenotypic abnormalities observed in oocytes carrying a *Parp-1* null mutation. Accordingly, PARP-1 deficiency during oogenesis predisposed the female gamete to genome instability through mechanisms involving abnormal homologous chromosome synapsis, the resolution of DNA double strand breaks as indicated by incomplete chromosome synapsis and persistence of γH2AX foci in fully synapsed chromosomes at the late pachytene stage during prophase-I of meiosis. In fully grown oocytes, PARP1 deficiency results in abnormal sister chromatid cohesion and failure to maintain metaphase-II arrest, and recruitment of the kinetochore-associated protein BUB3 to centromeric domains as well as premature anaphase-II onset upon removal from the oviductal environment (Yang et al., 2009). These observations indicate that poly(ADP ribosyl)ation is a critical epigenetic modification involved in the regulation of centromere structure and function and, hence, proper chromosome segregation in mammalian oocytes (Figure 7.5).

7.5 Environmental Effects Adversely Influencing the Female Gamete

Numerical chromosome abnormalities, or aneuploidies, in human conceptions commonly originate from errors in meiotic chromosome segregation. In the majority of cases aneuploidy is transmitted through the female gamete, although the underlying mechanisms that predispose oocytes to aneuploidy are not fully understood to date. Advanced maternal age is widely recognized as the single most important factor for an increased risk of chromosomal nondisjunction and premature sister chromatid separation (Hassold & Hunt, 2001; Vialard et al., 2006; Hunt & Hassold, 2008). Accordingly, the incidence of aneuploidy in clinically recognized pregnancies is about 2–3% in young women of fewer than 30 years, but increases to more than 35% of embryos in women in their 40s (Hassold & Hunt, 2001). Moreover, recent studies indicate that, in fact, the majority of human conceptuses present with some degree of chromosome abnormalities, most of which fail to implant or develop to term (Munné et al., 2007). In addition to chromosomal abnormalities, spindle defects, mitochondrial dysfunction, structural hardening of the zona pellucida, and a decrease in maturation-promoting factor (MPF) and mitogen-activated protein kinase (MAPK) activities have all been implicated in age-related declines in fertility and embryonic development (Kikuchi et al., 2000; Pellestor et al., 2003; Baird et al., 2005). In addition, programmed degradation of maternal mRNA stores during meiotic maturation occurs with significantly reduced efficiency in oocytes obtained from aged female mice. As degradation of transcripts in oocytes is not unbiased, this inefficiency may contribute to pathway-specific alterations in gene expression levels involved in cell cycle regulation and chromosome segregation, and may thus contribute to a detrimental impairment of meiotic progression (Pan et al., 2008). Microarray analyses of oocytes from young and old mice support the premise that aging is associated with a deregulation of mechanisms controlling the spindle assembly checkpoint, kinetochore function, and spindle assembly as well as chromatin remodeling processes (Pan et al., 2008). However, subsequent studies indicate that defects in the spindle assembly checkpoint of aged mouse oocytes are not the primary cause of abnormal chromosome segregation (Duncan et al., 2009). Additional studies have shed further light onto the underlying processes by establishing that deterioration of cohesion complexes is a major factor in the etiology of aneuploidy during reproductive senescence (Revenkova et al., 2004; Hodges et al., 2005; Chiang et al., 2010; Lister et al., 2010; Revenkova et al., 2010). For example, cohesins play an essential role as components of the cell cycle machinery in somatic and germ cells through the formation of physical associations between homologous chromosomes (metaphase-I) and sister centromeres (metaphase-II) (Suja & Barbero, 2009; Wood et al., 2010) (Figure 7.5). Proper meiotic chromosome segregation requires, therefore, the coordinate resolution of cohesion at different sub-chromosomal domains during the progression through meiosis-I and -II. However, as chromosomal cohesin loading in oocytes occurs during the S-phase of the cell cycle during fetal development, the time period between incorporation of cohesin subunits, Rec8, Stag3, and SMC1b, and their degradation during metaphase by separase can span up to five decades in human oocytes. Importantly, further studies indicate that reproductive aging in the female gamete leads to a reduction in the amount of the alpha-kleisin protein Rec8 and an increase in the inter-kinetochore distance, suggesting a gradual loss of cohesion (Chiang et al., 2010; Lister et al., 2010). Moreover, aged oocytes also exhibit a significant decrease in chromosomal Shugoshin 2 (Lister et al., 2010), a protein crucial to prevent cleavage of centromeric cohesins during metaphase-I and, hence, premature sister chromatid separation (Lee et al., 2008b; Llano et al., 2008).

Oocytes obtained from aged female mice show a significant reduction of Atrx mRNA and protein levels (Svoboda et al., 2001; Pan et al., 2008) suggesting that loss of ATRX function during reproductive senescence may contribute to the onset of aneuploidy in the female gamete. In addition, the most commonly observed chromosome segregation defects in ATRX-deficient oocytes, namely nondisjunction of whole chromosomes and premature separation of sister chromatids (Baumann et al., 2010), have also been identified as the major types of aneuploidy in oocytes from women of advanced reproductive age (Vialard et al., 2006). This underscores the importance of ATRX-deficient ova as a model to determine the epigenetic mechanisms involved in the onset of aneuploidy as a function of maternal age. Other chromatin-associated factors affected by changes in gene expression during reproductive aging include heterochromatin protein HP1 and the NAD-dependent histone deacetylase SIRT1 (Manosalva & Gonzalez, 2010), a known regulator of the histone methyltransferase Suv39h1 at heterochromatic domains.

Evidence from several model organisms including the mouse indicates that reproductive aging has the capability to alter the epigenetic composition of chromatin domains that are critical for a fast and efficient response to a developmental transition such as the onset of chromosome condensation upon germinal vesicle breakdown. For example, global histone deacetylation is seemingly impaired in oocytes obtained from older mice as indicated by the persistence of a high level of acetylated histone H4 on lysine 8 and lysine 12 at the metaphase-II stage of meiosis (Akiyama et al., 2006; Suo et al., 2010), whereas the lysine 14 and 16 residues in histone H3 (H3K14) and H4 (H4K16) present a similarly deacetylated chromatin status in both young and old oocytes. In addition, studies conducted on human oocytes obtained following standard IVF treatments show a significant correlation between defective deacetylation of the lysine 12 residue in histone H4 and chromosome misalignment in oocytes from women of advanced reproductive age (van den Berg et al., 2011). Moreover, aberrant chromatin configurations at the GV stage in aged mouse oocytes in the form of condensed, irregular, or clumped chromatin have been reported, which was also associated with a striking reduction in the patterns of global tri-methylation of histone H3 (H3K9me3) and di-methylation of histone H4 (H4K20me2) and histone H3 (H3K36me2 and HeK79me2) (Manosalva & Gonzalez, 2010). By contrast, the methylation patterns of H3K4 and H3K9 di-methylation remain unaltered in old GV and metaphase-II stage oocytes (Manosalva & Gonzalez, 2010). However, whether induced through the genetic manipulation of model organisms (ATRX), pharmacological inhibition of global histone deacetylation (TSA), or reproductive senescence, disruption of global histone modification patterns carry the potential to impair chromatin condensation and chromosome segregation, leading to aneuploidy and infertility in mammalian females.

In addition to changes in global histone acetylation and methylation levels, reproductive aging has also been linked to a significantly altered expression of several DNA methyltransferases (Hamatani et al., 2004). Interestingly, although the incidence of embryo resorption, morphological abnormalities, and delayed development were found to be increased during gestation in aged compared to young female mice, no significant changes in methylation patterns within the DMR's of Snrpn, Kcnq1ot1, U2af1-rs1, Peg1, Igf2r, and H19 were noted in blastocysts and mid-gestation embryos and placentas obtained from young or aged females (Lopes et al., 2009). However, whether the maternal age effect influences DNA methylation patterns at other imprinted loci in senescent germ cells remains yet to be determined.

Notably, it is evident that the establishment of epigenetic modifications during gametogenesis and the normal progression of meiotic transitions can be adversely influenced by environmental factors (Dolinoy et al., 2006; Jirtle & Skinner, 2007; Susiarjo et al., 2007). For example, studies suggest that high levels of exogenous gonadotropins are associated with increased nondisjunction in both human and mouse oocytes (Zuccotti et al., 1998; Roberts et al., 2005). Importantly, hormonal changes during the decade preceding reproductive senescence in women are characterized by increasing FSH serum concentrations. Ovarian hyperstimulation and suboptimal culture conditions during assisted reproductive technology may have additional effects on the quality of the female gamete as the establishment of epigenetic modifications in the oocyte genome is extremely susceptible during oocyte growth and maturation, potentially contributing to oocyte aneuploidy. In fact, a growing number of reports suggest that *in vitro* culture and embryo manipulation may have striking effects on DNA methylation of several DMRs and hence perturb the mono-allelic expression of imprinted genes in the preimplantation embryo. For instance, the culture of 2-cell embryos to the blastocyst stage in Whitten's medium resulted in illegitimate expression of the paternal H19 allele through loss of methylation within the upstream imprinting control region, whereas culture in the KSOM+AA medium conserved DNA methylation and transcriptional silencing of the paternal allele (Doherty et al., 2000). Similarly, a comparison of allelic expression and DNA methylation patterns in unmanipulated embryos as well as following embryo transfer and/or in vitro culture revealed mono-allelic expression of 10 imprinted genes at d9.5 of development in all tissues in the unmanipulated group. In contrast, both in vitro culture and embryo transfer induced aberrant expression of one or more imprinted genes in the yolk sac, in the placenta, and in embryonic tissues. Moreover, loss of methylation on the maternal allele of the KvDMR1 locus was associated with biallelic expression of the Kcnq1ot1 gene, a common characteristic of Beckwith-Wiedemann syndrome in human patients (Rivera et al., 2008). These studies strongly support clinical data, which suggest a link between an increased incidence of BWS and other imprinting-related disorders in children born following ART.

Lastly, a steadily growing list of environmental toxins has been associated with abnormal chromosome segregation, aneuploidy, and severe reproductive failure in the female germline. For example, Bisphenol A (BPA) is a component of a variety of polycarbonate plastic compounds and a ubiquitously found chemical in our daily life. BPA acts as an endocrine disruptor with various effects on the female reproductive tract at environmentally relevant doses and has the potential to induce changes in the genetic quality of the oocyte also during the final stages of oogenesis, leading to spindle abnormalities, abnormal meiotic chromosome congression at metaphase-I, and nondisjunction-associated aneuploidy (Hunt et al., 2003). Although the precise molecular pathways involved remain to be determined, studies provide some mechanistic insight by indicating that neonatal exposure to BPA results in hypomethylation of the imprinted genes Igf2r and Prg3 during the oocyte growth phase as well as an acceleration of primordial-to-primary follicle transitions leading to precocious depletion of the ovarian reservoir in mice (Chao et al., 2012). Other environmental toxins capable of interfering with meiotic chromosome segregation and the establishment of a euploid gamete include compounds such as 2-,ethoxyestradiol (Eichenlaub-Ritter et al., 2007), trichlorfon (Cukurcam et al., 2004), and phthalate monoesters (Tranfo et al., 2012). Additional studies are required to elucidate the precise molecular mechanism responsible for predisposing the mammalian oocyte to adverse effects of an array of environmental disruptors: insight that may potentially pave the way for the application of "epigenetic therapies" to reestablish proper chromatin marks to maintain chromosome stability during assisted reproductive technologies, and perhaps to identify reliable early detection markers for induced epigenetic changes with potential application to prevent aneuploidy in the female gamete.

7.6 Chromosome-microtubule Interactions in the Mammalian Oocyte

Accurate segregation of chromosomes during mitotic and meiotic division is crucially dependent on stable chromosome–microtubule attachments. Inappropriate interactions with spindle microtubules result in deleterious chromosome segregation errors (Figure 7.5), which can lead to aneuploidy and genomic instability in oocytes and developing embryos. Microtubules bind to the chromosomes specifically at the kinetochore, a specialized region of the centromere (Cheeseman & Desai, 2008). Proper kinetochore assembly involves the action of several epigenetic factors. For example, chromatin remodeling proteins of the SNF2 family of helicases have been implicated in the sequestration of the histone H3 variant centromere protein A (CENP-A) (reviewed in Prasad & Ekwall, 2011).

During mitosis, kinetochores from each sister chromatid normally attach to microtubules from opposing spindle poles, referred to as bi-oriented or amphitelic attachment. Yet, different types of attachment errors can occur. These include: (1) syntelic errors in which kinetochores from sister chromatids both attach to the same spindle pole; and (2) merotelic errors, whereby a single kinetochore is attached to both spindle poles simultaneously. Chromosome–microtubule interactions during meiosis-I are even more complex as it involves the separation of homologous chromosomes rather than sister chromatids. This requires the attachment and segregation of sister chromatids of each homologue to the same spindle pole (co-orientation). Studies are beginning to investigate the underlying mechanisms that regulate chromosome–microtubule interactions during meiosis;, however, our currently knowledge is largely based on functional analogies of mitotic division.

7.6.1 Regulation of Chromosome-Microtubule Interactions at the Kinetochore

The kinetochore functions as an essential interface between the chromosomes and spindle microtubules during cell division. It is a large multi-protein complex in which over 90 proteins assemble. Electron microscopy studies describe the kinetochore complex as having a tri-layer structure, composed of an inner and outer kinetochore plate as well as an outermost corona layer (Figure 7.6). The electron-dense inner kinetochore plate contacts centromeric heterochromatin, whereas the outer kinetochore plate contacts the spindle microtubules as well as the corona layer (Pluta et al., 1995; McEwen & Dong, 2010). Kinetochores only assemble at the centromeres upon entry into mitosis, when binding to spindle microtubules for chromosome segregation is required. Upon mitotic exit, the structure disassembles (Oegema et al., 2001; Cheeseman & Desai, 2008).

It is essential that a kinetochore assemble only at one location on each chromosome, and various properties of the centromere mark it as the exclusive site for kinetochore formation. This includes a combination of epigenetic marks such as nucleosomes that contain CENP-A, DNA sequence

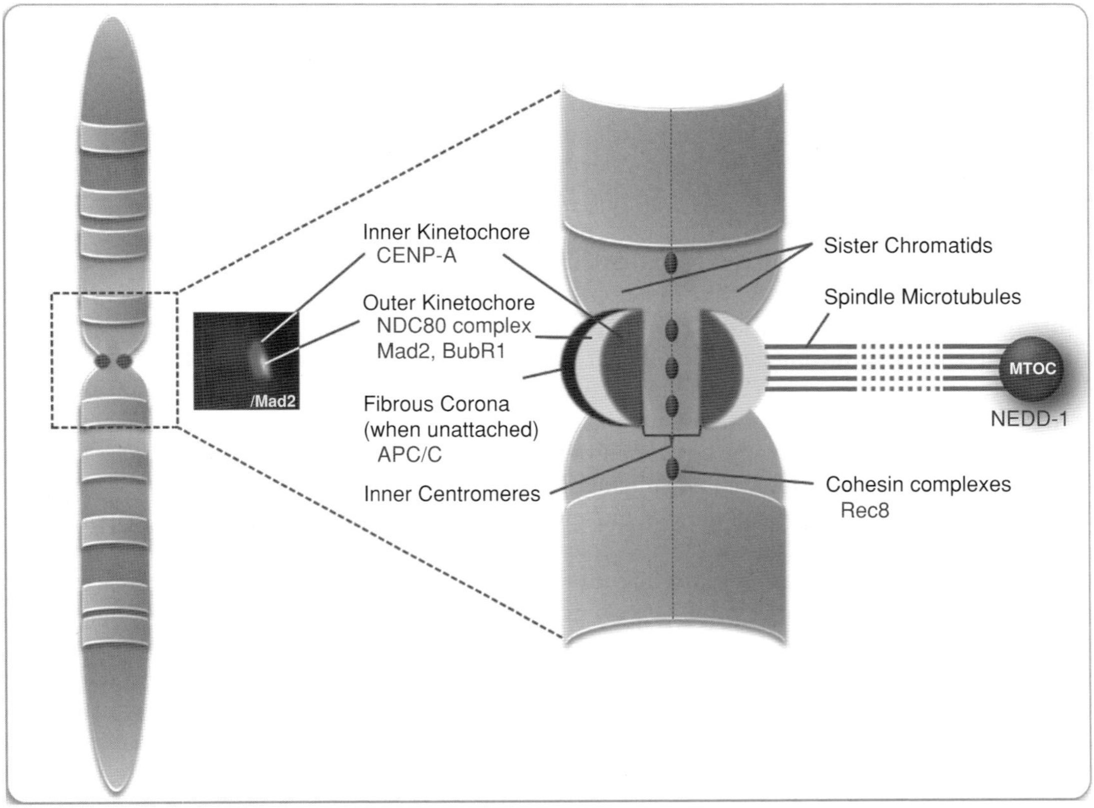

Figure 7.6 Schematic representation of the kinetochore complex assembled at the centromere. Key proteins shown in red localize to specific regions. (For color detail, see color plate section.)

properties, as well as chromatin structure. For example, incorporation of the histone H3 variant CENP-A occurs predominantly at the centromere and is required for kinetochore assembly (Howman et al., 2000; Oegema et al., 2001; Black et al., 2007). In contrast, alpha satellite DNA sequences normally associated with centromeres are not required for kinetochore assembly (Amor & Choo, 2002; Schuh et al., 2007; Bassett et al., 2010). Interestingly, studies in Drosophila reveal that incorporation of new CENP-A into centromeric chromatin does not occur until telophase and G1 (Jansen et al., 2007). This timing differs from the incorporation of typical histone molecules during DNA replication and may protect against potential mis-incorporation of new CENP-A at non-centromeric sites along the chromosome. Studies are also beginning to identify key regulatory mechanisms and demonstrate that the incorporation of the CENP-A into centromeric chromatin is dependent on the Mis 18 complex (Hayashi et al., 2004; Fujita et al., 2007) as well as the HJURP/Smc3 chaperone protein (Dunleavy et al., 2009; Foltz et al., 2009). Although CENP-A is sufficient to promote kinetochore assembly in Drosophila, additional proteins are required in human cells. A group of 15 proteins referred to as the Constitutive Centromere Associated Network (CCAN) have been identified, which are suggested to provide an appropriate environment for CENP-A as well as form a base for dynamic kinetochore assembly (Foltz et al., 2006; Hori et al., 2008).

While the inner kinetochore plate contacts centromeric heterochromatin, the outer kinetochore plate and corona layer contain proteins required for microtubule interaction. Identified proteins with microtubule binding activity include the CENP-E (Duesbery et al., 1997; Wood et al., 1997) and the Ska1 complex (Welburn et al., 2009) as well as the KMN network composed of KNL1, Mis12, and Ndc80 complex (Cheeseman & Desai, 2008). These pathways are likely conserved during meiotic division. Previous studies have shown that mouse oocytes express CENP-E at the kinetochores. Moreover, microinjection of specific antibodies to inhibit CENP-E function lead to MI-arrest likely due to disrupted chromosome–microtubule interactions (Duesbery et al., 1997). Elegant studies in mitotic cells demonstrate that the NDC80 microtubule interacting complex directly binds to microtubules and is essential for the formation of kinetochore–microtubule attachments (Cheeseman et al., 2006). This complex is composed of four proteins: Spc24, Spc25, Nuf2, and Hec1/Ndc80. The C-termini of Spc24 and Spc25 anchor the complex at the kinetochore, while the N-terminal domains of Nuf2 and Hec1 interact with the plus-ends of spindle microtubules (DeLuca et al., 2006). Depletion of the NDC80 complex component proteins leads to severe microtubule attachment defects during mitosis (Cheeseman & Desai, 2008). Studies are needed to assess whether Hec1 plays an analogous, crucial function in oocytes.

Another key group of proteins assemble at the outer kinetochore plate, which functions in the detection and correction of inappropriate chromosome–microtubule interactions. Studies indicate that accurate chromosome–microtubule interactions are promoted by cellular mechanisms that stabilize appropriate bi-oriented (amphitelic) attachments to opposing spindle poles, while inappropriate (ex. syntelic or merotelic) attachments are eliminated (Lampson & Cheeseman, 2011). Pioneering experiments by Bruce Nicklas provided direct evidence that tension across the centromere can stabilize chromosome–microtubule interactions (Nicklas & Ward, 1994; Nicklas et al., 2001). In contrast, the phosphorylation of kinetochore substrates by the conserved Ipl1/Aurora B kinase in mitotic cells selectively eliminates incorrect attachments by promoting the turnover of microtubule attachments in the absence of tension. Aurora B has been implicated in controlling kinetochore microtubule attachments by modulating the microtubule binding activity of key kinetochore proteins such as Hec1/Ndc8 (Cheeseman et al., 2002; DeLuca et al., 2006; Welburn et al., 2010). Importantly, studies in mouse oocytes identify Aurora C kinase, and not Aurora B, as playing a critical role in regulating kinetochore microtubules during meiotic division(Yang et al., 2010). This points to potentially unique regulatory mechanisms that govern chromosome–microtubule interactions during meiotic division in mammalian oocytes.

Inappropriate chromosome attachments activate the spindle assembly checkpoint (SAC), which inhibits APC/C^{cdc20} activity to prevent anaphase onset until kinetochores attach to spindle microtubules. SAC functions as a critical surveillance mechanism and delays anaphase, allowing for error correction mechanisms to promote bi-orientation of the chromosomes. Key component proteins of SAC include Mad2, BubR1, and Bub3, which localize to the outer kinetochore plate until stable chromosome–microtubule interactions are established. Studies have confirmed the expression and function of key checkpoint proteins, including MAD2, in mammalian oocytes (Homer et al., 2005; Niault et al., 2007; Leland et al., 2009; McGuinness et al., 2009), which activate the spindle assembly checkpoint (SAC) to prevent anaphase onset until all chromosomes are attached to the meiotic spindle apparatus.

7.6.2 Formation of Meiotic Spindle Microtubules

Accurate chromosome–microtubule interactions are dependent not only on kinetochore capture and function, but also on the formation of stable spindle microtubules. Studies are beginning to elucidate the underlying mechanisms that regulate spindle microtubule formation and organization in the mammalian oocyte. The primary organizing centers for the formation and anchorage of spindle microtubules (MTOCs) include centrosomes in higher eukaryotes and spindle pole bodies in fungal organisms. Centrosomes in animal cells are structurally complex, non–membrane-bound organelles that typically consist of a pair of centrioles surrounded by a protein matrix of pericentriolar material (PCM) (Raynaud-Messina & Merdes, 2007). Microtubule nucleation occurs in the PCM and is dependent on γ-tubulin (Oakley et al., 1990; Stearns & Kirschner, 1994). This unique tubulin superfamily member is highly conserved and interacts with specific gamma complex proteins (GCPs) to form large γ-tubulin ring complexes (γTuRCs) (Wiese & Zheng, 2006), which in turn bind dimers of α- and β-tubulin and catalyze their assembly into microtubules (Moritz & Agard, 2001). Microtubule minus-ends are anchored at the MTOCs, while the growing plus-ends extend outward for chromosome capture. Disruption of γ-tubulin in diverse systems perturbs mitotic spindle formation and structure (Job et al., 2003; Raynaud-Messina & Merdes, 2007). Moreover, targeted deletion of the γ-tubulin gene (*Tubg1*) in mice leads to significant mitotic errors and embryonic death by the blastocyst stage (Yuba-Kubo et al., 2005).

Mammalian oocytes lack classic centrosomes and analysis of meiotic spindle formation reveals microtubule nucleation from multiple microtubule organizing centers (MTOCs) around the condensing chromosomes, upon the resumption of meiosis (Combelles & Albertini, 2001; Schuh & Ellenberg, 2007; Ma et al., 2008). Multiple microtubule asters are evident in proximity to the chromosomes, which cluster into a single mass of microtubules. A current model proposes that these microtubules self-organize to form the anastral bipolar, barrel-shaped, meiotic spindle at MI (Schuh & Ellenberg, 2007). Studies have also indicated a role for Ran GTPase in promoting microtubule formation around chromatin in mouse oocytes (Dumont *et al.*, 2007; Schuh & Ellenberg, 2007). Interestingly, although disruption of Ran GTPase activity in mouse oocytes significantly perturbs spindle formation during meiosis-II, it does not inhibit assembly of functional meiotic spindles during meiosis-I (Dumont et al., 2007). This suggests key differences between MI and MII and, supports a critical role for microtubule organizing centers (MTOCs) in spindle assembly during meiosis-I. Mammalian oocytes contain unique MTOCs composed of key pericentriolar matrix (PCM) proteins including γ-tubulin, necessary for microtubule nucleation (Gueth-Hallonet et al., 1993; Combelles & Albertini, 2001; Ma et al., 2008), but lack centrioles (Szollosi et al., 1972; Schatten, 1994). Therefore, MTOCs in oocytes are often described as aggregates, or foci, of pericentriolar material that can functionally nucleate microtubules.

Fundamental aspects of meiotic spindle formation, such as γ-tubulin targeting to MTOCs, are likely regulated by conserved signaling pathways. Upon mitotic entry the levels of γ-tubulin at MTOCs and microtubule nucleation activity increase significantly. The recruitment of γ-tubulin to MTOCs in mitotic cells is attributed to a γTuRC associated protein, NEDD1 (Neuronal precursor cell Expressed Developmentally Down-regulated protein 1) (Haren et al., 2006; Luders et al., 2006). NEDD1 has been identified as a key component of acentriolar MTOCs in oocytes and is essential for microtubule nucleation and meiotic spindle stability (Ma et al., 2010). Depletion of NEDD1 in oocytes disrupts γ-tubulin targeting to MTOCs, leading to significant defects in meiotic spindle stability, loss of chromosome–microtubule attachments, and activation of SAC (Figure 7.5). Despite SAC activation, over 50% of NEDD1-depleted oocytes that escaped MI-arrest contained an abnormal chromosome number at MII. This suggests that potentially more subtle defects in meiotic spindle stability, which may be insufficient to promote effective SAC-mediated meiotic arrest, pose a significant risk in contributing to aneuploidy during meiotic division.

7.7 Conclusion

Mammalian germ cell maturation, fertilization, and preimplantation development are highly complex processes that rely on epigenetic chromatin remodeling to enable timely chromatin condensation, the formation of stable kinetochore–microtubule interactions for chromosome segregation, and the stringent regulation of gene expression in order to form a developmentally competent, euploid conceptus. Disturbances of these processes, as potentially conferred through disruptions of epigenetic states (i.e., maternal age, suboptimal culture conditions, environmental toxins, or hormonal imbalances) have the significant potential to cause chromosome segregation defects, congenital birth defects, and infertility. Clearly, the impact of a multitude of factors on a positive pregnancy outcome is only beginning to be elucidated, and additional research efforts are necessary to optimize ART protocols in both human and veterinary medicine. Most important, further insight into the affected pathways and mechanisms might pave the way for the development of "epigenetic therapy" approaches that could allow for the correction of acquired adverse epigenetic states through, for instance, nutritional supplements or customized embryo manipulation protocols in the future.

References

Abe, K.-i., Inoue, A., Suzuki, M.G., & Aoki, F. (2010). Global gene silencing is caused by the dissociation of RNA polymerase II from DNA in mouse oocytes. *The Journal of Reproduction & Development, 56*(5), 502–507.

Adenot, P. G., Mercier, Y., Renard, J. P. et al. (1997). Differential H4 acetylation of paternal and maternal chromatin precedes DNA replication and differential transcriptional activity in pronuclei of 1-cell mouse embryos. *Development, 124*(22), 4615–4625.

Adhikari, D., Zheng, W., Shen, Y. et al. (2010). Tsc/mTORC1 signaling in oocytes governs the quiescence and activation of primordial follicles. *Human Molecular Genetics, 19*(3), 397–410.

Akiyama, T., Nagata, M., & Aoki, F. (2006). Inadequate histone deacetylation during oocyte meiosis causes aneuploidy and embryo death in mice. *Proceedings of the Nationall Academy of Sciences of the United States of America, 103*(19), 7339–7344.

Ame, J., Spenlehauer, C., & de Murcia, G. (2004). The PARP superfamily. *Bioessays, 26*(8), 882–893.

Amor, D. J., & Choo, K. H. A. (2002). Neocentromeres: Role in human disease, evolution, and centromere Study. *The American Journal of Human Genetics, 71*(4), 695–714.

Andreu-Vieyra, C. V., Chen, R., Agno, J. E. et al. (2010). MLL2 is required in oocytes for bulk histone 3 lysine 4 trimethylation and transcriptional Silencing. *PLOS Biology, 8*(8), e1000453.

Arney, K. L., Bao, S., Bannister, A. J. et al. (2002). Histone methylation defines epigenetic asymmetry in the mouse zygote. *International Journal of Developmental Biology, 46*(3), 317–320.

Baird, D., Collins, J., Egozcue, J. et al. (2005). Fertility and ageing. *Human Reproduction Update, 11*(3), 261–276.

Bannister, A. J., Schneider, R., & Kouzarides, T. (2002). Histone methylation: Dynamic or static?. *Cell, 109*(7), 801–806.

Bannister, A. J., Zegerman, P., Partridge, J. F. et al. (2001). Selective recognition of methylated lysine 9 on histone H3 by the HP1 chromo domain. *Nature, 410*(6824), 120–124.

Barton S. C., Surani, M., & Norris, M. L. (1984). Role of paternal and maternal genomes in mouse development. *Nature, 311*(5984), 374–376.

Bassett, E. A., Wood, S., Salimian, K. J. et al. (2010). Epigenetic centromere specification directs aurora B accumulation but is insufficient to efficiently correct mitotic errors. *The Journal of Cell Biology, 190*(2), 177–185.

Baumann, C., & De La Fuente, R. (2009). ATRX marks the inactive X chromosome (Xi) in somatic cells and during imprinted X chromosome inactivation in trophoblast stem cells. *Chromosoma, 118*(2), 209–222.

Baumann, C., Viveiros, M., & De La Fuente, R. (2010). Loss of maternal ATRX results in centromere instability and aneuploidy in the mammalian oocyte and pre-implantation embryo. *PLOS Genetics, 6*(9), e1001137.

Bellone, M., Zuccotti, M., Redi, C. A., & Garagna, S. (2009). The position of the germinal vesicle and the chromatin organization together provide a marker of the developmental competence of mouse antral oocytes. *Reproduction, 138*(4), 639–643.

Berger, S., & Felsenfeld, G. (2001). Chromatin goes global. *Molecular Cell, 8*(2), 263–268.

Bernard, P., & Allshire, R. (2002). Centromeres become unstuck without heterochromatin. *Trends in Cell Biology, 12*(9), 419–424.

Bernard, P., Maure, J., Partridge, J. et al. (2001). Requirement of heterochromatin for cohesion at centromeres., *Science, 294*(5551), 2539–2542.

Bestor, T. H., & Bourc'his, D. (2004). Transposon silencing and imprint establishment in mammalian germ cells. *Cold Spring Harbor Symposium on Quantum Biology, 69*, 381–387.

Bird, A. (2007). Perceptions of epigenetics. *Nature, 447*(7143), 396–398.

Bird, A., & Wolffe, A. P. (1999). Methylation-induced repression – belts, braces, and chromatin. *Cell, 99*, 451–454.

Black, B. E., Brock, M. A., Bedard, S. et al. (2007). An epigenetic mark generated by the incorporation of CENP-A into centromeric nucleosomes. *Proceedings of the National Academy of Sciences USA, 104*(12), 5008–5013.

Bourc'his, D., & Bestor, T. H. (2004). Meiotic catastrophe and retrotransposon reactivation in male germ cells lacking Dnmt3L. *Nature, 431*(7004), 96–99.

Bourc'his, D., & Bestor, T. H. (2006). Origins of extreme sexual dimorphism in genomic imprinting. *Cytogenetic & Genome Research, 113*(1–4), 36–40.

Bourc'his, D., Xu, G. L., Lin, C. S. et al. (2001). Dnmt3L and the establishment of maternal genomic imprints. *Science, 294*(5551), 2536–2539.

Brunet, S., & Maro, B. (2007). Germinal vesicle position and meiotic maturation in mouse oocyte. *Reproduction, 133*(6), 1069–1072.

Celeste, A., Petersen, S., Romanienko, P. J. et al. (2002). Genomic instability in mice lacking histone H2AX, *Science, 296*(5569), 922–927.

Chao, H., Zhang, X., Chen, B. et al. (2012). Bisphenol A exposure modifies methylation of imprinted genes in mouse oocytes via the estrogen receptor signaling pathway. *Histochemistry & Cell Biology, 137*(2), 249–259.

Cheeseman, I. M., Anderson, S., Jwa, M. et al. (2002). Phospho-regulation of kinetochore-microtubule attachments by the Aurora Kinase Ipl1p. *Cell, 111*(2), 163–172.

Cheeseman, I. M., Chappie, J. S., Wilson-Kubalek, E. M., & Desai, A. (2006). The conserved KMN network constitutes the core microtubule-binding site of the kinetochore. *Cell, 127*(5), 983–997.

Cheeseman, I. M., & Desai, A. (2008). Molecular architecture of the kinetochore-microtubule interface. *Nature Reviews Molecular Cell Biology, 9*(1), 33–46.

Chen, R. Z., Pettersson, U., Beard, C. et al. (1998). DNA hypomethylation leads to elevated mutation rates. *Nature, 395*(6697), 89–93.

Chen, T., & Li, E. (2006). Establishment and maintenance of DNA methylation patterns in mammals. *Current Topics in Microbiology & Immunology, 301*, 179–201.

Cheung, P., Allis, C. D., & Sassone-Corsi, P. (2000). Signaling to chromatin through histone modifications. *Cell, 103*(2), 263–271.

Chiang, T., Duncan, F. E., Schindler, K. et al. (2010). Evidence that weakened centromere cohesion is a leading cause of age-related aneuploidy in oocytes. *Current Biology, 20*(17), 1522–1528.

Chotalia, M., Smallwood, S.A., Ruf, N. et al. (2009). Transcription is required for establishment of germline methylation marks at imprinted genes. *Genes & Development, 23*(1), 105–117.

Ciccone, D. N., Su, H., Hevi, S. et al. (2009). KDM1B is a histone H3K4 demethylase required to establish maternal genomic imprints. *Nature, 461*(7262), 415–418.

Cleveland, D., Mao, Y. & Sullivan, K. (2003). Centromeres and kinetochores: From epigenetics to mitotic checkpoint signaling. *Cell, 112*(4), 407–421.

Combelles, C. M., Gupta, S., & Agarwal, A. (2009). Could oxidative stress influence the in-vitro maturation of oocytes?. *Reproductive Biomedicine Online, 18*(6): 864–880.

Combelles, C. M. H., & Albertini, D. F. (2001). Microtubule patterning during meiotic maturation in mouse oocytes is determined by cell cycle-specific sorting and redistribution of g-tubulin. *Developmental Biology, 239*, 281–294.

Cowell, I. G., Aucott, R., Mahadevaiah, S. K. et al. (2002). Heterochromatin, HP1 and methylation at lysine 9 of histone H3 in animals. *Chromosoma, 111*(1), 22–36.

Cremer, T., & Cremer, C. (2001). Chromosome territories, nuclear architecture and gene regulation in mammalian cells. *Nature Reviews Genetics, 2*, 292–301.

Cukurcam, S., Sun, F., Betzendahl, I. et al. (2004). Trichlorfon predisposes to aneuploidy and interferes with spindle formation in in vitro maturing mouse oocytes. *Mutation Research, 465*(2);, 871–561.

D'Amours, D., Desnoyers, S., D'Silva, I., & Poirier, G. (1999). Poly(ADP-ribosyl)ation reactions in the regulation of nuclear functions. *Biochemistry Journal, 342*(2), 249–268.

Davis, P. K., & Brachmann. R. (2003). Chromatin remodeling and cancer. *Cancer Biology & Therapy, 2*(1), 24–31.

De La Fuente, R. (2006). Chromatin modifications in the germinal vesicle (GV) of mammalian oocytes. *Developmental Biology, 292*(1), 1–12.

De La Fuente, R., Baumann, C., Fan, T. et al. (2006). Lsh is required for meiotic chromosome synapsis and retrotransposon silencing in female germ cells. *Nature Cell Biology, 8*(12), 1448–1454.

De La Fuente, R., & Eppig, J. J. (2001). Transcriptional activity of the mouse oocyte genome: Companion granulosa cells modulate transcription and chromatin remodeling. *Developmental Biology, 229*(1), 224–236.

De La Fuente, R., Viveiros, M., Burns, K. et al. (2004a). Major chromatin remodeling in the germinal vesicle (GV) of mammalian oocytes is dispensable for global transcriptional silencing but required for centromeric heterochromatin function. *Developmental Biology, 275*(2), 447–458.

De La Fuente, R., Viveiros, M., Wigglesworth, K., & Eppig, J. (2004b). ATRX, a member of the SNF2 family of helicase /ATPases, is required for chromosome alignment and meiotic spindle organization in metaphase II stage mouse oocytes. *Developmental Biology, 272*, 1–14.

DeBaun, M. R., & Tucker, M. A. (1998). Risk of cancer during the first four years of life in children from The Beckwith-Wiedemann Syndrome Registry. *The Journal of Pediatrics, 132*(3), 398–400.

Debey, P., Szöllösi, M., Szöllösi, D. et al. (1993). Competent mouse oocytes isolated from antral follicles exhibit different chromatin organization and follow different maturation dynamics. *Molecular Reproduction & Development, 36*(1), 59–74.

Delaval, K., & Feil, R. (2004). Epigenetic regulation of mammalian genomic imprinting. *Current Opinion in Genetics & Development, 14*(2), 188–195.

DeLuca, J. G., Gall, W. E., Ciferri, C. et al. (2006). Kinetochore microtubule dynamics and attachment stability are regulated by Hec1. *Cell, 127*(5), 969–982.

Dennis, K., Fan, T., Geiman, T. et al. (2001). Lsh, a member of the SNF2 family, is required for genome-wide methylation. *Genes Development, 15*(22), 2940–2944.

Dillon, N., & Festenstein, R. (2002). Unravelling heterochromatin: competition between positive and negative factors regulates accessibility. *Trends in Genetics, 18*(5), 252–258.

Dodge, J. E., Kang, Y. K., Beppu, H. et al. (2004). Histone H3-K9 methyltransferase ESET is essential for early development. *Molecular and Cell Biology, 24*(6), 2478–2486.

Doherty, A. S., Mann, M. R. W., Tremblay, K. D. et al. (2000). Differential effects of culture on imprinted H19 expression in the preimplantation mouse embryo. *Biology of Reproduction, 62*(6), 1526–1535.

Dolinoy, D. C., Weidman, J. R., Waterland, R. A., & Jirtle, R. L. (2006). Maternal genistein alters coat color and protects Avy mouse offspring from obesity by modifying the fetal epigenome. *Environmental Health Perspectives, 114*(4), 567–572.

Duesbery, N. S., Choi, T., Brown, K. D. et al. (1997). CENP-E is an essential kinetochore motor in maturing oocytes and is masked during Mos-dependent, cell cycle arrest at metaphase-II. *Proceedings of the National Academy of Sciences of the United States of America, 94*(17), 9165–9170.

Dumont, J., Petri, S., Pellegrin, F. et al. (2007). A centriole- and RanGTP-independent spindle assembly pathway in meiosis I of vertebrate oocytes. *Journal of Cell Biology, 176*(3), 295–305.

Duncan, F. E., Chiang, T., Schultz, R. M., & Lampson, M. A. (2009). Evidence that a defective spindle assembly checkpoint is not the primary cause of maternal age-associated aneuploidy in mouse eggs. *Biology of Reproduction, 81*(4), 768–776.

Dundr, M., & Misteli, T. (2001). Functional architecture in the cell nucleus. *Biochemistry Journal, 356*(2), 297–310.

Dunleavy, E. M., Roche, D., Tagami, H. et al. (2009). HJURP is a cell-cycle-dependent maintenance and deposition factor of CENP-A at centromeres. *Cell, 137*(3), 485–497.

Eichenlaub-Ritter, U., Winterscheidt, U., Vogt, E. et al. (2007). 2-methoxyestradiol induces spindle aberrations, chromosome congression failure, and nondisjunction in mouse oocytes. *Biology of Reproduction, 76*(5), 784–793.

Eppig, J. J., Viveiros, M. M., Marin-Bivens, C., & De La Fuente, R. (2004). Regulation of mammalian oocyte maturation. In P. C. K. Leung & E. Y. Adashi (eds.), *The ovary* (2nd ed., pp. 113–129). Amsterdam: Eslevier.

Evsikov, A. V., Graber, J. H., Brockman, J. M. et al. (2006). Cracking the egg: Molecular dynamics and evolutionary aspects of the transition from the fully grown oocyte to embryo. *Genes & Development, 20*(19), 2713–2277.

Fan, T., Hagan, J., Kozlov, S. et al. (2005). Lsh controls silencing of the imprinted Cdkn1c gene. *Development, 132*, 635–644.

Faraone-Mennella, M. R. (2005). Chromatin architecture and functions: The role(s) of poly(ADP-RIBOSE) polymerase and poly(ADPribosyl)ation of nuclear proteins. *Biochemistry & Cell Biology, 83*(3), 396–404.

Fedoriw, A. M., Stein, P., Svoboda, P. et al. (2004). Transgenic RNAi reveals essential function for CTCF in H19 gene imprinting. *Science, 303*(5655), 238–240.

Ferguson-Smith, A. C., & Surani, M. A. (2001). Imprinting and the epigenetic asymmetry between parental genomes. *Science, 293*(5532), 1086–1089.

Flaus, A., Martin, D. M., Barton, G. J., & Owen-Hughes, T. (2006). Identification of multiple distinct Snf2 subfamilies with conserved structural motifs. *Nucleic Acids Research, 34*(10), 2887–2905.

Foltz, D. R., Jansen, L. E. T., Bailey, A. O. et al. (2009). Centromere-specific assembly of CENP-A nucleosomes is mediated by HJURP. *Cell, 137*(3), 472–484.

Foltz, D. R., Jansen, L. E. T., Black, B. E. et al. (2006). The human CENP-A centromeric nucleosome-associated complex. *Nature Cell Biology, 8*(5), 458–469.

Fry, C. J., & Peterson, C. L. (2001). Chromatin remodeling enzymes: Who's on first?. *Current Biology, 11*(5), R185–R197.

Fu, G., Ghadam, P., Sirotkin, A. et al. (2003). Mouse oocytes and early embryos express multiple histone H1 subtypes. *Biology of Reproduction, 68*(5), 1569–1576.

Fujita, Y., Hayashi, T., Kiyomitsu, T. et al. (2007). Priming of centromere for CENP-A recruitment by human hMis18[alpha], hMis18[beta], and M18BP1. *Developmental Cell, 12*(1), 17–30.

Geiman, T., Durum, S., & Muegge, K. (1998). Characterization of gene expression, genomic structure, and chromosomal localization of Hells (Lsh). *Genomics, 54*(3), 477–483.

Geiman, T., Tessarollo, L., Anver, M., Kopp, J., Ward, J., & Muegge, K. (2001). Lsh, a SNF2 family member, is required for normal murine development. *Biochimica et Biophysica Acta,* 1526(2), 211–220.

Gibbons, R., Bachoo, S., Picketts, D. et al. (1997). Mutations in transcriptional regulator ATRX establish the functional significance of a PHD-like domain. *Nature Genetics, 17*(2), 146–148

Gibbons, R. J., McDowell, T. L., Raman, S. et al. (2000). Mutations in ATRX, encoding a SWI/SNF-like protein, cause diverse changes in the pattern of DNA methylation. *Nature Genetics, 24*(4), 368–371.

Gill, G. (2004). SUMO and ubiquitin in the nucleus: Different functions, similar mechanisms?. *Genes & Development, 18*(17), 2046–2059.

Goldberg, A. D., Allis, C. D., & Bernstein, E. (2007). Epigenetics: A landscape takes shape. *Cell, 128*(4), 635–638.

Grunstein, M. (1997). Histone acetylation in chromatin structure and transcription. *Nature, 389*(6649), 349–352.

Gruzova, M. N., Parfenov, V. N. (1993). Karyosphere in oogenesis and intranuclear morphogenesis. *International Review of Cytology – A Survey of Cell Biology, 144*, 1–52.

Guenatri, M., Bailly, D., Maison, C., & Almouzni, G. (2004). Mouse centric and pericentric satellite repeats form distinct functional heterochromatin. *Journal of Cell Biology, 166*(4), 493–505.

Gueth-Hallonet, C., Antony, C., Aghion, J. et al. (1993). Gamma-Tubulin is present in acentriolar MTOCs during early mouse development. *Journal of Cell Biology, 105*, 157–166.

Guthrie, H. D., & Garrett, W. M. (2000). Changes in porcine oocyte germinal vesicle development as follicles approach preovulatory maturity. *Theriogenology, 54*(3), 389–399.

Hakimi, M., Bochar, D., Schmiesing, J. et al. (2002). A chromatin remodelling complex that loads cohesin onto human chromosomes. *Nature, 418*(6901), 994–998.

Hamatani, T., Falco, G., Carter, M. G. et al. (2004). Age-associated alteration of gene expression patterns in mouse oocytes. *Human Molecular Genetics, 13*(19), 2263–2278.

Haren, L., Remy, M.-H., Bazin, I. et al. (2006). NEDD1-dependent recruitment of the {gamma}-tubulin ring complex to the centrosome is necessary for centriole duplication and spindle assembly. *Journal of Cell Biology, 172*(4), 505–515.

Hashimshony, T., Zhang, J., Keshet, I. et al. (2003). The role of DNA methylation in setting up chromatin structure during development. *Nature Genetics, 34*(2), 187–192.

Hassold, T., & Hunt, P. (2001). To err (meiotically) is human: The genesis of human aneuploidy. *Nature Reviews Genetics, 2*, 280–291.

Hayashi, T., Fujita, Y., Iwasaki, O. et al. (2004). Mis16 and Mis18 are required for CENP-A loading and histone deacetylation at centromeres. *Cell, 118*(6), 715–729.

Heard, E. (2004). Recent advances in X-chromosome inactivation., *Current Opinion in Cell Biology, 16*(3), 247–255.

Henikoff, S., Ahmad, K., & Malik, H. S. (2001). The centromere paradox: Stable inheritance with rapidly evolving DNA. *Science, 293*(5532), 1098–1102.

Hinrichs, K., Choi, Y. H., Love, L. B. et al. (2005). Chromatin configuration within the germinal vesicle of horse oocytes: Changes post mortem and relationship to meiotic and developmental competence. *Biology of Reproduction, 72*(5), 1142–1150.

Hinrichs, K., & Williams, K. A. (1997). Relationships among oocyte-cumulus morphology, follicular atresia, initial chromatin configuration, and oocyte meiotic competence in the horse. *Biology of Reproduction, 57*(2), 377–384.

Hiura, H., Obata, Y., Komiyama, J. et al. (2006). Oocyte growth-dependent progression of maternal imprinting in mice. *Genes to Cells, 11*(4), 353–361.

Hodges, C., & Hunt, P. (2002). Simultaneous analysis of chromosomes and chromosome-associated proteins in mammalian oocytes and embryos. *Chromosoma, 111*(3), 165–169.

Hodges, C. A., Revenkova, E., Jessberger, R. et al. (2005). SMC1beta-deficient female mice provide evidence that cohesins are a missing link in age-related nondisjunction, *Nature Genetics, 37*(12), 1351–1355.

Homer, H. A., McDougall, A., Levasseur, M. et al. (2005). Mad2 prevents aneuploidy and premature proteolysis of cyclin B and securin during meiosis I in mouse oocytes. *Genes and Development, 19*, 202–207.

Hori, T., Amano, M., Suzuki, A. et al. (2008). CCAN makes multiple contacts with centromeric DNA to provide distinct pathways to the outer kinetochore. *Cell, 135*(6), 1039–1052.

Houlard, M., Berlivet, S., Probst, A. V. et al. (2006). CAF-1 is essential for heterochromatin organization in pluripotent embryonic cells. *PLOS Genetics, 2*(11), e181.

Howman, E. V., Fowler, K. J., Newson, A. J. et al. (2000). Early disruption of centromeric chromatin organization in centromere protein A (Cenpa) null mice. *Proceedings of the National Academy of Sciences of the United States of America, 97*(3), 1148–1153.

Huang, J., Fan, T., Yan, Q. et al. (2004). Lsh, an epigenetic guardian of repetitive elements. *Nucleic Acids Research, 32*(17), 5019–5028.

Huang, J. C., Yan, L. Y., Lei, Z. L. et al. (2007). Changes in histone acetylation during postovulatory aging of mouse oocyte. *Biology of Reproduction, 77*(4), 666–670.

Hunt, P. A., & Hassold, T. J. (2008). Human female meiosis: What makes a good egg go bad?. *Trends in Genetics, 24*(2), 86–93.

Hunt, P. A., Koehler, K. E., Susiarjo, M. et al. (2003). Bisphenol A exposure causes meiotic aneuploidy in the female mouse. *Current Biology, 13*(7), 546–553.

Inoue, A., Nakajima, R., Nagata, M., & Aoki, F. (2008). Contribution of the oocyte nucleus and cytoplasm to the determination of meiotic and developmental competence in mice. *Human Reproduction, 23*(6), 1377–1384.

Ivanovska, I., Khandan, T., Ito, T., & Orr-Weaver, T. L. (2005). A histone code in meiosis: The histone kinase, NHK-1, is required for proper chromosomal architecture in Drosophila oocytes. *Genes & Development, 19*(21), 2571–2582.

Ivanovska, I., & Orr-Weaver, T. L. (2006). Histone modifications and the chromatin scaffold for meiotic chromosome architecture. *Cell Cycle, 5*(18), 2064–2071.

Jablonka, E., Matzke, M., Thieffry, D., & Van Speybroeck, L. (2002). The genome in context: Biologists and philosophers on epigenetics. *Bioessays, 24*, 392–394.

Jaenisch, R., & Bird, A. (2003). Epigenetic regulation of gene expression: How the genome integrates intrinsic and environmental signals. *Nature Genetics, 33*(Suppl.), 245–254.

Jansen, L. E. T., Black, B. E., Foltz, D. R., & Cleveland, D. W. (2007). Propagation of centromeric chromatin requires exit from mitosis. *The Journal of Cell Biology, 176*(6), 795–805.

Jarvis, C., Geiman, T., Vila-Storm, M. et al. (1996). A novel putative helicase produced in early murine lymphocytes. *Gene, 169*(2), 203–207.

Jeffery, L., & Nakielny, S. (2004). Components of the DNA methylation system of chromatin control are RNA-binding proteins. *Journal of Biological Chemistry, 279*(47), 49479–49487.

Jenuwein, T., & Allis, C. D. (2001). Translating the histone code. *Science, 293*(5532), 1074–1080.

Jirtle, R. L., & Skinner, M. K. (2007). Environmental epigenomics and disease susceptibility. *Nature Reviews Genetics, 8*(4), 253–262.

Job, D., Valiron, O., & Oakley, B. (2003). Microtubule nucleation. *Current Opinion in Cell Biology, 15*(1), 111–117.

Kageyama, S., Liu, H., Nagata, M., & Aoki, F. (2006). Stage specific expression of histone deacetylase 4 (HDAC4) during oogenesis and early preimplantation development in mice. *Journal of Reproductive Development, 52*(1), 99–106.

Kageyama, S.-i., Liu, H., Kaneko, N. et al. (2007). Alterations in epigenetic modifications during oocyte growth in mice. *Reproduction, 133*(1), 85–94.

Kaneda, M., Hirasawa, R., Chiba, H. et al. (2009). Genetic evidence for Dnmt3a-dependent imprinting during oocyte growth obtained by conditional knockout with Zp3-Cre and complete exclusion of Dnmt3b by chimera formation. *Genes to Cells, 15*(3), 169–179.

Kaneda, M., Okano, M., Hata, K. et al. (2004). Essential role for de novo DNA methyltransferase Dnmt3a in paternal and maternal imprinting. *Nature, 429*(6994), 900–903.

Karpen, G. H., & Allshire, R. C. (1997). The case for epigenetic effects on centromere identity and function. *Trends in Genetics, 13*(12), 489–496.

Kelly, T. L., & Trasler, J. M. (2004). Reproductive epigenetics. *Clinical Genetics, 65*(4), 247–260.

Kikuchi, K., Naito, K., Noguchi, J. et al. (2000). Maturation/M-phase promoting factor: A regulator of aging in porcine cocytes. *Biology of Reproduction, 63*(3), 715–722.

Kim, J., Liu, H., Tazaki, M. et al. (2003). Changes in histone acetylation during mouse oocyte meiosis. *Journal of Cell Biology, 162*(1), 37–46.

Kim, M., Mauro, S., Gevry, N. et al. (2004). NAD(+)-dependent modulation of chromatin structure and transcription by nucleosome binding properties of PARP-1.' *Cell, 119*(6), 803–814.

Kimmins, S., & Sassone-Corsi, P. (2005). Chromatin remodelling and epigenetic features of germ cells. *Nature, 434*(7033), 583–589.

Kimura, H., Tada, M., Nakatsuji, N., & Tada, T. (2004). Histone code modifications on pluripotential nuclei of reprogrammed somatic cells. *Molecular and Cellular Biology, 24*(13), 5710–5720.

Kota, S. K., & Feil, R. (2010). Epigenetic transitions in germ cell development and meiosis. *Developmental Cell, 19*(5), 675–686.

Kourmouli, N., Jeppesen, P., Mahadevhaiah, S. et al. (2004). Heterochromatin and tri-methylated lysine 20 of histone H4 in animals. *Journal of Cell Science, 117*(Pt 12), 2491–2501.

Kourmouli, N., Sun, Y.M., van der Sar, S. et al. (2005). Epigenetic regulation of mammalian pericentric heterochromatin *in vivo* by HP1. *Biochemical and Biophysical Research Communications, 337*(3), 901–907.

Kouzarides, T. (2007). Chromatin modifications and their function. *Cell, 128*(4), 693–705.

Kruhlak, M. J., Hendzel, M. J., Fischle, W. et al. (2001). Regulation of global acetylation in mitosis through loss of histone acetyltransferases and deacetylases from chromatin. *Journal of Biological Chemistry, 276*(41), 38307–38319.

Lachner, M., O'Carroll, D., Rea, S. et al. (2001). Methylation of histone H3 lysine 9 creates a binding site for HP1 proteins. *Nature, 410*(6824), 116–120.

Lampson, M. A., & Cheeseman, I. M. (2011). Sensing centromere tension: Aurora B and the regulation of kinetochore function. *Trends in Cell Biology, 21*(3), 133–140.

Langst, G., & Becker P. B. (2001). Nucleosome mobilization and positioning by ISWI-containing chromatin-remodeling factors. *Journal of Cell Science, 114*(Pt 14), 2561–2568.

Lee, H., Yin, X., Jin, Y. et al. (2008a). Germinal vesicle chromatin configuration and meiotic competence is related to the oocyte source in canine. *Animal Reproduction Science, 103*, 336–347.

Lee, J., Kitajima, T. S., Tanno, Y. et al. (2008b). Unified mode of centromeric protection by shugoshin in mammalian oocytes and somatic cells. *Nature Cell Biology, 10*(1), 42–52.

Lees-Murdock, D. J., Lau, H.-T., Castrillon, D. H. et al. (2008). DNA methyltransferase loading, but not de novo methylation, is an oocyte-autonomous process stimulated by SCF signalling. *Developmental Biology, 321*(1), 238–250.

Lefevre, B., Gougeon, A., Nome, F. & Testart, J. (1989). *In vivo* changes in oocyte germinal vesicle related to follicular quality and size at mid-follicular phase during stimulated cycles in the cynomolgus monkey. *Reproduction Nutrition Development, 29*(5), 523–531.

Leland, S., Nagarajan, P., Polyzos, A. et al. (2009). Heterozygosity for a Bub1 mutation causes female-specific germ cell aneuploidy in mice. *Proceedings of the National Academy of Sciences USA, 106*(31), 12776–12781.

Li, X., Ito, M., Zhou, F. et al. (2008). A maternal-zygotic effect gene, zfp57, maintains both maternal and paternal imprints, *Developmental Cell, 15*(4), 547–557.

Lister, L. M., Kouznetsova, A., Hyslop, L. A. et al. (2010). Age-related meiotic segregation errors in mammalian oocytes are preceded by depletion of cohesin and Sgo2. *Current Biology, 20*(17), 1511–1521.

Liu, H., & Aoki, F. (2002). Transcriptional activity associated with meiotic competence in fully grown mouse GV oocytes. *Zygote, 10*(4), 327–332.

Liu, H., Kim, J., & Aoki, F. (2004). Regulation of histone H3 lysine 9 methylation in oocytes and early pre-implantation embryos. *Development, 131*(10), 2269–2280.

Liu, K., Rajareddy, S., Liu, L. et al. (2006a). Control of mammalian oocyte growth and early follicular development by the oocyte PI3 kinase pathway: New roles for an old timer. *Developmental Biology, 299*(1), 1–11.

Liu, Y., Sui, H.-S., Wang, H.-L. et al. (2006b). Germinal vesicle chromatin configurations of bovine oocytes. *Microscopy Research and Technique, 69*(10), 799–807.

Llano, E., Gomez, R., Gutierrez-Caballero, C. et al. (2008). Shugoshin-2 is essential for the completion of meiosis but not for mitotic cell division in mice. *Genes & Development, 22*(17), 2400–2413.

Lopes, F. L., Fortier, A. L., Darricarr're, N. et al. (2009). Reproductive and epigenetic outcomes associated with aging mouse oocytes, *Human Molecular Genetics, 18*(11), 2032–2044.

Lucchesi, J. C., Kelly, W. G., & Panning, B. (2005). Chromatin remodeling in dosage compensation. *Annual Review of Genetics, 39*, 615–651.

Lucifero, D., Mann, M. R., Bartolomei, M. S., & Trasler, J. M. (2004). Gene-specific timing and epigenetic memory in oocyte imprinting. *Humam Molecular Genetics, 13*(8), 839–849.

Luders, J., Patel, U. K., & Stearns, T. (2006). GCP-WD is a [gamma]-tubulin targeting factor required for centrosomal and chromatin-mediated microtubule nucleation. *Nature Cell Biology, 8*(2), 137–147.

Luger, K. (2003). Structure and dynamic behavior of nucleosomes. *Current Opinion in Genetics & Development, 13*(2), 127–135.

Ma, W., Baumann, C., & Viveiros, M. M. (2010). NEDD1 is crucial for meiotic spindle stabilty and accurate chromosome segregation in mammalian oocytes. *Developmental Biology, 339*, 439–450.

Ma, W., Koch, J. A., & Viveiros, M. M. (2008). Protein kinase C delta (PKCδ) interacts with microtubule organizing center (MTOC)-associated proteins and participates in meiotic spindle organization. *Developmental Biology, 320*, 414–425.

Mager, J., & Bartolomei, M. S. (2005). Strategies for dissecting epigenetic mechanisms in the mouse. *Nature Genetics, 37*(11), 1194–1200.

Maison, C., & Almouzni, G. (2004). HP1 and the dynamics of heterochromatin maintenance. *Nature Reviews Mollecular Cell Biology, 5*(4), 296–304.

Maison, C., Bailly, D., Peters, A. et al. (2002). Higher-order structure in pericentric heterochromatin involves a distinct pattern of histone modification and RNA component. *Nature Genetics, 30*, 329–334.

Mandl, A. M. (1962). Preovulatory changes in the oocyte of the adult rat. *Proceedings of the Royal Society of London, 158*, 105–118.

Manosalva, I., & Gonzalez, A. (2010). Aging changes the chromatin configuration and histone methylation of mouse oocytes at germinal vesicle stage. *Theriogenology, 74*(9), 1539–1547.

Margueron, R., Trojer, P., & Reinberg, D. (2011). The key to development: Interpreting the histone code?. *Current Opinion in Genetics & Development, 15*(2), 163–176.

Matsui, Y., & Hayashi, K. (2007). Epigenetic regulation for the induction of meiosis. *Cellular and Molecular Life Sciences, 64*(3), 257–262.

Mattson, B., & Albertini, D. (1990). Oogenesis: Chromatin and microtubule dynamics during meiotic prophase. *Molecular Reproduction & Development, 25*(4), 374–383.

Matzuk, M. M., Burns, K. H., Viveiros, M. M., & Eppig, J. J. (2002). Intercellular communication in the mammalian ovary: Oocytes carry the conversation. *Science, 296*(5576), 2178–2180.

McDowell, T. L., Gibbons, R. J., Sutherland, H. et al. (1999). Localization of a putative transcriptional regulator (ATRX) at pericentromeric heterochromatin and the short arms of acrocentric chromosomes. *Proceedings of the National Academy of Sciences USA, 96*(24), 13983–13988.

McEwen, B., & Dong, Y. (2010). Contrasting models for kinetochore microtubule attachment in mammalian cells. *Cellular and Molecular Life Sciences, 67*(13), 2163–2172.

McGrath, J., & Solter, D. (1984). Completion of mouse embryogenesis requires both the maternal and paternal genomes. *Cell, 37*(1): 179–183.

McGuinness, B. E., Anger, M., Kouznetsova, A. et al. (2009). Regulation of APC/C activity in oocytes by a Bub1-dependent spindle assembly checkpoint. *Current Biology, 19*(5), 369–380.

Meglicki, M., Zientarski, M., & Borsuk, E. (2008). Constitutive heterochromatin during mouse oogenesis: The pattern of histone H3 modifications and localization of HP1alpha and HP1beta proteins. *Molecular Reproduction & Development, 75*(2), 414–428.

Moore, T. (2001). Genetic conflict, genomic imprinting and establishment of the epigenotype in relation to growth. *Reproduction, 122*(2), 185–193.

Morgan, H. D., Santos, F., Green, K. et al. (2005). Epigenetic reprogramming in mammals. *Human Molecular Genetics, 14*(Spec. No. 1), R47–R58.

Moritz, M., & Agard, D. A. (2001). Gamma-tubulin complexes and microtubule nucleation. *Current Opinion in Structural Biology, 11*(2), 174–181.

Motlik, J., & Fulka, J. (1976). Breakdown of the germinal vesicle in pig oocytes *in vivo* and *in vitro*. *Journal of Experimental Zoology, 198*(2), 155–162.

Muegge, K. (2005). Lsh, a guardian of heterochromatin at repeat elements. *Biochemistry & Cell Biology, 83*(4), 548–554.

Munné, S., Chen, S., Colls, P. et al. (2007). Maternal age, morphology, development and chromosome abnormalities in over 6000 cleavage-stage embryos. *Reproductive Biomedicine Online, 14*(5), 628–634.

Murphy, T. D., & Karpen, G. H. (1998). Centromeres take flight: Alpha satellite and the quest for the human centromere. *Cell, 93*(3), 317–320.

Niault, T. O, Hached, K., Sotillo, R. O. et al. (2007). Changing Mad2 levels affects chromosome segregation and spindle assembly checkpoint control in female mouse meiosis I. *PLoS One, 2*(11), e1165.

Nicklas, R. B., & Ward, S. C. (1994). Elements of error correction in mitosis: Microtubule capture, release, and tension. *The Journal of Cell Biology, 126*(5), 1241–1253.

Nicklas, R. B., Waters, J. C., Salmon, E. D., & Ward, S. C. (2001). Checkpoint signals in grasshopper meiosis are sensitive to microtubule attachment, but tension is still essential. *Journal of Cell Science, 114*(23), 4173–4183.

Nimura, K., Ishida, C., Koriyama, H. et al. (2006). Dnmt3a2 targets endogenous Dnmt3L to ES cell chromatin and induces regional DNA methylation. *Genes Cells, 11*(10), 1225–1237.

Oakley, B. R., Oakley, C. E., Yoon, Y., & Jung, M. K. (1990). [Gamma]-tubulin is a component of the spindle pole body that is essential for microtubule function in Aspergillus nidulans. *Cell, 61*(7), 1289–1301.

Obata, Y., Kaneko-Ishino, T., Koide T. et al. (1998). Disruption of primary imprinting during oocyte growth leads to the modified expression of imprinted genes during embryogenesis. *Development, 125*(8), 1553–1560.

Obata. Y., & Kono, T. (2002). Maternal primary imprinting is established at a specific time for each gene throughout oocyte growth. *Journal of Biological Chemistry, 277*(7), 5285–5289.

Oegema, K., Desai, A., Rybina, S. et al. (2001). Functional analysis of kinetochore assembly in Caenorhabditis elegans. *The Journal of Cell Biology, 153*(6), 1209–1226.

Ooga, M., Inoue, A., Kageyama, S. et al. (2008). Changes in H3K79 methylation during preimplantation development in mice. *Biology of Reproduction, 78*(3), 413–424.

Page, S., & Hawley, R. (2003). Chromosome choreography: The meiotic ballet. *Science, 301*(5634), 785–789.

Pan, H., Ma, P., Zhu, W., & Schultz, R. M. (2008). Age-associated increase in aneuploidy and changes in gene expression in mouse eggs. *Developmental Biology, 316*(2), 397–407.

Parfenov, V., Potchukalina, G., Dudina, L. et al. (1989). Human antral follicles: Oocyte nucleus and the karyosphere formation (electron microscopic and autoradiographic data). *Gamete Research, 22*(2), 219–231.

Pellestor, F., Andréo, B., Arnal, F. et al. (2003). Maternal aging and chromosomal abnormalities: New data drawn from *in vitro* unfertilized human oocytes. *Human Genetics, 112*(2), 195–203.

Peters, A. H., O'Carroll, D., Scherthan, H. et al. (2001). Loss of the Suv39h histone methyltransferases impairs mammalian heterochromatin and genome stability. *Cell, 107*(3), 323–337.

Peterson, C. L., & Laniel, M. A. (2004). Histones and histone modifications, *Current Biology, 14*(14), R546–R551.

Petronczki, M., Siomos, M., & Nasmyth, K. (2003). Un menage a quatre: The molecular biology of chromosome segregation in meiosis. *Cell, 112*(4), 423–440

Picketts, D. J., Tastan, A. O., Higgs, D. R., & Gibbons, R. J. (1998). Comparison of the human and murine ATRX gene identifies highly conserved, functionally important domains. *Mammalian Genome, 9*, 400–403.

Pluta, A. F., Mackay, A. M., Ainsztein, A. M. et al. (1995). The centromere: Hub of chromosomal activities. *Science, 270*(5242), 1591–1594.

Polo, S. E., & Almouzni, G. (2006). Chromatin assembly: A basic recipe with various flavours. *Current Opinion in Genetics and Development, 16*(2), 104–111.

Prasad, P., & Ekwall, K. (2011). New insights into how chromatin remodellers direct CENP-A to centromeres. *EMBO Journal, 30*(10), 1875–1876.

Racki, W. J., & Richter, J. D. (2006). CPEB controls oocyte growth and follicle development in the mouse. *Development, 133*(22), 4527–4537.

Raynaud-Messina, B., & Merdes, A. (2007). [Gamma]-tubulin complexes and microtubule organization. *Current Opinion in Cell Biology, 19*(1), 24–30.

Rea, S., Eisenhaber, F., O'Carroll, D. et al. (2000). Regulation of chromatin structure by site-specific histone H3 methyltransferases. *Nature, 406*(6796), 593–599.

Reik, W., Dean, W., & Walter, J. (2001). Epigenetic reprogramming in mammalian development. *Science, 293*(5532), 1089–1093.

Revenkova, E., Eijpe, M., Heyting, C. et al. (2004). Cohesin SMC1 beta is required for meiotic chromosome dynamics, sister chromatid cohesion and DNA recombination. *Nature Cell Biology, 6*(6), 555–562.

Revenkova, E., Herrmann, K., Adelfalk, C., & Jessberger, R. (2010). Oocyte cohesin expression restricted to predictyate stages provides full fertility and prevents aneuploidy. *Current Biology, 20*(17), 1529–1533.

Richter, J. (2001). Think globally, translate locally: What mitotic spindles and neuronal synapses have in common. *Proceedings of the National Academy of Sciences of the United States of America, 98*(13), 7069–7071.

Rivera, R. M., Stein, P., Weaver, J. R. et al. (2008). Manipulations of mouse embryos prior to implantation result in aberrant expression of imprinted genes on day 9.5 of development. *Human Molecular Genetics, 17*(1), 1–14.

Roberts, R., Iatropoulou, A., Ciantar, D. et al. (2005). Follicle-stimulating hormone affects metaphase I chromosome alignment and increases aneuploidy in mouse oocytes matured *in vitro*. *Biology of Reproduction, 72*(1), 107–118.

Robertson, K. D., & Wolffe, A. P. (2000). DNA methylation in health and disease. *Nature Reviews Genetics, 1*(1), 11–19.

Rollins, R. A., Haghighi, F., Edwards, J. R. et al. (2006). Large-scale structure of genomic methylation patterns. *Genome Research, 16*(2), 157–163.

Russo, V., Martelli, A., Berardinelli, P. et al. (2007). Modifications in chromatin morphology and organization during sheep oogenesis. *Microscopy Research and Technique, 70*(8), 733–744.

Sarma, K., & Reinberg, D. (2005). Histone variants meet their match. *Nature Reviews Molecular Cell Biology, 6*(2), 139–149.

Sarmento, O. F., Digilio, L. C., Wang, Y. et al. (2004). Dynamic alterations of specific histone modifications during early murine development. *Journal of Cell Science, 117*(Pt 19), 4449–4459.

Sasaki, H., & Matsui, Y. (2008). Epigenetic events in mammalian germ-cell development: Reprogramming and beyond. *Nature Reviews Genetics, 9*(2), 129–140.

Sassone-Corsi, P. (2002). Unique chromatin remodeling and transcriptional regulation in spermatogenesis. *Science, 296*(5576), 2176–2178.

Schatten, G. (1994). The centrosome and its mode of inheritance: The reduction of the centrosome during gametogenesis and its restoration during fertilization. *Developmental Biology, 165*(2), 299–335.

Schober, C. S., Aydiner, F., Booth, C. J. et al. (2011). The kinase VRK1 is required for normal meiotic progression in mammalian oogenesis. *Mechanisms of Development, 128*(3–4), 178–190.

Schotta, G., Lachner, M., Sarma, K. et al. (2004). A silencing pathway to induce H3-K9 and H4-K20 trimethylation at constitutive heterochromatin. *Genes & Development, 18*(11), 1251–1262.

Schramm, R. D., Tennier, M. T., Boatman, D. E., & Bavister, B. D. (1993). Chromatin configurations and meiotic competence of oocytes are related to follicular diameter in nonstimulated rhesus monkeys. *Biology of Reproduction, 48*(2), 349–356.

Schreiber, V., Dantzer, F., Ame, J. C., & de Murcia, G. (2006). Poly(ADP-ribose): Novel functions for an old molecule, *Nature Reviews Molecular Cell Biology, 7*(7), 517–528.

Schuh, M., & Ellenberg, J. (2007). Self-organization of MTOCs replaces centrosome function during acentrosomal spindle assembly in live mouse oocytes. *Cell, 130*(3), 484–498.

Schuh, M., Lehner, C. F., & Heidmann, S. (2007). Incorporation of Drosophila CID/CENP-A and CENP-C into centromeres during early embryonic anaphase. *Current Biology, 17*(3), 237–243.

Spinaci, M., Seren, E., & Mattioli, M. (2004). Maternal chromatin remodeling during maturation and after fertilization in mouse oocytes. *Molecular Reproduction and Development, 69*(2), 215–221.

Stearns, T., & Kirschner, M. (1994). *In vitro* reconstitution of centrosome assembly and function: The central role of [gamma]-tubulin. *Cell, 76*(4), 623–637.

Sui, H. S., Liu, Y., Miao, D. Q. et al. (2005). Configurations of germinal vesicle (GV) chromatin in the goat differ from those of other species. *Molecular Reproduction & Development, 71*(2), 227–236.

Suja, J., & Barbero, J. (2009). Cohesin complexes and sister chromatid cohesion in mammalian meiosis. *Genome Dynamics, 5*, 94–116.

Sullivan, B. A., & Karpen, G. H. (2004). Centromeric chromatin exhibits a histone modification pattern that is distinct from both euchromatin and heterochromatin. *Nature Structural & Molecular Biology, 11*(11), 1076–1083.

Sun, L., Lee, D., Zhang, Q. et al. (2004a). Growth retardation and premature aging phenotypes in mice with disruption of the SNF2-like gene, PASG. *Genes & Development, 18*(9), 1035–1046.

Sun, X.-S., Liu, Y., Yue, K.-Z. et al. (2004b). Changes in germinal vesicle (GV) chromatin configurations during growth and maturation of porcine oocytes. *Molecular Reproduction and Development, 69*(2), 228–234.

Suo, L., Meng, Q.-G., Pei, Y. et al. (2010). Changes in acetylation on lysine 12 of histone H4 (acH4K12) of murine oocytes during maternal aging may affect fertilization and subsequent embryo development. *Fertility & Sterility, 93*(3), 945–951.

Surani, M. (2001). Reprogramming of genome function through epigenetic inheritance. *Nature, 414*(6859), 122–128.

Surani, M., Barton, S. C., & Norris, M. L. (1984). Development of reconstituted mouse eggs suggests imprinting of the genome during gametogenesis. *Nature, 308*(5959), 548–550.

Surani, M. A., Hayashi, K., & Hajkova, P. (2007). Genetic and epigenetic regulators of pluripotency. *Cell, 128*(4), 747–762.

Susiarjo, M., Hassold, T. J., Freeman, E., & Hunt, P. A. (2007). Bisphenol A exposure in utero disrupts early oogenesis in the mouse. *PLoS Genetics, 3*(1), e5.

Svoboda, P., Stein, P. & Schultz, R. (2001). RNAi in mouse oocytes and preimplantation embryos: Effectiveness of hairpin dsRNA. *Biochemical & Biophysical Research Communications, 287*(5), 1099–1104.

Swain, J. E., Ding, J., Brautigan, D. L. et al. (2007). Proper chromatin condensation and maintenance of histone H3 phosphorylation during mouse oocyte meiosis requires protein phosphatase activity. *Biology of Reproduction, 76*(4), 628–638.

Szollosi, D., Calarco, P., & Donahue, R. P. (1972). Absence of centrioles in the first and second meiotic spindles of mouse oocytes. *Journal of Cell Science, 11*(2), 521–541.

Tachibana, M., Nozaki, M., Takeda, N., & Shinkai, Y. (2007). Functional dynamics of H3K9 methylation during meiotic prophase progression. *EMBO Journal, 26*(14), 3346–3359.

Tilghman, S. (1999). The sins of the fathers and mothers: Genomic imprinting in mammalian development. *Cell, 96*(2), 185–193.

Tranfo, G., Caporossi, L., Paci, E. et al. (2012). Urinary phthalate monoesters concentration in couples with infertility problems. *Toxicology Letters* (epub ahead of print).

Tsukiyama, T. (2002). The *in vivo* functions of ATP-dependent chromatin-remodelling factors. *Nature Reviews Mollecular Cell Biology, 3*(6), 422–429.

van den Berg, I. M., Eleveld, C., van der Hoeven, M. et al. (2011). Defective deacetylation of histone 4 K12 in human oocytes is associated with advanced maternal age and chromosome misalignment. *Human Reproduction* (epub ahead of print).

Varga-Weisz, P. (2001). ATP-dependent chromatin remodeling factors: Nucleosome shufflers with many missions. *Oncogene, 20*(24), 3076–3085.

Vialard, F., Petit, C., Bergere, M. et al. (2006). Evidence of a high proportion of premature unbalanced separation of sister chromatids in the first polar bodies of women of advanced age. *Human Reproduction, 21*(5), 1172–1178.

Wang, H.-L., Sui, H.-S., Liu, Y. et al. (2009a). Dynamic changes of germinal vesicle chromatin configuration and transcriptional activity during maturation of rabbit follicles. *Fertility & Sterility, 91*(4, Supp. 1), 1589–1594.

Wang, J., Hevi, S., Kurash, J. K. et al. (2009b). The lysine demethylase LSD1 (KDM1) is required for maintenance of global DNA methylation. *Nature Genetics, 41*(1), 125–129.

Wang, Q., Wang, C. M., Ai, J. S. et al. (2006). Histone phosphorylation and pericentromeric histone modifications in oocyte meiosis. *Cell Cycle, 5*(17), 1974–1982.

Weber, M., & Schuebeler, D. (2007). Genomic patterns of DNA methylation: Targets and function of an epigenetic mark. *Current Opinion in Cell Biology, 19*(3), 273–280.

Webster, K., O'Bryan, M., Fletcher, S. et al. (2005). Meiotic and epigenetic defects in Dnmt3L-knockout mouse spermatogenesis. *Proceedings of the National Academy of Sciences of the United States of America, 102*(11), 4068–4073.

Welburn, J. P. I., Grishchuk, E. L., Backer, C. B. et al. (2009). The human kinetochore ska1 complex facilitates microtubule depolymerization-coupled motility. *Developmental Cell, 16*(3), 374–385.

Welburn, J. P. I., Vleugel, M., Liu, D. et al. (2010). Aurora B phosphorylates spatially distinct targets to differentially regulate the kinetochore-microtubule interface. *Molecular Cell, 38*(3), 383–392.

Wickramasinghe, D., Ebert, K. M., & Albertini, D. F. (1991). Meiotic competence acquisition is associated with the appearance of M-phase characteristics in growing mouse oocytes. *Developmental Biology, 143*(1), 162–172.

Wiebe, M. S., Nichols, R. J., Molitor, T. P. et al. (2010). Mice deficient in the serine/threonine protein kinase VRK1 are infertile due to a progressive loss of spermatogonia. *Biology of Reproduction, 82*(1), 182–193.

Wiens, G. R., & Sorger, P. K. (1998). Centromeric chromatin and epigenetic effects in kinetochore assembly. *Cell, 93*(3), 313–316.

Wiese, C., & Zheng, Y. (2006). Microtubule nucleation: {Gamma}-tubulin and beyond. *Journal of Cell Science, 119*(20), 4143–4153.

Wood, A., Severson, A., & Meyer, B. (2010). Condensin and cohesin complexity: The expanding repertoire of functions. *Nature Reviews Genetics, 11*(6), 391–404.

Wood, K. W., Sakowicz, R., Goldstein, L. S. B., & Cleveland, D. W. (1997). CENP-E is a plus end-directed kinetochore motor required for metaphase chromosome alignment. *Cell, 91*(3), 357–366.

Xu, G. L., Bestor, T. H., Bourc'his, D. et al. (1999). Chromosome instability and immunodeficiency syndrome caused by mutations in a DNA methyltransferase gene. *Nature, 402*, 187–191.

Yan, Q., Cho, E., Lockett, S., & Muegge, K. (2003a). Association of Lsh, a regulator of DNA methylation, with pericentromeric heterochromatin is dependent on intact heterochromatin. *Molecular & Cellular Biology, 23*(23), 8416–8428.

Yan, Q., Huang, J., Fan, T. et al. (2003b). Lsh, a modulator of CpG methylation, is crucial for normal histone methylation. *EMBO Journal, 22*(19), 5154–5162.

Yang, F., Baumann, C., & De La Fuente, R. (2009). Persistence of histone H2AX phosphorylation after meiotic chromosome synapsis and abnormal centromere cohesion in poly (ADP-ribose) polymerase (Parp-1) null oocytes. *Developmental Biology, 331*, 326–338.

Yang, K.-T., Li, S.-K., Chang, C.-C. et al. (2010). Aurora-C kinase deficiency causes cytokinesis failure in meiosis I and production of large polyploid oocytes in mice. *Molecular Biology of the Cell, 21*(14), 2371–2383.

Yuba-Kubo, A., Kubo, A., Hata, M., & Tsukita, S. (2005). Gene knockout analysis of two gamma-tubulin isoforms in mice. *Developmental Biology, 282*(2), 361–373.

Zeng, W., Baumann, C., Schmidtmann, A. et al. (2011). Lymphoid-specific helicase (HELLS) is essential for meiotic progression in mouse spermatocytes. *Biology of Reproduction* (epub ahead of print).

Zhang, Y. (2003). Transcriptional regulation by histone ubiquitination and deubiquitination. *Genes & Development, 17*(22), 2733–2740.

Zuccotti, M., Giorgi Rossi, P., Martinez, A., et al. (1998). Meiotic and developmental competence of mouse antral oocytes. *Biology of Reproduction, 58*(3), 700–704.

Zuccotti, M., Piccinelli, A., Rossi, P., et al. (1995). Chromatin organization during mouse oocyte growth. *Molecular Reproduction & Development, 4*, 479–485.

Zucker, R. M., Keshaviah, A. P., Price, O. T., & Goldman, J. M. (2000). Confocal laser scanning microscopy of rat follicle development. *Journal of Histochemistry & Cytochemistry, 48*(6), 781–791.

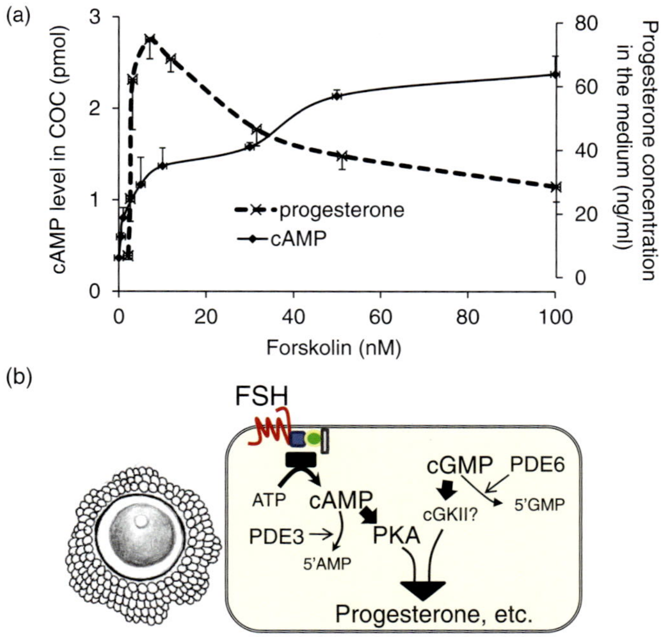

Plate 3.3 The relationship between cAMP level and progesterone production in cumulus cells of COC cultured with different doses of forskolin for 20 hr. (a) Porcine COC were cultured with 0 to 100 nM forskolin for 20 hr. The condition medium was recovered and then used for progesterone assay by EIA, and the cultured COC were used for cAMP analysis as described in the Figure 3.1 legend. (b) A schematic diagram with regard to the regulation of cAMP and cGMP levels in cumulus cells to induce differentiation. FSH produces cAMP to activate the PKA pathway. cGMP is also produced and accumulated in cumulus cells, which may activate the cGKII pathway. Both pathways are required to produce progesterone. The level of cAMP or cGMP is negatively regulated by specific enzyme PDE3 or PDE6 in cumulus cells.

Oocyte Physiology and Development in Domestic Animals, First Edition. Edited by Rebecca L. Krisher.

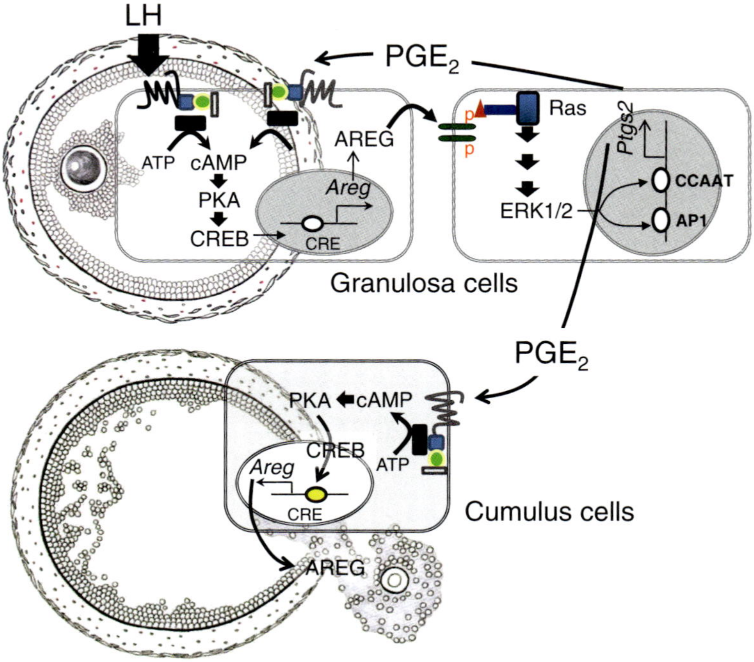

Plate 3.7 Schematic showing potential paracrine and autocrine pathways by which expression of amphiregulin (*Areg*) and *Ptgs2* is regulated in granulosa cells and cumulus cells of ovulating follicles. The proposed sequence is as follows: 1) LH binds to its cognate receptor localized to granulosa cells stimulating cAMP production to activate PKA-CREB pathway. Phosphorylated CREB binds to the cAMP responsible element (CRE) in the promoter region of the *Areg* gene. 2) As a consequence, *Areg* mRNAs are induced rapidly providing ligand that 3) bind EGF-receptors on granulosa cells (autocrine) leading to activation of ERK1/2 and induced expression of *Ptgs2*. In the promoter region of the *Ptgs2* gene, AP1 site and CCAAT site are essential transcription factor binding sites to upregulate gene expression. 4) The resulting increased production of prostaglandins (PGs; PGE) then provides ligands that bind EP2 on granulosa cells and cumulus cells (paracrine and autocrine) that (like the FSH and LH receptors) activate the cAMP-PKA-CREB pathway to increase the expression level of *Areg* mRNA. Thus, by the PGE/EP2 pathway, *Areg* mRNA can be induced in both cell types.

(a)

```
sh   DQESASSELK KPLVPASVNL SEYFKQFLFP QNECELHDFR LSFSQLKWDN WIVAPHKYNP RYCKGDCPRA
ca   DQESVSSELK KPLVPASFNL SEYFKQFLFP QNECELHDFR LSFSQLKWDN WIVAPHKYNP RYCKGDCPRA
go   DQESVSSELK KPLVPASVNL SEYFKQFLFP QNECELHDFR LSFSQLKWDN WIVAPHKYNP RYCKGDCPRA
pi   AQDTVSSELK KPLVPASFNL SEYFKQFLFP QNECELHDFR LSFSQLKWDN WIVAPHKYNP RYCKGDCPRA

hu   GQETVSSELK KPLGPASFNL SEYFRQFLLP QNECELHDFR LSFSQLKWDN WIVAPHRYNP RYCKGDCPRA
mo   GQKAIRSEAK GPLLTASFNL SEYFKQFLFP QNECELHDFR LSFSQLKWDN WIVAPHRYNP RYCKGDCPRA
ra   GQKTLSSETK KPL-TASFNL SEYFRQFLFP QNECELHDFR LSFSQLKWDN WIVAPHRYNP RYCKGDCPRA
---------------------------------------------------------------------------------
sh   VGHRYGSPVH TMVQNIIHEK LDSSVPRPSC VPAKYSPLSV LAIEPDGSIA YKEYEDMIAT KCTCR
ca   VGHRYGSPVH TMVMNIIHEK LDSSVPRPSC VPAKYSPLSV LAIEPDGSIA YKEYEDMIAT KCTCR
go   VGHRYGSPVH TMVQNIIHEK LDSSVPRPSC VPAKYSPLSV LAIEPDGSIA YKEYEDMIAT KCTCR
pi   VGHRYGSPVH TMVQNIIHEK LDSSVPRPSC VPAKYSPLSV LAIEPDGSIA YKEYEDMIAT KCTCR

hu   VGHRYGSPVH TMVQNIIYEK LDSSVPRPSC VPAKYSPLSV LTIEPDGSIA YKEYEDMIAT KCTCR
mo   VRHRYGSPVH TMVQNIIYEK LDPSVPRPSC VPGKYSPLSV LTIEPDGSIA YKEYEDMIAT RCTCR
ra   VRHRYGSPVH TMVQNIIYEK LDPSVPRPSC VPGKYSPLSV LTIEPDGSIA YKEYEDMIAT RCTCR
```

(b)

```
sh   QAGSIASEVP GPSREHDGPE SNQCSLHPFQ VSFQQLGWDH WIIAPHLYTP NYCKGVCPRV LHYGLNSPNH
ca   QAGSIASEVP GPSREHDGPE SNLCSLHPFQ VSFQQLGWDH WIIAPHLYTP NYCKGVCPRV LHYGLNSPNH
go   QAGSIASEVP GPSREHDGPE SNQCSLHPFQ VSFQQLGWDH WIIAPHLYTP NYCKGVCPRV LYYGLNSPNH
pi   QAGSIASEVL GPSREHDGPE SNQCSLHPFQ VSFHQLGWDH WIIAPHFYTP NYCKGVCPRV LHYGLNSPNH

hu   QADGISAEVT ASSSKHSGPE NNQCSLHPFQ ISFRQLGWDH WIIAPPFYTP NYCKGTCLRV LRDGLNSPNH
mo   QACSIESDAS CPSQEHDGSV NNQCSLHPYK VSFHQLGWDH WIIAPRLYTP NYCKGICTRV LPYGLNSPNH
ra   QTCSIASDVP CPSQEQDRSV NNQCSLHPYK VSFHQLGWDH WIIAPRLYTP NYCKGICTGV LPYGLNSPNH
---------------------------------------------------------------------------------
sh   AIIQNLVSEL VDQNVPQPSC VPYKYVPISI LLIEANGSIL YKEYEGMIAQ SCTCR
ca   AIIQNLVNEL VDQSVPQPSC VPYKYVPISI LLIEANGSIL YKEYEGMIAQ SCTCR
go   AIIQNLVNEL VDQNVPQPSC VPYKYVPISI LLIEANGSIL YKEYEGMIAQ SCTCR
pi   AIIQNLVNEL VDQSVPQPSC VPYKYVPISI LLIEANGSIL YKEYEDMIAQ PCTCR

hu   AIIQNLINQL VDQSVPRPSC VPYKYVPISV LMIEANGSIL YKEYEGMIAE SCTCR
mo   AIIQSLVNEL VNHSVPQPSC VPYNFLPMSI LLIETNGSIL YKEYEGMIAQ SCTCR
ra   AIIQSLVNEL VNRSVPQLSC VPYKFLPMSI LLIEANGSIL YKEYEGMIAQ SCTCR
```

Plate 4.2 Alignment of GDF9 and BMP15 mature region amino acid sequences. Multiple sequence alignment was carried out using the T-Coffee software (www.tcoffee.org). **(a)** GDF9 sequences. **(b)** BMP15 sequences. sh: sheep, ca: cattle, go: goat, pi: pig, hu: human, mo: mouse, ra: rat. The six conserved cysteines that make up the "cysteine knot" are shown in yellow. Sequence differences between the four domestic species are highlighted in blue. Sequence differences between the domestic species and the human/rodent sequences are highlighted in green.

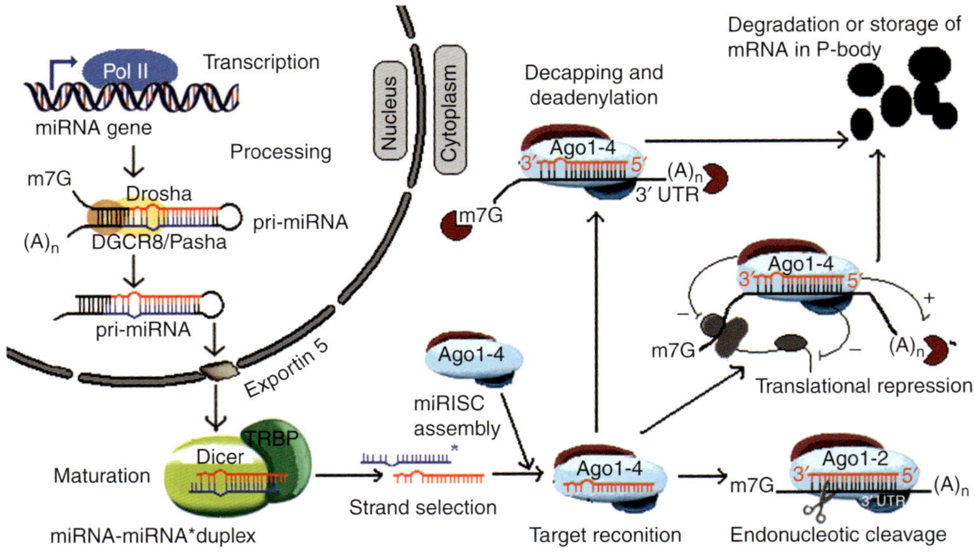

Plate 5.1 Biogenesis of miRNAs and their mechanism of gene regulation. MiRNAs are transcribed intragenic or intergenic chromosomal DNA regions by RNA polymerase II (Pol II) into primary transcripts of primary miRNA (pri-miRNA), which are then subjected to two RNase III endonucleases (Drosha and DGCR8/Pasha) to result in a stem-loop hairpin double stranded structure pre-miRNA. After being transported to the cytoplasm by the enzyme exportin 5, the pre-miRNA will be further processed to a 20–22 nt long miRNA duplex by the enzyme Dicer. Depnding on the sequence similarity with the target mRNA, the mature miRNA stand coupled with the RISC complex lead to the translational repression or endonucleatic cleavage of the target mRNA.

Argonaute proteins 1-4 (Ago1-4), inhibition of initiation of translation (−), Promoting deadenylation (+), 7-methyl-G cap (m7G), the passenger strand (miRNA*).

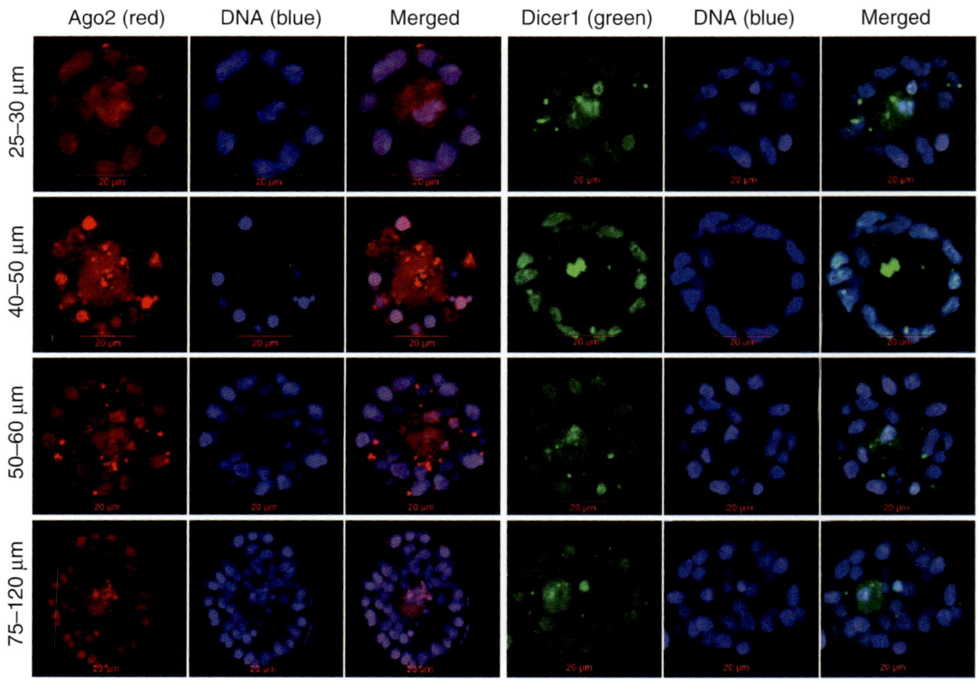

Plate 5.2 Expression of Dicer1 (green) and Ago2 (red) at different stages of bovine ovarian follicles categorized based on their size. Both Ago2 and Dicer are found to be expressed in both granulose and theca cells of the follicle at different stages of follicular development.

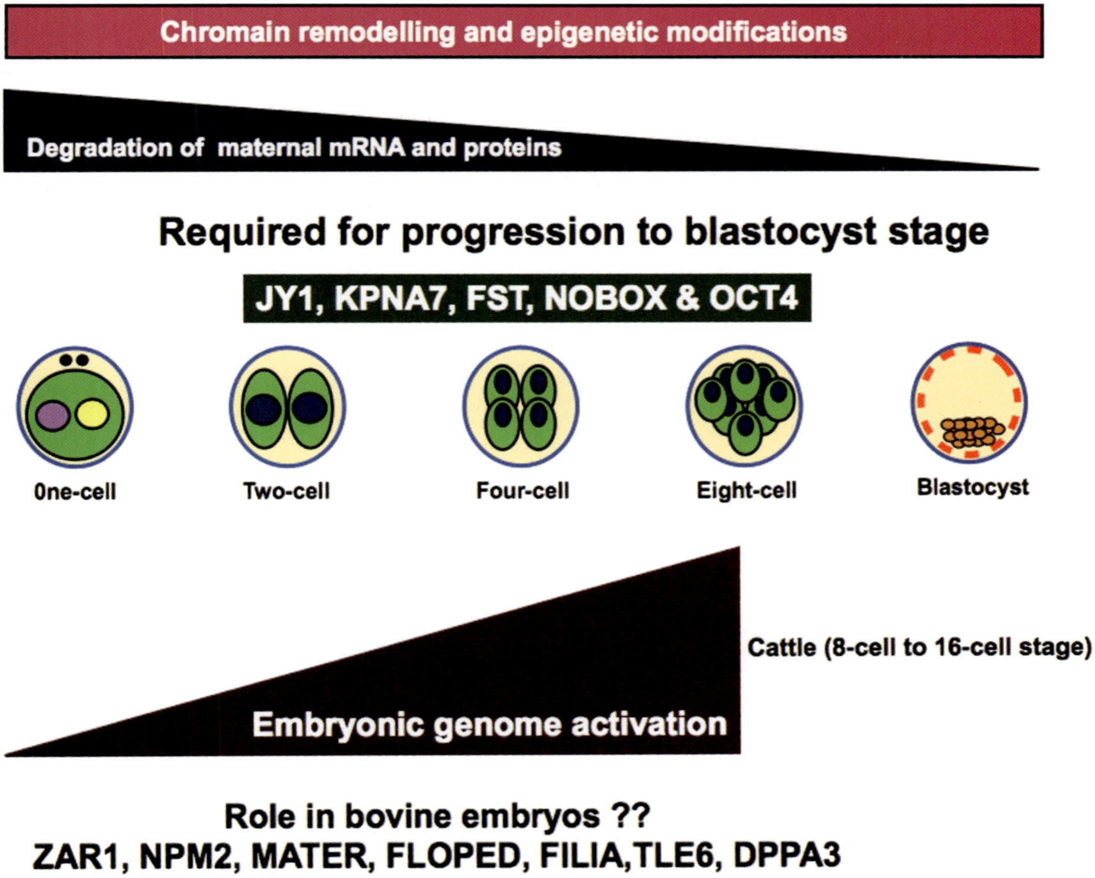

Plate 6.2 Oocyte-derived regulators and regulatory events during the bovine maternal-to-embryonic transition and subsequent stages of early embryogenesis.

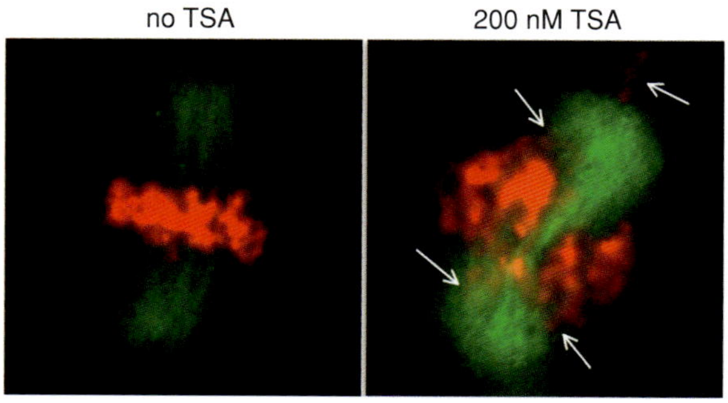

Plate 7.3 Treatment with TSA disrupts metaphase chromosome configurations. Pharmacological interference with histone deacetylation (200nM TSA) during the resumption of meiosis is associated with decondensed chromatin and severe chromosome misalignment (red, arrows) at the metaphase spindle (green), and ultimately leads to the formation of aneuploid gametes.

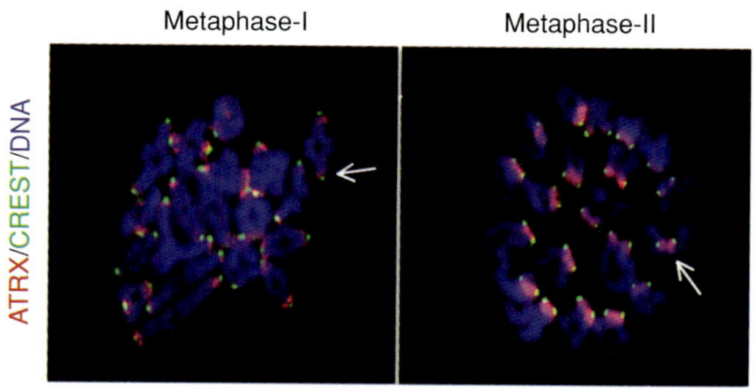

Plate 7.4 Chromatin remodeling proteins, such as ATRX, play an essential role in the functional differentiation of chromatin domains. ATRX (red) is a constitutive component of pericentric heterochromatin during metaphase-I as well as metaphase-II in mouse oocytes (arrows). Chromosomes are shown in blue, and the kinetochore marker CREST is shown in green.

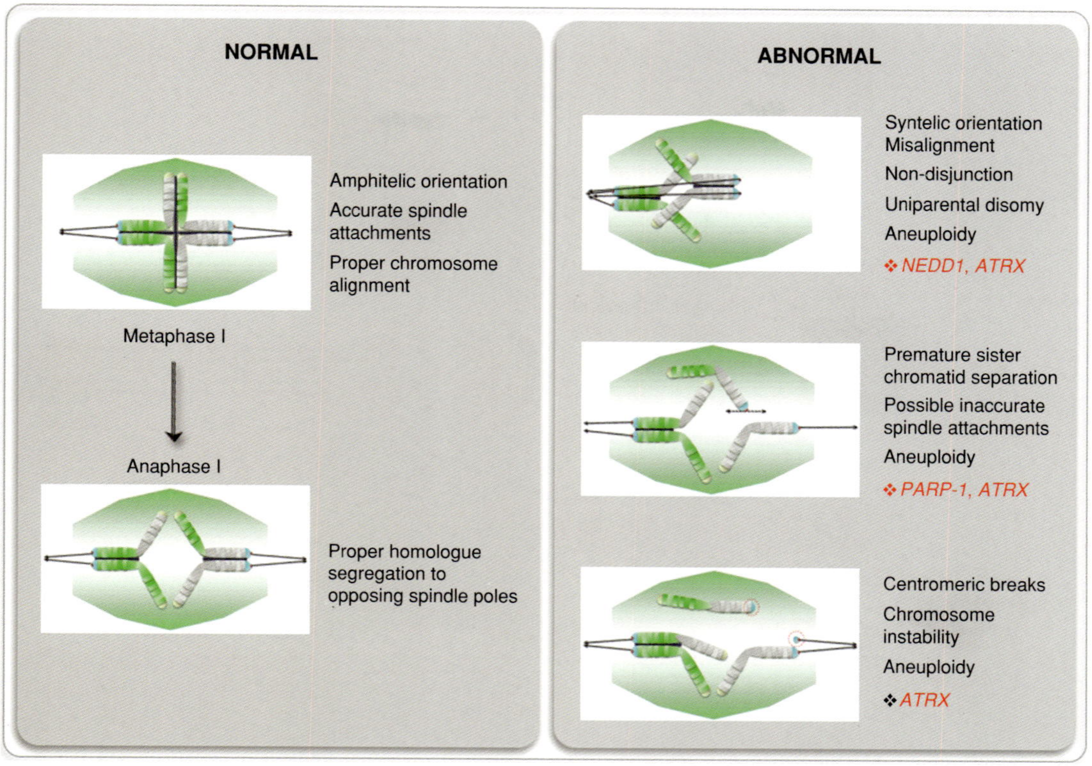

Plate 7.5 Schematic representation of normal (bi-oriented) as well as inappropriate chromosome-spindle microtubule interactions. Proper chromosome alignment at the metaphase spindle depends critically on the establishment of a proper epigenetic chromatin environment as well as on the formation of accurate and stable attachments between kinetochore domains and spindle microtubules. Loss-of-function analyses of key proteins (shown in red) lead to chromosome-microtubule attachment errors.

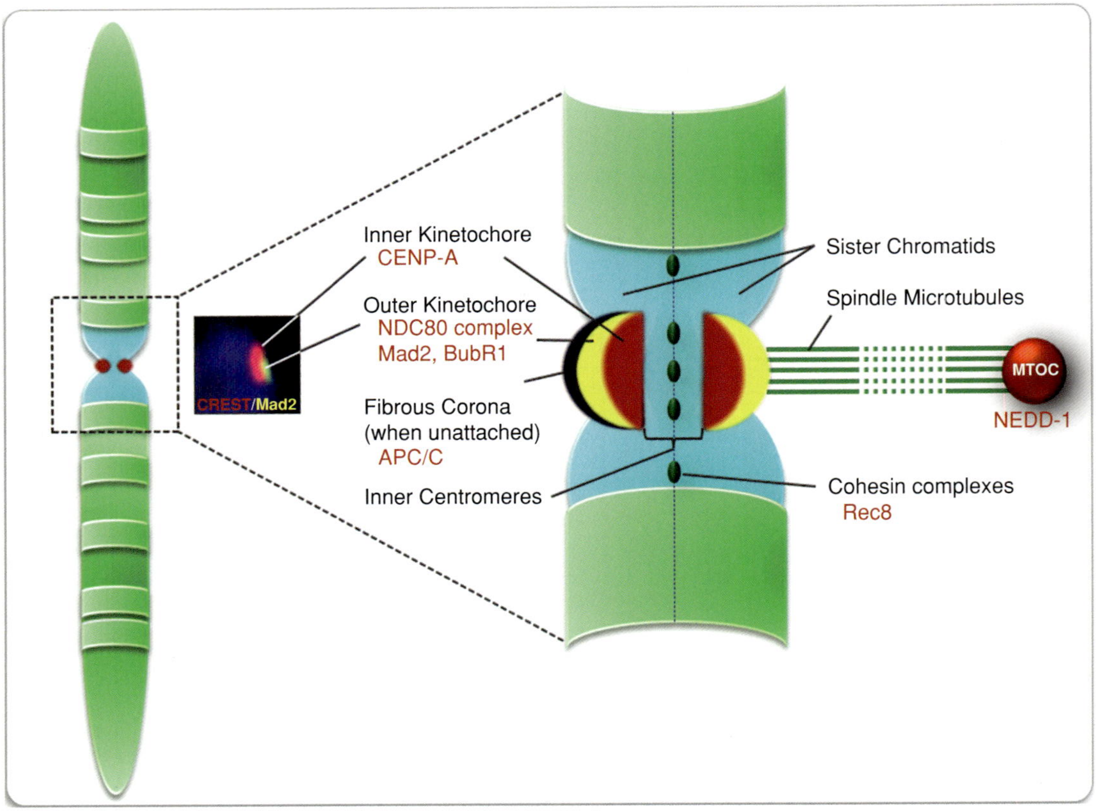

Plate 7.6 Schematic representation of the kinetochore complex assembled at the centromere. Key proteins shown in red localize to specific regions.

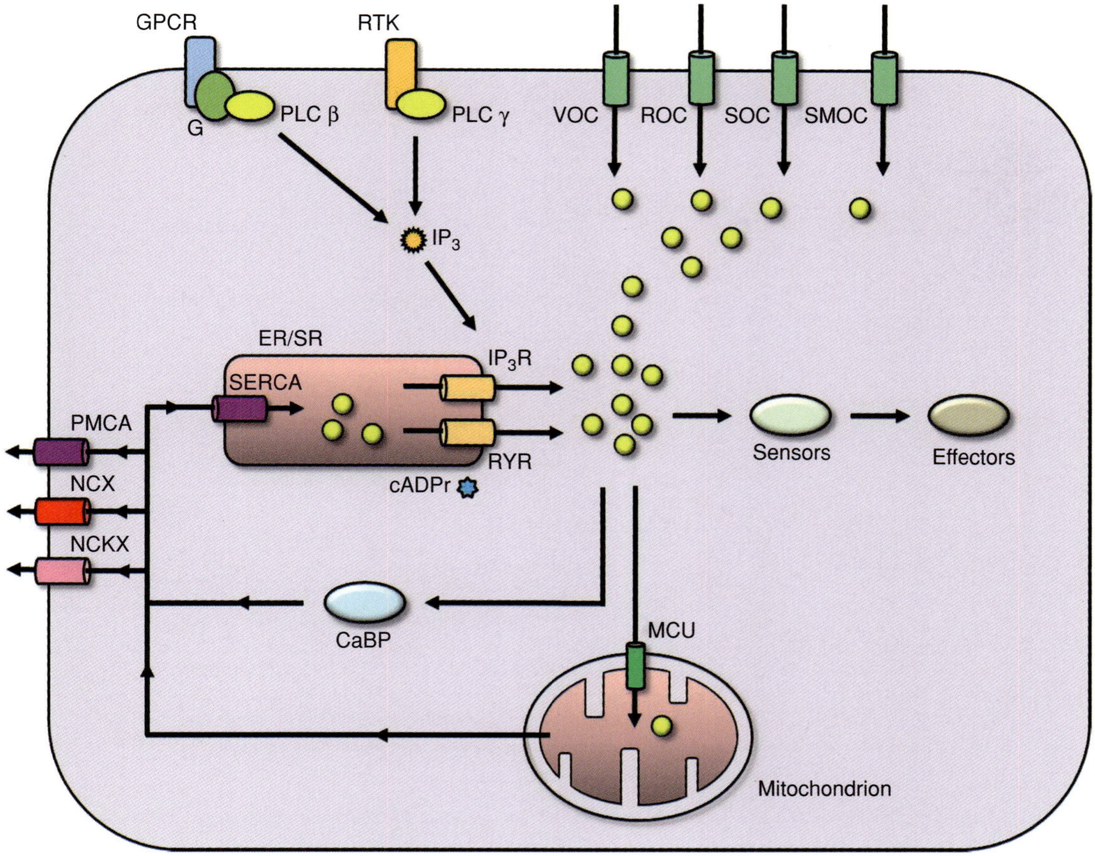

Plate 8.1 Schematic illustration of the Ca^{2+} signaling toolkit in cells. Abbreviations are as follows: GPCR, G protein-coupled receptor; G, G protein; PLCβ, phospholipase Cβ; RTK, receptor tyrosine kinase; PLCγ, phospholipase Cγ; IP_3, inositol 1,4,5-trisphosphate; IP_3R, inositol 1,4,5-trisphosphate receptor; RYR, ryanodine receptor; cADPr, cyclic ADP-ribose; VOC, voltage-operated channel; ROC, receptor-operated channel; SOC, store-operated channel; SMOC, second messenger-operated channel; ER/SR, endoplasmic/sarcoplasmic reticulum; MCU, mitochondrial Ca^{2+} uniporter; CaBP, Ca^{2+}-binding protein; SERCA, sarcoplasmic/endoplasmic reticulum Ca^{2+} ATPase; PMCA, plasma membrane Ca^{2+} ATPase; NCX, Na^+/Ca^{2+} exchanger; NCKX, Na^+/Ca^{2+}-K^+ exchanger. Yellow circles represent Ca^{2+}; the arrows indicate the movement of Ca^{2+} or the transmission of the signal. Based on Berridge, 2007.

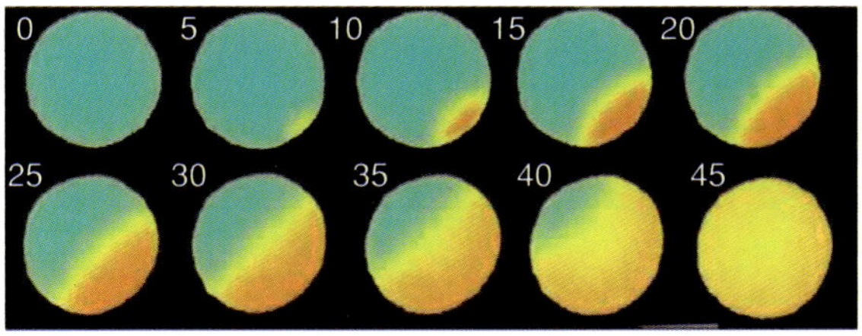

Plate 8.2 A fertilization Ca^{2+} wave observed in an ascidian oocyte. The outer chorion layer of the oocyte was removed; the oocyte was then injected with the Ca^{2+} indicator dye and inseminated. The figure shows the first Ca^{2+} wave that originates at the site of sperm entry. After the initial Ca^{2+} increase, in ascidian oocytes, a second set of waves is also generated whose point of origin is the vegetal hemisphere. Numbers denote time in seconds; different colors indicate different Ca^{2+} concentrations. Courtesy of Carroll et al. (2003); with permission.

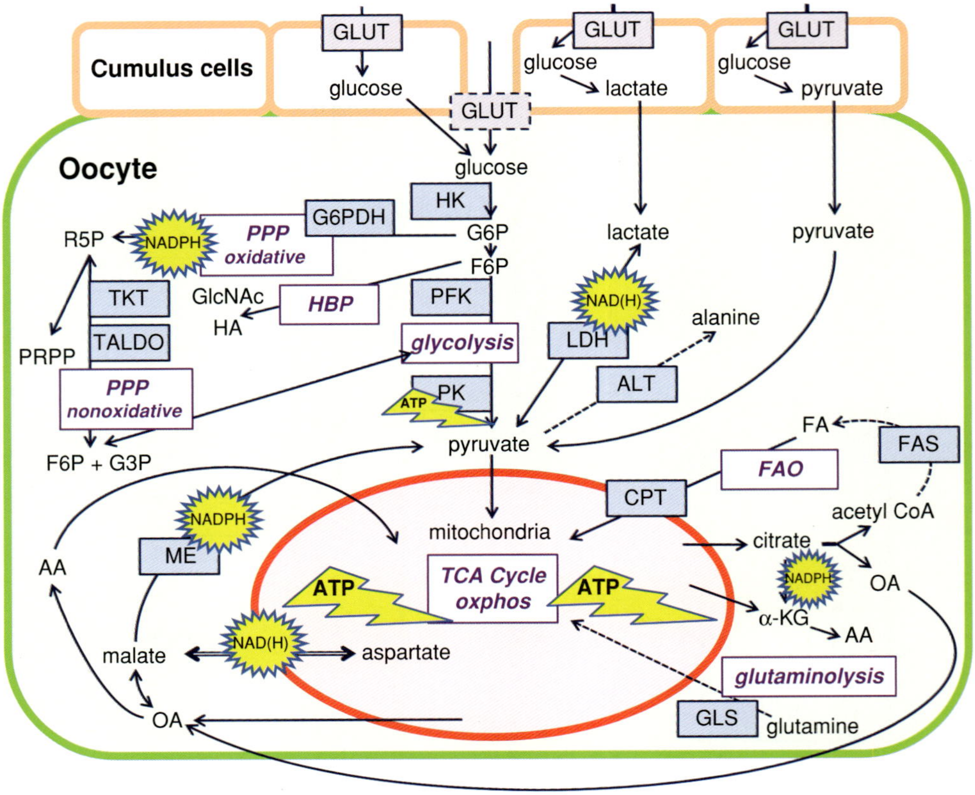

Plate 9.1 Known and potential (dashed lines) metabolic pathways operating in the mammalian oocyte. Pathways names are shown in white boxes, key regulatory enzymes are shown in blue boxes, and important end products are shown in yellow. Abbreviations: α-KG: alpha ketoglutarate; AA: amino acids; ALT: alanine transaminase; ATP: adenosine tri phosphate; CPT: carnitine palmitoyltransferase; F6P: fructose 6 phosphate; FA: fatty acids; FAO: fatty acid β-oxidation; FAS: fatty acid synthase; G3P: glyceraldehyde 3 phosphate; G6P: glucose 6 phosphate; G6PDH: glucose 6 phosphate dehydrogenase; GLS: glutaminase; GLUT: glucose transporters; HK: hexokinase; LDH: lactate dehydrogenase; ME: malic enzyme; OA: oxaloacetate; Oxphos: oxidative phosphorylation; PFK: phosphofructokinase; PK: pyruvate kinase; PRPP: phosphoribosyl pyrophosphate; R5P: ribose 5 phosphate; TALDO: transaldolase; TCA: tricarboxylic acid; TKT: transketolase.

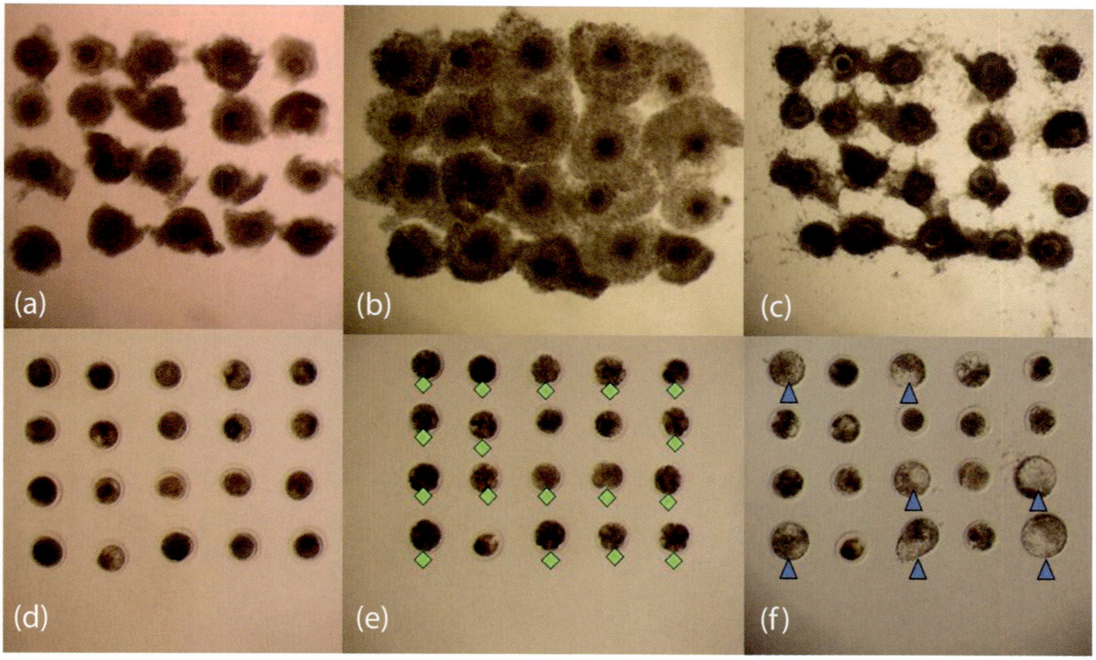

Plate 11.3 Representative images of the same batch of immature bovine oocytes cultured to Day 7 on Cell-Tak (Matoba et al., 2010). Twenty oocytes/embryos are placed on the bottom of the dish on Cell-Tak in a 5 × 4 grid formation in a 100 μl droplet of medium under mineral oil. For post-fertilization culture, embryos are placed at a maximum of 160 μm apart (d–f). (a) immature cumulus oocyte complexes (COCs), (b) *in vitro*–matured COCs, (c) presumptive zygotes 18–20 h after IVF, (d) denuded Day 1 zygotes, (e) Day 2 cleaved embryos (indicated by a diamond), and (f) Day 7 embryos.

8 Oocyte Calcium Homeostasis

Zoltan Machaty

8.1 Significance of Ca^{2+}

Calcium is the fifth-most abundant element in the earth's crust and the single-most abundant mineral in many animals, including humans (Baird, 2011). It is a soft gray metal that reacts violently with water to liberate hydrogen. When ionized, it becomes less reactive, and in the body it is present exclusively in an ionic (Ca^{2+}) form. Most of the Ca^{2+} in the body is incorporated into bones; only a small fraction exists as soluble Ca^{2+}. Some soluble Ca^{2+} is bound to molecules such as negatively charged proteins or other organic compounds, and some exists as free cations. Soluble free Ca^{2+} is essential for various biological functions, perhaps the most prominent being signal transduction.

Cells generate signals to adapt to the changing environment, and Ca^{2+} is the most universal carrier of signals. Changes in the cell's intracellular free Ca^{2+} concentration regulate a range of cell functions as diverse as cell proliferation, muscle contraction, synaptic plasticity, secretion, fertilization, and cell death. Ca^{2+} is properly suited for the job of a signaling messenger because it can bind to proteins and by doing so it can change the shape and charge of these molecules, thus altering their function (Clapham, 2007). It can accommodate 4–12 oxygen atoms in its primary coordination sphere; in most cases it binds to proteins through ~7 oxygen atoms provided by the glutamate and aspartate residues of the proteins (Krebs & Heizmann, 2007). The first indication that Ca^{2+} is not merely a structural element came from studying heart contraction more than a century ago (Ringer, 1883). Isolated frog hearts continued beating when suspended in a saline medium that was made using hard tap water, but stopped if distilled water was used to prepare the bathing solution. To sustain the beat, Ca^{2+} had to be added, and this observation laid the foundation for the notion that Ca^{2+} is a carrier of signals.

Ca^{2+} is the most tightly regulated ion in cells. Organisms keep their intracellular Ca^{2+} concentrations low for two major reasons. Ca^{2+} precipitates phosphates, the universal energy currency of cells; prolonged high Ca^{2+} levels in the cytoplasm lead to cell death. In addition, at low cytosolic levels, a relatively small increase in the Ca^{2+} concentration can trigger a distinct signal. Hence, although in the extracellular space the Ca^{2+} concentration is approximately 2 mM, cells maintain a ~20,000-fold gradient across their plasma membrane and keep their intracellular free Ca^{2+} concentration at ~100 nM. To achieve this, Ca^{2+} is either chelated or compartmentalized inside the cell, or is extruded into the extracellular space (Clapham, 2007). Specialized transporters located in the plasma membrane or membranes of the cellular organelles are used to move Ca^{2+} in and out of the cytosol in order to maintain the cellular Ca^{2+} homeostasis.

Oocyte Physiology and Development in Domestic Animals, First Edition. Edited by Rebecca L. Krisher.

8.2 Signaling by Ca^{2+}

8.2.1 The Signaling Apparatus

Ca^{2+} is an extremely versatile signaling agent. It is traditionally considered as an intracellular second messenger mediating the effects of first messengers that arrive at the cell surface. However, it also has the ability to recognize specific plasma membrane receptors and thus act on the outside of cells working as a first messenger (Brown et al., 1993). Despite the versatility, the toolkit utilized to generate the signal is startlingly similar in most cell types. Seminal findings in muscle and nerve cells led to the establishment of the concept that Ca^{2+} signaling involves a transient elevation of the intracellular free Ca^{2+} concentration by either the release of Ca^{2+} from cellular compartments or by Ca^{2+} entry across the plasma membrane, and that the signal is terminated by the resequestration or extrusion of Ca^{2+} by specialized Ca^{2+} pumps (reviewed by Whitaker, 2008).

The endoplasmic reticulum (ER) is the compartment that stores Ca^{2+} inside of the cell (in smooth and striated muscle, its somewhat modified version, the sarcoplasmic reticulum, serves this purpose). It is a multifunctional organelle that performs two major functions: in addition to its role in Ca^{2+} homeostasis, the ER is involved in synthesizing proteins and lipids (Baumann & Walz, 2001). The ER architecture is extremely diverse and reflects its multiple functions. It takes the shape of flattened cisternae when the organelle's primary role is protein synthesis, and it appears as an interconnected meshwork of membrane tubules when involved in signaling. In addition, the ER surface may appear rough or smooth; the rough ER with its associated ribosomes is the site of protein synthesis whereas the smooth ER is the primary depot for Ca^{2+}.

Two types of Ca^{2+} release channel receptors, the inositol 1,4,5-trisphosphate (IP_3) receptors and ryanodine receptors, span the ER membrane and are responsible for discharging Ca^{2+} from the store (Figure 8.1). The IP_3 receptor is a massive protein complex; its pore is formed by four ~310 kDa subunits (Mikoshiba, 1993). It allows Ca^{2+} to flow out of the ER when opened by its ligand IP_3 or by Ca^{2+} itself. The ryanodine receptor is composed of four ~560 kDa tetramers. It is gated by electrochemical coupling to the dihydropyridine receptor of the plasma membrane in skeletal muscle, by Ca^{2+}, or by cyclic ADP ribose (Coronado et al., 1994).

Ca^{2+} is loaded into the ER lumen by sarcoplasmic/endoplasmic reticulum Ca^{2+} ATPases (SERCA pumps). In addition, SERCA pumps are also located in the nuclear envelope and in the membrane of the Golgi network where, together with secretory pathway Ca^{2+} ATPases (SPCA pumps), they regulate Golgi function, thereby contributing to the regulation of Ca^{2+} homeostasis (Brini & Carafoli, 2009). Their work requires energy: SERCA pumps load two Ca^{2+} into the intracellular store at the expense of one ATP molecule (Inesi et al., 1980). As a result of their action, the net Ca^{2+} concentration in the ER reaches ~1 mM. Once in the store, Ca^{2+} is attached to luminal Ca^{2+}-binding proteins. Such proteins are usually classified as either buffers or chaperones, although there is a substantial overlap between the two groups. Buffer proteins such as calsequestrin and calreticulin simply bind Ca^{2+} with high capacity and thus regulate storage capacity of the ER. Chaperones (e.g., calnexin, GRP78, and GRP94) assist in protein processing while they can also modulate Ca^{2+} signaling (Berridge, 2002).

8.2.2 Elevation of Cytosolic Ca^{2+} Levels

To generate a Ca^{2+} signal, the cell must transiently elevate the Ca^{2+} concentration in the cytoplasm. In general, Ca^{2+} can enter the cytosol from intracellular compartments or the extracellular

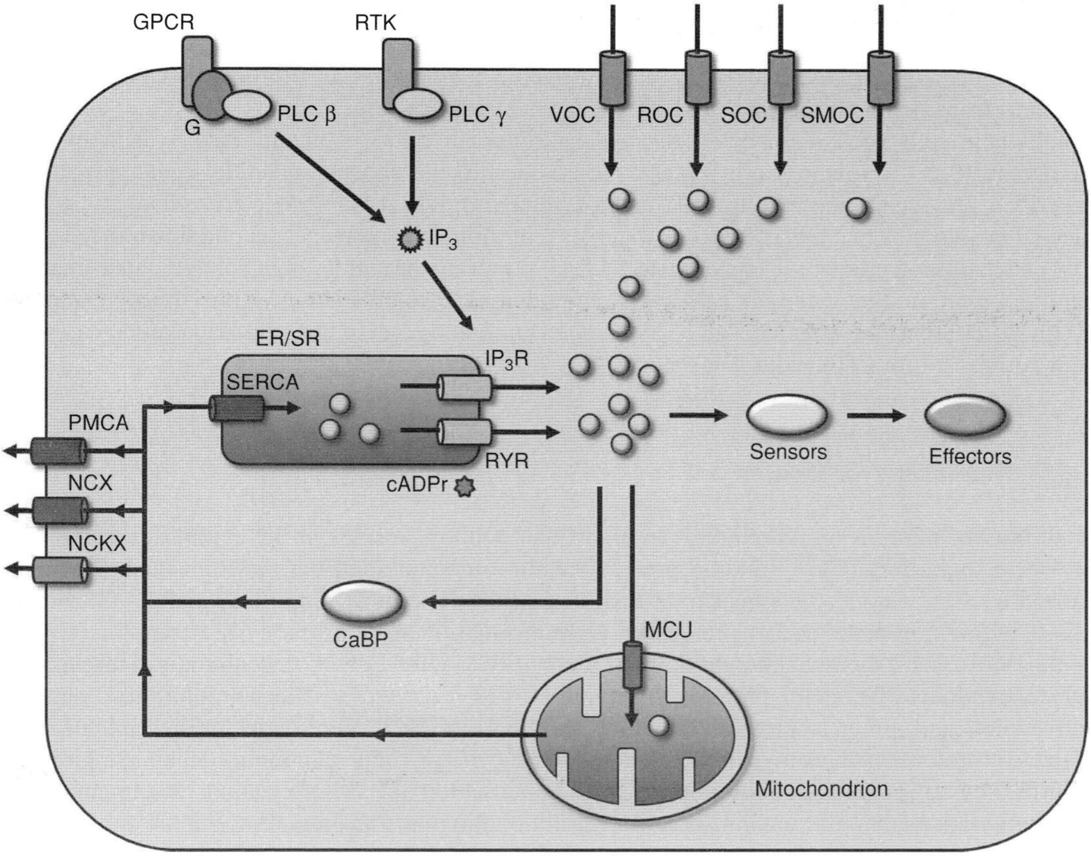

Figure 8.1 Schematic illustration of the Ca^{2+} signaling toolkit in cells. Abbreviations are as follows: GPCR, G protein-coupled receptor; G, G protein; PLCβ, phospholipase Cβ; RTK, receptor tyrosine kinase; PLCγ, phospholipase Cγ; IP_3, inositol 1,4,5-trisphosphate; IP_3R, inositol 1,4,5-trisphosphate receptor; RYR, ryanodine receptor; cADPr, cyclic ADP-ribose; VOC, voltage-operated channel; ROC, receptor-operated channel; SOC, store-operated channel; SMOC, second messenger-operated channel; ER/SR, endoplasmic/sarcoplasmic reticulum; MCU, mitochondrial Ca^{2+} uniporter; CaBP, Ca^{2+}-binding protein; SERCA, sarcoplasmic/endoplasmic reticulum Ca^{2+} ATPase; PMCA, plasma membrane Ca^{2+} ATPase; NCX, Na^+/Ca^{2+} exchanger; NCKX, Na^+/Ca^{2+}-K^+ exchanger. Yellow circles represent Ca^{2+}; the arrows indicate the movement of Ca^{2+} or the transmission of the signal. Based on Berridge, 2007. (For color detail, see color plate section.)

space. In nonexcitable cells, signaling predominantly involves the liberation of stored Ca^{2+} via the IP_3-mediated signaling pathway. Ligand binding to two receptor classes located in the plasma membrane is able to trigger this pathway. G protein-coupled receptors activate the enzyme phospholipase Cβ (PLCβ) via an intermediate G protein; receptor tyrosine kinases on the other hand directly stimulate another PLC isoform, PLCγ. Activated PLC hydrolyzes phosphatidylinositol 4,5-bisphosphate (PIP_2) in the plasma membrane and generates two important second messengers, IP_3 and diacylglycerol. IP_3 binds to its receptor on the surface of the ER (Furuichi et al., 1989); binding allows diffusion of Ca^{2+} from the ER to increase intracellular free Ca^{2+} levels from the resting ~100 nM to ~1 μM. Opening of the IP_3 receptor requires the simultaneous presence of both IP_3 and Ca^{2+}. At low and high Ca^{2+} levels, the receptor is relatively insensitive to IP_3 (Bezprozvanny et al., 1991). At the same time, the receptor's sensitivity to IP_3 is also biphasic and is greatest at physiological IP_3 levels (between 0.5 and 1 μM). Ca^{2+} can also seep out of the ER along a leak pathway (Lomax et al., 2002), and in some cell types another second messenger, nicotinic acid adenine dinucleotide phosphate (NAADP), is also able to mobilize Ca^{2+} through NAADP-sensitive channels from stores other than the ER (Berridge et al., 2002).

Excitable cells such as muscle cells and neurons can utilize a much faster system in order to increase Ca^{2+} levels in the cytoplasm. In addition to the phosphoinositide signaling pathway described previously, these cells make use of voltage-operated Ca^{2+} selective channels in their plasma membranes that can conduct as much as a million Ca^{2+} ions per second down the 20,000-fold concentration gradient (Clapham, 1995). Ca^{2+} entering the cell through voltage-gated channels can bind to ryanodine receptors and induce the release of stored Ca^{2+} (McPherson & Campbell, 1993). Additional entry channels gated by receptors or second messengers are also available in some cells for Ca^{2+} influx (Berridge, 2007).

8.2.3 Transmitting the Signal

A number of proteins are specialized to bind Ca^{2+} while it resides in the cytoplasm; some simply serve as buffers (Ca^{2+}-binding proteins) while others are sensors that transduce the signal carried by Ca^{2+} along the signal transduction cascade. The most important of them is calmodulin, a highly conserved sensor protein whose conformation changes upon Ca^{2+} binding (Copley et al., 1999). Ca^{2+}-bound calmodulin gains the ability to interact with the calmodulin recruitment site of hundreds of effector proteins; by doing so it can relieve protein autoinhibition, remodel active sites, and dimerize proteins (Hoeflich & Ikura, 2002). It often wraps around amphipathic helices of proteins such as myosin light chain kinase (MLCK) and calmodulin-dependent kinase II (CaMKII). Ca^{2+}/calmodulin binding causes multimerization of CaMKII, leading to auto- and inter-phosphorylations that prolong kinase activity. Calmodulin thus translates transient Ca^{2+} elevations into more extended signals. Additional Ca^{2+} sensing proteins are also available to bind Ca^{2+} and attach to a great number of downstream targets (reviewed by Krebs & Heizmann, 2007).

8.2.4 Removal of Ca^{2+} from the Cytosol

After Ca^{2+} has carried out its mission as a signal, the low resting cytosolic Ca^{2+} level must be restored. As described earlier, SERCA pumps on the surface of the ER are responsible for sequestering Ca^{2+} into the intracellular store. In addition, plasma membrane Ca^{2+} ATPases (PMCA pumps) located in the plasma membrane work against the large concentration gradient to move Ca^{2+} out of the cell. PMCA pumps push one Ca^{2+} out at the cost of one ATP hydrolyzed (Niggli et al., 1982). Exchange mechanisms also play a significant role in lowering elevated cytosolic Ca^{2+} levels. Na^+/Ca^{2+} exchangers located in the plasma membrane remove one Ca^{2+} in exchange for three Na^+ (Blaustein & Lederer, 1999) and the Na^+/Ca^{2+}-K^+ exchangers cotransport one K^+ with one Ca^{2+} in exchange for four Na^+ (Altimimi & Schnetkamp, 2007).

Mitochondria also contribute to the Ca^{2+} homeostasis of cells. Their function is intimately linked to that of the ER; the interaction between the two organelles shapes the Ca^{2+} signal (Duchen, 2000). Mitochondria can accumulate up to 0.5 mM Ca^{2+} in the mitochondrial matrix due to the electrochemical gradient generated by mitochondrial hydrogen exchange. Ca^{2+} uptake is carried out by the mitochondrial Ca^{2+} uniporter (MCU) located in the organelle's inner membrane. MCUs have lower affinity for Ca^{2+} than SERCA pumps, and they probably play a role only when intracytoplasmic Ca^{2+} levels climb beyond ~0.5 μM (Pozzan et al., 1994). Thus mitochondria sequester Ca^{2+} during the recovery phase of the Ca^{2+} transients and later return it to the ER.

8.2.5 Store-operated Ca^{2+} Entry

The PMCA pump and the exchange mechanisms may remove Ca^{2+} from the cell so effectively that a Ca^{2+} influx across the plasma membrane becomes necessary to avoid a shortage of Ca^{2+} in the stores. In many nonexcitable cells a Ca^{2+} entry mechanism called store-operated Ca^{2+} entry is activated in response to store depletion (Putney et al., 2001). This can be demonstrated by using thapsigargin or cyclopiazonic acid (CPA) that specifically inhibits the SERCA pumps. Ca^{2+} constantly leaks out of the ER, and in the absence of functional SERCA pumps, the stores soon become empty; this leads to the generation of a Ca^{2+} influx. The Ca^{2+} influx induced by store depletion is believed to be important for prolonging cytoplasmic Ca^{2+} increases, accelerating Ca^{2+} waves, and refilling the ER. In some somatic cell types, the Ca^{2+} release-activated Ca^{2+} current is well characterized (Hoth & Penner, 1992); nevertheless, the molecular components of the cascade remained unknown for a long time. The first hint came when Stim1 was identified as a protein essential for the onset of the inward Ca^{2+} current generated after store depletion (Liou et al., 2005). With its single transmembrane-spanning domain, Stim1 is localized primarily in the ER and its Ca^{2+} binding EF hand (a characteristic motif of Ca^{2+}-binding proteins arranged as the thumb and forefinger of a right hand in which the E and F helices are connected by a loop of amino acids) serves as a sensor of luminal Ca^{2+} content. Another protein termed Orai1 was soon identified in the plasma membrane (Feske et al., 2006). Orai1 is a four-transmembrane domain protein, and mutagenesis of its pore has indicated that it is the channel-forming subunit (Yeromin et al., 2006). It is now believed that upon store depletion, Stim1 forms clusters, moves to plasma membrane-adjacent regions of the ER, and opens Orai1, thereby stimulating a Ca^{2+} influx to refill the stores.

8.2.6 Spatial and Temporal Organization of the Ca^{2+} Signal

As mentioned earlier, Ca^{2+} signaling is highly versatile and regulates many cellular processes. The versatility is due primarily to the signal being organized in both time (oscillations) and space (elementary events and waves). Ca^{2+} oscillations and waves are complex signals that are believed to carry significantly more information than simple static intracellular Ca^{2+} rises.

In many cell types, Ca^{2+} waves are initiated in a discrete region of the cell and then propagated throughout the cytoplasm. In the globalization of the signal, a regenerative Ca^{2+} release mechanism known as Ca^{2+}-induced Ca^{2+} release (CICR) plays a central role by linking together the individual Ca^{2+} release sites. It enables the Ca^{2+} released at one site to stimulate neighboring receptors, thus igniting a regenerative wave (Berridge, 1997). In excitable cells, Ca^{2+} enters the cytoplasm through voltage-gated Ca^{2+} channels and stimulates ryanodine receptors to release Ca^{2+} from the stores. Imaging techniques revealed the existence of elementary events called sparks in cardiac cells; sparks are small bursts of Ca^{2+} released from a localized group of ryanodine receptors (Cheng et al., 1993). The Ca^{2+} released from the ER diffuses to adjacent sites, increases the sensitivity of ryanodine receptors, and induces further Ca^{2+} release (Endo et al., 1970). This creates a Ca^{2+} wavefront. The

Ca^{2+} concentration then drops behind the wavefront as pumps remove Ca^{2+} from the cytosol. There are indications that the IP_3/Ca^{2+} signaling system is also organized into elementary events; such events generated by the IP_3 receptor are known as puffs (Yao et al., 1995). In addition, sensitivity of IP_3 receptors to Ca^{2+} indicates that this receptor type is also responsive to CICR (Bezprozvanny et al., 1991). The key elements of Ca^{2+} waves generated by the IP_3/Ca^{2+} signaling system are well characterized in Xenopus oocytes (Lechleiter & Clapham, 1992). In such cells, IP_3 can be generated by the stimulation of plasma membrane receptors; it then diffuses throughout the cell and binds to IP_3 receptors to induce the discharge of Ca^{2+}. The released Ca^{2+} moves to adjacent sites, increases the sensitivity of the IP_3 receptors, and induces further Ca^{2+} release via CICR. Local Ca^{2+} release then produces high Ca^{2+} concentration at the mouth of the IP_3 receptor that eventually inhibits the channel. Again, in both receptor types, the role of CICR in signal globalization depends on the state of excitability of the receptors, which is determined by the concentration of Ca^{2+} mobilizing ligands (cyclic ADP-ribose or IP_3, respectively) present at the receptor (Berridge, 2002).

CICR is also critical for the temporal regulation of the signal as it provides the positive feedback mechanism during the upstroke of each Ca^{2+} spike of the repetitive signal. In addition, Ca^{2+} release can occur through both types of release channels when the stores become overloaded with Ca^{2+}. It has been suggested that the loading state of the store sets the sensitivity of the IP_3 receptor and determines when the next Ca^{2+} spike will occur (Berridge, 1993). The frequency of Ca^{2+} oscillations thus depends on how fast the store can be reloaded, and this in turn is determined by the rate at which Ca^{2+} enters the cell.

8.3 Ca^{2+} Signaling in Oocytes

8.3.1 Development of the Signaling Mechanism

The Ca^{2+} signals in oocytes rely on the intracellular stores. Several lines of evidence indicate that the stores are located primarily within the ER (reviewed by Stricker, 2006). As in somatic cells, SERCA pumps sequester Ca^{2+} into the ER lumen where it is stored bound to specialized buffer proteins. The ER plays a fundamental role in shaping the Ca^{2+} signal, and its architecture changes during maturation as the oocytes gain the ability to generate the signal. Typically, in immature oocytes, the ER spreads relatively uniformly throughout the ooplasm (Kline, 2000). Maturation in most species stimulates a major reorganization of the ER with the development of distinct ER clusters in the cortical cytoplasm. At the same time the amount of Ca^{2+} that can be mobilized from the ER increases progressively (Jones et al., 1995; Machaty et al., 1997a). The IP_3 receptor also undergoes significant changes: its mass and sensitivity (the amount of Ca^{2+} released upon stimulation) increases markedly during maturation, and this is associated with the formation of IP_3 receptor clusters in the oocyte cortex (Mehlmann et al., 1996). These changes are responsible for the evolvement of the Ca^{2+} release mechanism, and as a result, the oocyte acquires the ability to generate the explosive Ca^{2+} signal upon the encounter with the sperm. This has been demonstrated experimentally: when injected with a sperm-derived Ca^{2+}-releasing factor, mature oocytes produce repetitive Ca^{2+} transients that are about twofold larger and longer compared to those measured in immature oocytes. Also, Ca^{2+} waves are generated in mature, but not in immature, oocytes when treated with the sperm extract (Carroll et al., 1994).

Interestingly, after fertilization, the ER architecture may reorganize again. The clusters disappear in species that display a single Ca^{2+} rise in response to the sperm, whereas in oocytes that generate repetitive Ca^{2+} transients they persist during oscillations. It is possible that ER clustering may be needed to maintain the repetitive signal, and the absence of the clusters probably plays a role in preventing the generation of additional Ca^{2+} spikes (Kline, 2000).

8.3.2 *Role of Ca^{2+} in Oocyte Maturation*

Oocytes are produced by meiosis, which consists of two rounds of nuclear divisions. The first meiotic division separates the homologous chromosomes. It is then followed by the second meiotic division without DNA replication, leading to the formation of haploid gametes. Unlike most cells that either divide continuously or remain quiescent in interphase, oocytes stop in the middle of the cell cycle (Whitaker, 1996). In fact, in most animal species they stop twice during meiosis. The first arrest usually takes place as oocytes enter the first meiotic prophase and is maintained by low activity of the M-phase Promoting Factor (MPF), a complex of cyclin-dependent kinase 1 (cdk1) and cyclin B. Low MPF activity prevents germinal vesicle breakdown (GVBD) and keeps the oocytes arrested for an extended period of time. In order to become capable of fertilization, the oocytes undergo maturation. During maturation, meiosis progresses until it stops at a species-specific stage; mammalian oocytes complete the first meiotic division and arrest at second meiotic metaphase. The metaphase arrest is maintained by high activity of MPF (which in turn is sustained by the cytostatic factor); high MPF activity keeps the chromatin in a condensed state and stabilizes the meiotic spindle.

Limited information is available about the role of Ca^{2+} in the maturation process. Although Ca^{2+} signals were suggested to have a role in the initiation of maturation (Guerrier et al., 1982), it has become generally accepted that in most species it is a decrease in the cytosolic cAMP levels induced by hormone stimulation that is responsible for triggering the signaling cascade at this time. It seems probable that hormone signaling does not include the Ca^{2+} signaling pathway, and Ca^{2+} may have a role later, at the time of GVBD. Nevertheless, it was also suggested that Ca^{2+} may play a role in the regulation of the first meiotic arrest by modulating cAMP levels in the oocyte through Ca^{2+}-sensitive adenylate cyclases (Silvestre et al., 2011). However, it is not known in detail how the Ca^{2+} signal is linked to the decrease in cAMP levels or how MPF activation, which leads to the release from the first meiotic arrest, is triggered (Whitaker, 1996). Despite these observations, Ca^{2+} signals do have a role in the acquisition of maturation. Inhibition of the phosphoinositide pathway in immature mouse oocytes delays GVBD, which can be circumvented by the injection of IP_3 into the ooplasm (Pesty et al., 1994). A series of Ca^{2+} spikes are generated at GVBD that stop as the cell cycle reaches the first metaphase stage of meiosis (Carroll et al., 1994). Finally, the fact that GVBD can be blocked with Ca^{2+} chelators in some mammalian oocytes also implicates Ca^{2+} in the regulation of oocyte maturation (Homa, 1995).

8.3.3 *Role of Ca^{2+} in Oocyte Activation*

Ca^{2+} Release Through IP_3 Receptors

Under physiological conditions it is the fertilizing sperm that restarts the cell cycle machinery in mature oocytes. By fusing to the oocyte, the sperm induces a series of events that are collectively known as activation. These events primarily include a transient elevation in the intracellular free Ca^{2+} concentration, the release of the content of the cortical granules, resumption of meiosis, recruitment of maternal mRNAs, pronuclear formation, initiation of DNA synthesis, and cleavage (Schultz & Kopf, 1995). Much evidence indicates that the key trigger during oocyte activation is the

Ca^{2+} signal: the rise in the oocyte's intracellular Ca^{2+} level is responsible for all additional events associated with fertilization.

The idea that Ca^{2+} might be essential for oocyte activation dates back almost a century (Lillie, 1919). Direct evidence regarding the importance of Ca^{2+} during oocyte activation came later, when it was demonstrated that oocytes of various species displayed events similar to those after fertilization when exposed to the Ca^{2+} ionophore A23187 (Steinhardt et al., 1974). Subsequently, an increase in the intracellular free Ca^{2+} concentration was measured in medaka oocytes (Ridgway et al., 1977). The final proof that an elevation in the cytosolic free Ca^{2+} concentration is the activating signal was provided when it was shown that preventing the Ca^{2+} rise inhibited fertilization in sea urchins (Zucker & Steinhardt, 1978). These observations established that Ca^{2+} is an important signaling messenger in oocytes and that it plays a central role during fertilization in almost every species tested.

It was realized early on that the origin of Ca^{2+} that activates the oocyte is the intracellular store (Steinhardt & Epel, 1974). The stored Ca^{2+} is released from the ER when IP_3 binds to its receptor; this was first indicated by the observation that polyphosphoinositide turnover increased significantly in sea urchin oocytes at fertilization (Turner et al., 1984). This was followed by the discovery that activation could be induced artificially by injecting IP_3, the product of polyphosphoinositide hydrolysis into sea urchin oocytes (Whitaker & Irvine, 1984). The findings that IP_3 levels increase during fertilization, that microinjection of IP_3 induces oocyte activation, and that IP_3 antagonists inhibit sperm-induced activation in a great number of species indicate that IP_3 is the second messenger that triggers Ca^{2+} release in response to the fertilizing sperm (Stricker, 1999). In mammalian oocytes, IP_3 binds predominantly to the type 1 IP_3 receptor (Kurokawa et al., 2004).

How is IP_3 Generated?

Three major hypotheses have been formulated to explain the mechanism by which the fertilizing sperm triggers the Ca^{2+} signal in the oocyte. The Ca^{2+} conduit hypothesis proposed that Ca^{2+} from the extracellular medium enters the oocyte via channels of the fused sperm (Jaffe, 1991). Once in the oocyte cytoplasm, Ca^{2+} was suggested to cause a CICR similar to that observed in muscle cells. However, in sea urchins, the Ca^{2+} signal is successfully generated and the oocytes are activated even in the absence of extracellular Ca^{2+} (Schmidt et al., 1982). In addition, sustained injection of Ca^{2+} into hamster and mouse oocytes does not elicit repetitive Ca^{2+} signals that normally accompany fertilization (Igusa & Miyazaki, 1983; Swann, 1992). These observations argue against the Ca^{2+} conduit model.

According to the receptor hypothesis, the sperm-induced Ca^{2+} signal is initiated via a receptor-linked signaling cascade. The fertilizing sperm is suggested to bind a receptor (either a G protein-coupled receptor or a receptor tyrosine kinase) on the surface of the oocyte plasma membrane. Receptor binding then leads to the activation of a PLC and thus the generation of the Ca^{2+} mobilizing second messenger IP_3. The idea originated from the observation that G proteins played a role in fertilization: GDPβS, a G protein antagonist, blocked cortical granule exocytosis in sea urchin oocytes during fertilization (Turner et al., 1986). This idea has been supported by a number of observations. Introducing IP_3 into the cytoplasm induces Ca^{2+} release, sustained injection of IP_3 triggers repetitive Ca^{2+} transients, and GTPγS, a non-hydrolyzable analogue of GTP (a regulator of G protein activity), also causes Ca^{2+} oscillations in oocytes of various species (reviewed by Hogben et al., 1998). In addition, the stimulation of endogenous cell surface receptors or those expressed exogenously by mRNA microinjection also leads to oocyte activation (see, e.g., Miyazaki et al., 1990; Machaty et al., 1997b). However, these observations simply indicate the presence of a signaling pathway in the oocytes; they do not necessarily mean that the particular pathway operates at fertilization. Most important, although many cell surface molecules in the oocyte's plasma membrane have been identified, none of them seem to be involved in the generation of the Ca^{2+} signal (Schultz & Kopf, 1995).

The third hypothesis explains the initiation of the Ca^{2+} signal by the transfer of an activating factor from the sperm to the oocyte following gamete fusion. The initial indication that the sperm may act as a vehicle and deliver the activating factor came from a study in the sea urchin where a sperm extract induced activation after being injected into the oocyte cytoplasm (Dale et al., 1985). Since then, an extract isolated from sperm has been shown to induce Ca^{2+} increase in oocytes of many species. In mammalian oocytes, increasing the quantity of the sperm extract during microinjection raised the frequency of the Ca^{2+} transients without affecting their amplitude, which was consistent with the expectations from a true activating factor (Swann, 1990). Another clue that a sperm-derived factor may be responsible for triggering oscillations in the oocyte cytoplasm came from intracytoplasmic sperm injection (ICSI). Initially, activation of the oocyte during human ICSI was believed to be a side effect (i.e., the result of Ca^{2+} contamination from the external medium during the injection process). Later it was demonstrated that activation was triggered by Ca^{2+} oscillations that started hours after injection of the sperm (Tesarik et al., 1994). (By contrast, Ca^{2+} contamination during microinjection causes only a single Ca^{2+} increase immediately after the injection pipette penetrates the plasma membrane). This finding is consistent with the sperm factor model and cannot be explained by the conduit or receptor hypothesis.

The sperm extract showed high PLC activity and caused a large increase in IP_3 levels in sea urchin oocyte homogenates, indicating that the active factor may be a PLC (Jones et al., 1998). Injection of the extract into frog oocytes also led to an elevation in IP_3 (Wu et al., 2001). However, none of the known PLC isoforms induced repetitive Ca^{2+} transients when injected at physiological levels into the oocyte cytoplasm. The breakthrough in the search for the elusive sperm factor came when a novel PLC isoform was discovered in mammalian sperm. In the mouse, PLCζ (PLCzeta) was found to be expressed exclusively in the testis, and when its cRNA was microinjected into oocytes, a repetitive Ca^{2+} signal was generated that was indistinguishable from that measured during fertilization (Saunders et al., 2002). Microinjection of a recombinant PLCζ protein also caused repetitive Ca^{2+} transients in unfertilized mouse oocytes (Kouchi et al., 2004), and it was shown that the amount of PLCζ needed to generate the Ca^{2+} signal is present in a single sperm (Saunders et al., 2002). The involvement of PLCζ in oocyte activation was further demonstrated when PLCζ expression in the ovary was found to cause parthenogenetic oocyte activation and, subsequently, the formation of ovarian tumors (Yoshida et al., 2007). Finally, data obtained using an RNAi approach to downregulate expression indicated that sperm with reduced PLCζ levels were unable to fertilize the oocyte and induce embryo development (Knott et al., 2005). Although the exact mechanism by which PLCζ binds and hydrolyzes PIP_2 is yet to be clarified, it is now generally accepted that PLCζ is the sperm-derived factor that causes Ca^{2+} oscillations in mammalian oocytes (Swann & Yu, 2008).

The Fertilization Ca^{2+} Wave

By applying image intensification techniques after fertilization in medaka oocytes, the Ca^{2+} elevation was shown to take the form of a wave starting at the point of sperm entry (Gilkey et al., 1978). Since then it was demonstrated that in most deuterostome species, the sperm-induced Ca^{2+} signal takes the form of a propagating wave (Figure 8.2). The wave initiates at the point of sperm-oocyte fusion and moves across the oocyte to the opposite pole; it crosses mammalian oocytes in $\sim$2 sec.

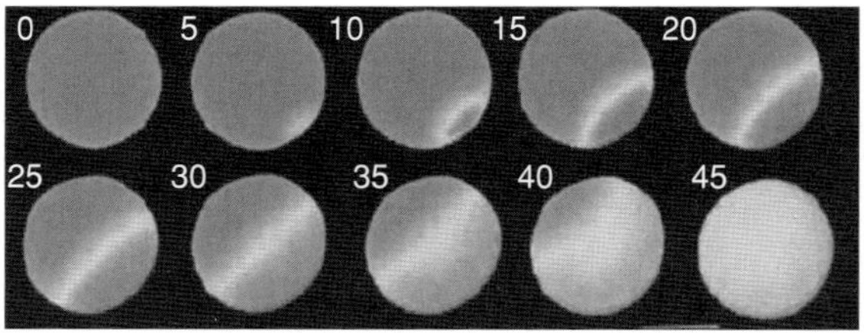

Figure 8.2 A fertilization Ca^{2+} wave observed in an ascidian oocyte. The outer chorion layer of the oocyte was removed; the oocyte was then injected with the Ca^{2+} indicator dye and inseminated. The figure shows the first Ca^{2+} wave that originates at the site of sperm entry. After the initial Ca^{2+} increase, in ascidian oocytes, a second set of waves is also generated whose point of origin is the vegetal hemisphere. Numbers denote time in seconds; different colors indicate different Ca^{2+} concentrations. Courtesy of Carroll et al. (2003); with permission. (For color detail, see color plate section.)

In the sea urchin, the wave travels for ~20 sec until it reaches the antipode, whereas in frog oocytes the process takes approximately 10 minutes (reviewed by Whitaker, 2006).

Typically, fertilization Ca^{2+} waves are conveyed by IP_3 receptors (Stricker, 1999). They are built up from many separate Ca^{2+} release events, each of which is produced by a limited number of Ca^{2+} release channels. The IP_3 receptor is sensitive to Ca^{2+} levels in the cytosol: Ca^{2+} stimulates Ca^{2+} release by CICR and the discharged Ca^{2+} then diffuses to the neighboring receptors (Solovyova et al., 2002). The production and diffusion of IP_3 also contributes to the propagation of the wave. It sensitizes the receptor to CICR; in the absence of IP_3, cytosolic Ca^{2+} blocks the receptor, but when IP_3 is present, Ca^{2+} triggers the release of additional Ca^{2+} from the ER (Adkins & Taylor, 1999). Although ryanodine receptors are present in oocytes of some mammalian species, they do not seem to have a major signaling role during fertilization. It is possible that the ryanodine receptors located at the cortex of mammalian oocytes release Ca^{2+} that play a contributory role in the stimulation of cortical granule exocytosis (Kline & Kline, 1994); they may also provide Ca^{2+} that helps sustain the train of Ca^{2+} transients (Swann, 1992).

Repetitive Ca^{2+} Oscillations

In most species, the sperm induces a single elevation in the intracellular Ca^{2+} concentration. Nemertean and annelid worms, ascidians, and mammals, on the other hand, display repetitive Ca^{2+} rises (Stricker, 1999). Such Ca^{2+} oscillations were detected for the first time in fertilized mouse oocytes (Cuthbertson et al., 1981); a prolonged oscillating Ca^{2+} signal measured in a pig oocyte is shown in Figure 8.3.

It is not entirely clear what makes the signal oscillate. Most Ca^{2+} signals generated by Ca^{2+} mobilization from intracellular stores have a predisposition for oscillation (Berridge & Galione, 1988). Both IP_3 and ryanodine receptors show CICR that is the basis of oscillatory activity in many cell types. High cytosolic Ca^{2+} inhibits further Ca^{2+} release through both receptor types, and they are both controlled by positive and negative feedback mechanisms. These features are sufficient to elicit the oscillatory pattern. Nevertheless, store depletion adds an additional negative feedback constituent (Berridge & Galione, 1988) and, because PLCζ shows very high sensitivity to Ca^{2+} (Kouchi et al., 2004), the Ca^{2+}-promoted production of IP_3 provides one more positive feedback for the IP_3 receptor.

The oscillations are probably controlled by the basic feedback properties of the IP_3 receptor (Adkins & Taylor, 1999). According to one popular model of Ca^{2+} oscillations, Ca^{2+}-dependent IP_3 production by PLCζ leads to oscillating IP_3 levels, which is accountable for the repetitive nature of the fertilization Ca^{2+} signal (Dupont & Dumollard, 2004). However, injection of the

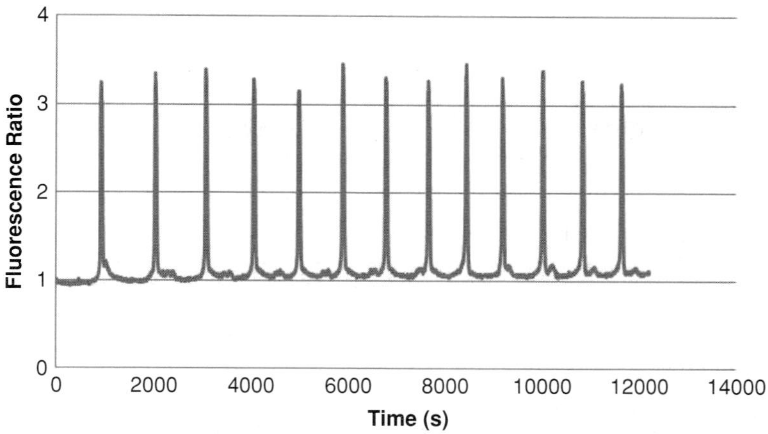

Figure 8.3 Repetitive Ca^{2+} transients in a pig oocyte at fertilization. Mature oocytes were treated with pronase to remove the zona pellucida, then loaded with fura-2 and inseminated. The fluorescence ratio on the Y axis indicates Ca^{2+} concentration (Chunmin Wang & Zoltan Machaty, unpublished).

mammalian sperm factor into frog oocytes causes only one Ca^{2+} transient (Wu et al., 2001), indicating that the nonlinear feedback of Ca^{2+} on PLCζ cannot in itself explain oscillations. Another plausible model argues that instead of controlling PLCζ activity, Ca^{2+} may act directly on the IP_3 receptors (De Young & Keizer, 1992). In this version, IP_3 concentrations do not oscillate, but instead IP_3, at a constant level, provides continuous stimulation to its receptor, which opens when intracellular Ca^{2+} is low and closes when Ca^{2+} concentration at the receptor rises above a threshold level. Observations that IP_3 levels do not oscillate in HeLa cells during the repetitive Ca^{2+} signal induced by metabotropic glutamate receptor stimulation (Matsu-ura et al., 2006), and that providing a sustained IP_3 supply can trigger Ca^{2+} oscillations in mouse oocytes (Jones & Nixon, 2000), support this model. Measuring IP_3 at fertilization could help in distinguishing between the different models; unfortunately, due to inherent difficulties, the data available can be interpreted in various ways (Shirakawa et al., 2006). It has also been suggested that in mammalian oocytes, the two mechanisms may coexist. In unfertilized oocytes, the IP_3 receptor alone is responsible for the Ca^{2+} oscillations seen after sustained injection of IP_3. Then, at gamete fusion, the sperm introduces PLCζ into the ooplasm after which a new mechanism, the regenerative IP_3 production, maintains the oscillatory Ca^{2+} signal (Swann & Yu, 2008).

During oscillations, the initiation site of the individual transients changes with time. The first Ca^{2+} wave in all species studied starts at the point where the sperm fuses with the oocyte plasma membrane. The site of origination in ascidians moves gradually toward the vegetal pole of the oocyte (Carroll et al., 2003). In a similar fashion, the Ca^{2+} wave initiation site of subsequent waves in mouse oocytes also translocates from the point of sperm entry to the cortex of the vegetal hemisphere (Deguchi et al., 2000).

In the mouse, the oscillations stop at the time of pronuclear formation; if exit from meiosis is blocked, the oscillations go on indefinitely (Marangos et al., 2003). In addition, the transfer of nuclei of 1- and 2-cell fertilized mouse embryos to unfertilized oocytes triggers Ca^{2+} oscillations that initiate at the time of nuclear envelope breakdown (Kono et al., 1995). These findings led to the conclusion that the train of Ca^{2+} spikes comes to an end when PLCζ, the messenger responsible for the generation of the signal, is sequestered into the pronucleus (Larman et al., 2004). However, in nucleate and anucleate halves of murine zygotes obtained by bisection after fertilization, the Ca^{2+} oscillations cease at about the same time irrespective of the presence of the pronuclei (Day et al., 2000). In addition, in fertilized bovine and rabbit oocytes, the oscillations persist well beyond pronuclear formation (Fissore et al., 1992; Fissore & Robl, 1993). This indicates that additional factors may also control the cessation of the Ca^{2+} oscillations.

Store-operated Ca^{2+} Entry in Oocytes

After each Ca^{2+} rise, resting intracellular Ca^{2+} levels are restored by various mechanisms. SERCA pumps on the surface of the ER load Ca^{2+} back into the store (Kline & Kline, 1992) and Ca^{2+} uptake by mitochondria is also significant (see below). Due to the absence of specific inhibitors, little is known about the role of PMCA pumps in oocytes, but another system, the Na^{+}/Ca^{2+} exchange mechanism, is available for Ca^{2+} removal (Carroll, 2000). Although the specific contribution of these extrusion mechanisms has yet to be determined, it was shown in sea urchin, frog, and mouse oocytes that a substantial efflux of Ca^{2+} occurs after the mobilization of the stored Ca^{2+} (Steinhardt & Epel, 1974; Shapira et al., 1996; Pepperell et al., 1999). The outward Ca^{2+} current may be so significant that an influx of Ca^{2+} across the plasma membrane becomes necessary to compensate for the loss. This may seem counterproductive, and the reason for it is not entirely clear. Long-lasting high intracellular Ca^{2+} levels can be detrimental; the extrusion mechanisms are programmed to remove Ca^{2+} from the cytosol immediately, and this apparently leads to a net loss of Ca^{2+} that needs to be compensated for in order to regenerate the signaling system.

In hamster and mouse oocytes, an influx of Ca^{2+} is necessary to maintain oscillations; the sperm-induced Ca^{2+} transients stop in the absence of extracellular Ca^{2+} or in the presence of Ca^{2+} influx channel antagonists (Igusa & Miyazaki, 1983; Shiina et al., 1993). A Ca^{2+} influx pathway is present in mouse oocytes (Kline & Kline, 1992), and it has also been demonstrated that during the sperm-induced Ca^{2+} oscillations, a Ca^{2+} influx is stimulated after each Ca^{2+} rise (McGuinness et al., 1996). A store-operated Ca^{2+} entry mechanism also exists in pig oocytes (Machaty et al., 2002), and evidence has been provided that the Stim1 protein is responsible for mediating the Ca^{2+} entry after store depletion (Koh et al., 2009). Under resting conditions, Stim1 is distributed fairly evenly in the cytoplasm. Following a treatment with thapsigargin in Ca^{2+}-free medium, Stim1 moves to the cortical cytoplasm and rearranges into clusters. Thapsigargin is known to induce store depletion by inhibiting SERCA pumps; Stim1 translocation under such conditions implies that the protein may be involved in transducing the empty signal from the store to the Ca^{2+} influx channels. Downregulation of Stim1 using RNA interference not only suppresses Ca^{2+} influx after Ca^{2+} add-back in thapsigargin-pretreated oocytes, but also prevents Ca^{2+} oscillations after fertilization. The oocytes pre-injected with Stim1 siRNA display only a few Ca^{2+} rises; additional Ca^{2+} transients cannot be generated in the absence of Stim1 (Kiho Lee, Chunmin Wang, Zoltan Machaty, unpublished). The situation is similar when the putative channel component of the store-operated Ca^{2+} entry pathway is inhibited. In somatic cells, Orai1 is believed to serve as a Ca^{2+} influx channel that is activated by Stim1 after store depletion. Downregulation of Orai1 with siRNA abolishes the Ca^{2+} oscillations associated with fertilization; oocytes without functional Orai1 display only the initial Ca^{2+} spike(s) induced by the fertilizing sperm (Wang et al., 2012). This indicates that store-operated Ca^{2+} entry mediated by an interaction between Stim1 and Orai1 is essential for refilling the intracellular stores and sustaining the repetitive Ca^{2+} signal.

Events Downstream of the Ca^{2+} Signal

The primary function of the fertilization Ca^{2+} signal is targeting the cell cycle machinery to induce the resumption of meiosis. In mammalian oocytes, which await fertilization in metaphase, the cell cycle arrest is maintained by MPF that keeps the chromatin in a condensed state and stabilizes the meiotic spindle (Whitaker, 1996). At metaphase, MPF shows high activity because elevated levels of its regulatory subunit, cyclin B maintains the kinase activity of the catalytic subunit, cdk1. For the cell cycle to proceed, MPF activity must be downregulated. Although MPF activity is guarded by CSF, the Ca^{2+} signal bypasses this checkpoint and takes a different route to inactivate MPF. Experimental data suggest that at fertilization Ca^{2+} binds calmodulin; Ca^{2+}-calmodulin in turn stimulates calmodulin-dependent protein kinase II (CaMKII), which is the major transducer of the Ca^{2+} signal (Lorca et al., 1993). In mammals, CaMKII activity increases after sperm-oocyte fusion; in fact it oscillates in step with the fertilization Ca^{2+} signal (Markoulaki et al., 2004). Activated CaMKII can phosphorylate Emi2; Emi2 is known to contribute to the maintenance of metaphase II arrest by inhibiting the anaphase-promoting complex/cyclosome ([APC/C). Phosphorylation of Emi2 leads to its destruction and subsequently, to the activation of APC/C (Jones, 2007). In mouse oocytes Emi2 is destroyed after parthenogenetic activation (Madgwick et al., 2006), and a constitutively active form of CaMKII causes resumption of meiosis and blastocyst formation (Knott et al., 2006). Once the APC/C is activated it stimulates the proteolysis of cyclin B: in mouse oocytes, cyclin B is ubiquitinated and destroyed by the proteasome soon after the initiation of the fertilization Ca^{2+} transients (Nixon et al., 2002). Cyclin B degradation leads to the decrease of MPF activity, and meiosis resumes as the oocyte enters anaphase.

The Significance of the Repetitive Pattern

Oocytes, including those of mammals, can be artificially activated by a one-time elevation of the Ca^{2+} concentration in their cytoplasm, so the question arises: What is the oscillatory signal needed for? Most agents used to stimulate parthenogenetic activation can only generate a single rise in cytosolic Ca^{2+} levels, which lasts for 5–10 minutes. It was observed that a single Ca^{2+} transient is effective in activating mostly aged oocytes; freshly ovulated oocytes show poor development after such a treatment (Ozil & Swann, 1995). As mentioned previously, in order for meiosis to resume in metaphase-arrested mammalian oocytes, MPF levels in the ooplasm must drop. This is achieved when Ca^{2+} triggers the destruction of cyclin B. However, mature mammalian oocytes continue to synthesize proteins and, although a single Ca^{2+} transient can downregulate cyclin B, continued cyclin production in freshly ovulated oocytes can restore MPF activity leading to arrest at a so-called third metaphase stage (Kubiak, 1989). A single Ca^{2+} rise ensures exit from meiosis in aged oocytes only in which MPF levels have already declined (Nixon et al., 2000). In contrast, repetitive Ca^{2+} transients cause a sustained destruction of cyclin B and thus a progressive decrease in MPF activity leading to the completion of meiosis even in the presence of ongoing cyclin synthesis in fresh oocytes.

By inducing various numbers of Ca^{2+} transients in oocytes through controlled electrical field pulses, it was shown that the duration and frequency of the Ca^{2+} rises determined the rate of pronuclear formation in mouse oocytes (Ozil & Swann, 1995) and affected the development of parthenogenetic rabbit conceptuses (Ozil & Huneau, 2001). In addition, the repetitive Ca^{2+} transients are also important to ensure complete activation; each Ca^{2+} rise pushes the process progressively forward. Cortical reaction takes place in a stepwise manner; each consequential transient stimulates the exocytosis of additional cortical granules. At the same time, early events such as resumption of meiosis are stimulated by a single Ca^{2+} transient whereas others including the downregulation of the activity of cell cycle regulatory proteins require several Ca^{2+} spikes (Ducibella et al., 2002). These data clearly indicate the importance of the oscillating Ca^{2+} signal at fertilization.

It came as something of a surprise, then, when it was reported that a single Ca^{2+} elevation of sufficiently long duration can induce oocyte activation effectively. In response to a cytosolic Ca^{2+}

rise that lasted for more than 20 minutes, freshly ovulated mouse oocytes underwent complete activation (Ozil et al., 2005). It was concluded that it is the sum of the elevated Ca^{2+} in the cytoplasm rather than the pattern of Ca^{2+} oscillations that is relevant in terms of inducing activation (Tóth et al., 2006). This finding seemingly argues against the importance of the oscillating fertilization signal. However, in the same study it was also noticed that post-implantation development of oocytes that were activated by the prolonged monotonic Ca^{2+} rise was compromised compared to that of oocytes receiving repetitive Ca^{2+} transient. The temporal arrangement of the repetitive transients also affected gene expression after implantation (Ozil et al., 2006). Although the mechanism by which the pattern of the Ca^{2+} signal influences gene expression in the developing embryo still needs to be clarified, it is now believed that the impact of Ca^{2+} oscillations at fertilization reaches farther than just a few hours after gamete fusion.

The existence of the oscillatory signal also implies that there is a constituent in the signaling cascade downstream of Ca^{2+} that decodes the amplitude, number, or frequency of the repetitive spikes and translates it from a digital signal into an analog response (Dupont & Goldbeter, 1998). Due to its ability to undergo auto-activation, CaMKII can remain active after Ca^{2+} concentrations drop and, in neurons, it is believed to act as a decoder of Ca^{2+} oscillations (De Konick & Schulman, 1998). However, in mouse oocytes, CaMKII shows activity that changes in phase with the Ca^{2+} oscillations and does not seem to have any "long-term memory" (Markoulaki et al., 2004). This suggests that the factor in oocytes that possesses the ability to interpret the information encoded in the repetitive Ca^{2+} signal lies downstream of CaMKII in the signaling cascade.

Protein kinase C (PKC) has also been implicated as a factor that may convert Ca^{2+} signals into cellular responses. Mammalian oocytes express conventional PKC isoforms that can be activated by Ca^{2+} and diacylglycerol (Halet, 2004). PLCζ, the sperm-derived oscillogen, is also believed to generate diacylglycerol that may contribute to the stimulation of PKC. Data indicate that PKC activity increases during fertilization, with activated PKC undergoing translocation to the plasma membrane or the meiotic spindle (Tatone et al., 2003). One particular PKC isoform, PKCγ was suggested to play the role of a decoder of Ca^{2+} signals (Oancea & Meyer, 1998). However, experimental data have not supported this idea. In mouse oocytes, PKCα and PKCγ are activated primarily and move to the plasma membrane in response to the first Ca^{2+} spike. After the Ca^{2+} transient, the PKC returns to the cytosol and the subsequent Ca^{2+} transients are associated with translocations of lesser amplitude with no apparent sign of incremental translocation (Halet et al., 2004).

Mitochondria and the Ca^{2+} Signal

Mitochondria also play a major role in the regulation of intracellular Ca^{2+} homeostasis. During the fertilization Ca^{2+} signal in sea urchin and ascidian oocytes, mitochondria actively sequester Ca^{2+} (Eisen & Reynolds, 1985; Dumollard et al., 2003). This has two main consequences. Mitochondrial Ca^{2+} uptake buffers cytosolic Ca^{2+} and thereby regulates Ca^{2+} release from the stores. In addition, Ca^{2+} in the mitochondrial matrix stimulates mitochondrial ATP synthesis (Duchen, 2000). Mammalian oocytes also contain $\sim$100,000 mitochondria that are responsible for ATP production via oxidative phosphorylation during fertilization and early development (Pikó & Matsumoto, 1976). In mouse oocytes, increased cytoplasmic Ca^{2+} levels at fertilization lead to Ca^{2+} elevations in the mitochondrial matrix, which in turn stimulates mitochondrial dehydrogenases of the Kreb's cycle and the electron transport chain. This results in oscillatory increases in FAD and NAD reduction and, eventually, an increase in ATP levels (Campbell & Swann, 2006). Inhibition of mitochondrial respiration blocks the Ca^{2+} transients whereas local uncaging of ATP generates supernumerary spikes (Dumollard et al., 2003, 2004). ATP is believed to be important for providing energy for

Ca^{2+} pumping during Ca^{2+} oscillations; suboptimal ATP supply in mouse oocytes leads to poor development to the expanded blastocyst stage (Van Blerkom et al., 1995). Because Ca^{2+} uptake by mitochondria controls respiration (Rizutto et al., 2004), the stimulation of metabolism in the fertilized oocyte may be an additional function of the oscillatory Ca^{2+} signal (Whitaker, 2006).

8.4 Summary

The significance of Ca^{2+} as an essential ion in oocytes with a central role in signal transduction is well established. Nevertheless, important questions remain to be answered in order for us to gain a better insight into the mechanisms that regulate proper function of the female gamete. One area that will benefit from such research is human-assisted reproduction. A more complete understanding of the role of PLCζ, the sperm-derived oscillogen, may provide an effective diagnostic biomarker of oocyte fertilizability, and may also make it possible to establish PLCζ as a possible mode of treatment for certain types of male infertility. In addition, defining the link between Ca^{2+} signals and downstream signaling molecules (i.e., how the Ca^{2+} signals are interpreted by specific proteins, and how these proteins control different events during oocyte maturation and activation) will improve the efficiency of various assisted reproductive technologies as well as that of somatic cell nuclear transfer.

References

Adkins, C. E., & Taylor, C. W. (1999). Lateral inhibition of inositol 1,4,5-trisphosphate receptors by cytosolic Ca^{2+}. *Current Biology, 9*(19), 1115–1118.

Altimimi, H. F., & Schnetkamp, P. P. (2007). Na^{+}/Ca^{2+}-K^{+} exchangers (NCKX): Functional properties and physiological roles. *Channels (Austin), 1*(2), 62–69.

Baird, G. S. (2011). Ionized calcium. *Clinica Chimica Acta, 412*(9–10), 696–701.

Baumann, O., & Walz, B. (2001). Endoplasmic reticulum of animal cells and its organization into structural and functional domains. *International Review of Cytology, 205*, 149–214.

Berridge, M. J. (1993). Inositol trisphosphate and calcium signaling. *Nature, 361*(6410), 315–325.

Berridge, M. J. (1997). Elementary and global aspects of calcium signalling. *The Journal of Physiology, 499*(Pt 2), 291–306.

Berridge, M. J. (2002). The endoplasmic reticulum: A multifunctional signaling organelle. *Cell Calcium, 32*(5–6), 235–249.

Berridge, M. J. (2007). Calcium signalling, a spatiotemporal phenomenon. In J. Krebs and M. Michalak (eds.), *Calcium: A Matter of Life or Death* (pp. 485–502). Oxford: Elsevier.

Berridge, M. J., & Galione, A. (1988). Cytosolic calcium oscillators. *The FASEB Journal, 2*(15), 3074–3082.

Berridge, G., Dickinson, G., Parrington, J., & Galione, A., & Patel, S. (2002). Solubilization of receptors for the novel Ca^{2+}-mobilizing messenger, nicotinic acid adenine dinucleotide phosphate. *The Journal of Biological Chemistry, 277*(46), 43717–43723.

Bezprozvanny, I., Watras, J., & Ehrlich, B. E. (1991). Bell-shaped calcium-response curves of Ins(1,4,5)P_3- and calcium-gated channels from endoplasmic reticulum of cerebellum. *Nature, 351*(6329), 751–754.

Blaustein, M. P., & Lederer, W. J. (1999). Sodium/calcium exchange: Its physiological implications. *Physiological Reviews, 79*(3), 763–854.

Brini, M., & Carafoli, E. (2009). Calcium pumps in health and disease. *Physiological Reviews, 89*(4), 1341–1378.

Brown, E. M., Gamba, G., Riccardi, D., Lombardi, M., Butters, R. et al. (1993). Cloning and characterization of an extracellular Ca^{2+}-sensing receptor from bovine parathyroid. *Nature, 366*(6455), 575–580.

Campbell, K., & Swann, K. (2006). Ca^{2+} oscillations stimulate an ATP increase during fertilization of mouse eggs. *Developmental Biology, 298*(1), 225–233.

Carroll, J. (2000). Na^{+}-Ca^{2+} exchange in mouse oocytes: Modifications in the regulation of intracellular free Ca^{2+} during oocyte maturation. *Journal of Reproduction and Fertility, 118*(2), 337–342.

Carroll, J., Swann, K., Whittingham, D., & Whitaker, M. (1994). Spatiotemporal dynamics of intracellular $[Ca^{2+}]_i$ oscillations during the growth and meiotic maturation of mouse oocytes. *Development, 120*(12), 3507–3517.

Carroll, M., Levasseur, M., Wood, C., Whitaker, M., Jones, K. T., McDougall, A. (2003). Exploring the mechanism of action of the sperm-triggered calcium-wave pacemaker in ascidian zygotes. *Journal of Cell Science, 116*(Pt 24), 4997–5004.

Cheng, H., Lederer, W. J., & Cannell, M. B. (1993). Calcium sparks – elementary events underlying excitation-contraction coupling in heart-muscle. *Science, 262*(5134), 740–744.

Clapham, D. E. (1995). Calcium signaling. *Cell, 80*(2), 259–268.

Clapham, D. E. (2007). Calcium signaling. *Cell, 131*(6), 1047–1058.

Copley, R. R., Schultz, J., Ponting, C. P., & Bork, P. (1999). Protein families in multicellular organisms. *Current Opinion in Structural Biology, 9*(3), 408–415.

Coronado, R., Morrissette, J., Sukhareva, M., & Vaughan, D. M. (1994). Structure and function of ryanodine receptors. *American Journal of Physiology, 266*(6), 1485–1504.

Cuthbertson, K. S., Whittingham, D. G., & Cobbold, P. H. (1981). Free Ca^{2+} increases in exponential phases during mouse oocyte activation. *Nature, 294*(5843), 754–757.

Dale, B., DeFelice, L. J., & Ehrenstein, G. (1985). Injection of a soluble sperm fraction into sea-urchin eggs triggers the cortical reaction. *Experientia, 41*(8), 1068–1070.

Day, M. L., McGuinness, O. M., Berridge, M. J., & Johnson, M. H. (2000). Regulation of fertilization-induced Ca^{2+} spiking in the mouse zygote. *Cell Calcium, 28*(1), 47–54.

Deguchi, R., Shirakawa, H., Oda, S., Mohri, T., & Miyazaki, S. (2000). Spatiotemporal analysis of Ca^{2+} waves in relation to the sperm entry site and animal-vegetal axis during Ca^{2+} oscillations in fertilized mouse eggs. *Developmental Biology, 218*(2), 299–313.

De Koninck, P., & Schulman, H. (1998). Sensitivity of CaM kinase II to the frequency of Ca^{2+} oscillations. *Science, 279*(5348), 227–230.

De Young, G. W., & Keizer, J. (1992). A single-pool inositol 1,4,5-trisphosphate-receptor-based model for agonist-stimulated oscillations in Ca^{2+} concentration. *Proceedings of the National Academy of Sciences of the United States of America, 89*(20), 9895–9899.

Duchen, M. (2000). Mitochondria and calcium: From cell signalling to cell death. *The Journal of Physiology, 529*(Pt 1), 57–68.

Ducibella, T., Huneau, D., Angelichio, E., Xu, Z., Schultz, R. M. et al. (2002). Egg-to-embryo transition is driven by differential responses to Ca^{2+} oscillation number. *Developmental Biology, 250*(2), 280–291.

Dumollard, R., Hammar, K., Porterfield, M., Smith, P. J., Cibert, C. et al. (2003). Mitochondrial respiration and Ca^{2+} waves are linked during fertilisation and meiosis completion. *Development, 130*(4), 683–692.

Dumollard, R., Marangos, P., Fitzharris, G., Swann, K., Duchen, M., & Carroll, J. (2004). Sperm-triggered Ca^{2+} oscillations and Ca^{2+} homeostasis in the mouse egg have an absolute requirement for mitochondrial ATP production. *Development, 131*(13), 3057–3067.

Dupont, G., & Dumollard, R. (2004). Simulation of calcium waves in ascidian eggs: Insights into the origin of the pacemaker sites and the possible nature of the sperm factor. *Journal of Cell Science, 117*(Pt 18), 4313–4323.

Dupont, G., & Goldbeter, A. (1998). CaM kinase II as frequency decoder of Ca^{2+} oscillations. *Bioessays, 20*(8), 607–610.

Eisen, A., & Reynolds, G. T. (1985). Source and sinks for the calcium released during fertilisation of single sea urchin eggs. *The Journal of Cell Biology, 100*(5), 1522–1527.

Endo, M., Tanaka, M., & Ogawa, Y. (1970). Calcium induced release of calcium from the sarcoplasmic reticulum of skinned skeletal muscle fibres. *Nature, 228*(5266), 34–36.

Feske, S., Gwack, Y., Prakriya, M., Srikanth, S., Puppel, S. H. et al. (2006). A mutation in Orai1 causes immune deficiency by abrogating CRAC channel function. *Nature, 441*(7090), 179–185.

Fissore, R. A., Dobrinsky, J. R., Balise, J. J., Duby, R. T., & Robl, J. M. (1992). Patterns of intracellular Ca^{2+} concentrations in fertilized bovine eggs. *Biology of Reproduction, 47*(6), 960–969.

Fissore, R. A., & Robl, J. M. (1993). Sperm, inositol trisphosphate, and thimerosal-induced intracellular Ca^{2+} elevations in rabbit eggs. *Developmental Biology, 159*(1), 122–130.

Furuichi, T., Yoshikawa, S., Miyawaki, A., Wada, K., Maeda, N., & Mikoshiba, K. (1989). Primary structure and functional expression of the inositol 1,4,5-trisphosphate-binding protein P400. *Nature, 342*(6245), 32–38.

Gilkey, J. C., Jaffe, L. F., Ridgway, E. B., & Reynolds, G. T. (1978). A free calcium wave traverses the activating egg of the medaka, Oryzias latipes. *The Journal of Cell Biology, 76*(2), 448–466.

Guerrier, P., Moreau, M., Meijer, L., Mazzei, G., Vilain, J. P., & Dubé, F. (1982). The role of calcium in meiosis reinitiation. *Progress in Clinical Biological Research, 91*, 139–155.

Halet, G. (2004). PKC signaling at fertilization in mammalian eggs. *Biochimica et Biophysica Acta, 1742*(1–3), 185–189.

Halet, G., Tunwell, R., Parkinson, S. J., & Carroll, J. (2004). Conventional PKCs regulate the temporal pattern of Ca^{2+} oscillations at fertilization in mouse eggs. *The Journal of Cell Biology, 164*(7), 1033–1044.

Hoeflich, K. P., & Ikura, M. (2002). Calmodulin in action: Diversity in target recognition and activation mechanisms. *Cell, 108*(6), 739–742.

Hogben, M., Parrington, J., Shevchenko, V., Swann, K., & Lai, F. A. (1998). Calcium oscillations, sperm factors and egg activation at fertilisation. *Journal of Molecular Medicine, 76*(8), 548–554.

Homa, S. T. (1995). Calcium and meiotic maturation of the mammalian oocyte. *Molecular Reproduction and Development, 40*(1), 122–134.

Hoth, M., & Penner, R. (1992). Depletion of intracellular calcium stores activates a calcium current in mast cells. *Nature, 355*(6358), 353–356.

Igusa, Y., & Miyazaki, S. (1983). Effects of altered extracelluar and intracellular calcium concentration on hyperpolarizing responses of hamster egg. *The Journal of Physiology, 340*, 611–632.

Inesi, G., Kurzmack, M., Coan, C., & Lewis, D. E. (1980). Cooperative calcium binding and ATPase activation in sarcoplasmic reticulum vesicles. *The Journal of Biological Chemistry, 255*(7), 3025–3031.

Jaffe, L. F. (1991). The path of calcium in cytosolic calcium oscillations: A unifying hypothesis. *Proceedings of the National Academy of Sciences of the United States of America, 88*(21), 9883–9887.

Jones, K. T. (2007). Intracellular calcium in the fertilization and development of mammalian eggs. *Clinical and Experimental Pharmacology and Physiology, 34*(10), 1084–1089.

Jones, K. T., Carroll, J., & Whittingham, D. G. (1995). Ionomycin, thapsigargin, ryanodine, and sperm induced Ca^{2+} release increase during meiotic maturation of mouse oocytes. *The Journal of Biological Chemistry, 270*(12), 6671–6677.

Jones, K. T., Cruttwell, C., Parrington, J., & Swann, K. (1998). A mammalian sperm cytosolic phospholipase C activity generates inositol trisphosphate and causes Ca^{2+} release in sea urchin egg homogenates. *FEBS Letters, 437*(3), 297–300.

Jones, K. T., & Nixon, V. L. (2000). Sperm-induced Ca^{2+} oscillations in mouse oocytes and eggs can be mimicked by photolysis of caged inositol 1,4,5-trisphosphate: Evidence to support a continuous low level production of inositol 1,4,5-trisphosphate during mammalian fertilization. *Developmental Biology, 225*(1), 1–12.

Kline, D. (2000). Attributes and dynamics of the endoplasmic reticulum in mammalian eggs. *Current Topics in Developmental Biology, 50*, 125–154.

Kline, D., & Kline, J. T. (1992). Thapsigargin activates a calcium influx pathway in the unfertilized mouse egg and suppresses repetitive calcium transients in the fertilized egg. *The Journal of Biological Chemistry, 267*(25), 17624–17630.

Kline, J. T., & Kline, D. (1994). Regulation of intracellular calcium in the mouse egg: Evidence for inositol trisphosphate-induced calcium release, but not calcium-induced calcium release. *Biology of Reproduction, 50*(1), 193–203.

Knott, J. G., Gardner, A. J., Madgwick, S., Jones, K. T., Williams, C. J., & Schultz, R. M. (2006). Calmodulin-dependent protein kinase II triggers mouse egg activation and embryo development in the absence of Ca^{2+} oscillations. *Developmental Biology, 296*(2), 388–395.

Knott, J. G., Kurokawa, M., Fissore, R. A., Schultz, R. M., & Williams, C. J. (2005). Transgenic RNA interference reveals role for mouse sperm phospholipase Czeta in triggering Ca^{2+} oscillations during fertilization. *Biology of Reproduction, 72*(4), 992–996.

Koh, S., Lee, K., Wang, C., Cabot, R. A., & Machaty, Z. (2009). STIM1 regulates store-operated Ca^{2+} entry in oocytes. *Developmental Biology, 330*(2), 368–376.

Kono, T., Carroll, J., Swann, K., & Whittingham, D. G. (1995). Nuclei from fertilized mouse embryos have calcium-releasing activity. *Development, 121*(4), 1123–1128.

Kouchi, Z., Fukami, K., Shikano, T., Oda, S., Nakamura, Y. et al. (2004). Recombinant phospholipase Czeta has high Ca^{2+} sensitivity and induces Ca^{2+} oscillations in mouse eggs. *The Journal of Biological Chemistry, 279*(11), 10408–10412.

Krebs, J., & Heizmann, C. W. (2007). Calcium-binding proteins and the EF-hand principle. In J. Krebs & M. Michalak (eds.), *Calcium: A matter of life or death* (pp. 51–93). Oxford: Elsevier.

Kubiak, J. Z. (1989). Mouse oocytes gradually develop the capacity for activation during the metaphase II arrest. *Developmental Biology, 136*(2), 537–545.

Kurokawa, M., Sato, K., & Fissore, R. A. (2004). Mammalian fertilization: From sperm factor to phospholipase Czeta. *Biology of the Cell, 96*(1), 37–45.

Larman, M. G., Saunders, C. M., Carroll, J., Lai, F. A., & Swann, K. (2004). Cell cycle-dependent Ca^{2+} oscillations in mouse embryos are regulated by nuclear targeting of PLCzeta. *Journal of Cell Science, 117*(Pt 12), 2513–2521.

Lechleiter, J. D., & Clapham, D. E. (1992). Molecular mechanisms of intracellular calcium excitability in X. laevis oocytes. *Cell, 69*(2), 283–294.

Lillie, F. R. (1919). *Problems of fertilization*. Chicago: University of Chicago Press.

Liou, J., Kim, M. L., Heo, W. D., Jones, J. T., Myers, J. W. et al. (2005). STIM is a Ca^{2+} sensor essential for Ca^{2+}-store-depletion-triggered Ca^{2+} influx. *Current Biology, 15*(13), 1235–1241.

Lomax, R. B., Camello, C., Van Coppenolle, F., Petersen, O. H., & Tepikin, A. V. (2002). Basal and physiological Ca^{2+} leak from the endoplasmic reticulum of pancreatic acinar cells. Second messenger-activated channels and translocons. *The Journal of Biological Chemistry, 277*(29), 26479–26485.

Lorca, T., Cruzalegui, F. H., Fesquet, D., Cavadore, J. C., Méry, J. et al. (1993). Calmodulin-dependent protein kinase II mediates inactivation of MPF and CSF upon fertilization of Xenopus eggs. *Nature, 366*(6452), 270–273.

Machaty, Z., Funahashi, H., Day, B. N., & Prather, R. S. (1997a). Developmental changes in the intracellular Ca^{2+} release mechanisms in porcine oocytes. *Biology of Reproduction, 56*(4), 921–930.

Machaty, Z., Mayes, M. A., Kovács, L. G., Balatti, P. A., Kim, J. H., & Prather, R. S. (1997b). Activation of porcine oocytes via an exogenously introduced rat muscarinic M1 receptor. *Biology of Reproduction, 57*(1), 85–91.

Machaty, Z., Ramsoondar, J. J., Bonk, A. J., Bondioli, K. R., & Prather, R. S. (2002). Capacitative calcium entry mechanism in porcine oocytes. *Biology of Reproduction, 66*(3), 667–674.

Madgwick, S., Hansen, D. V., Levasseur, M., Jackson, P. K., & Jones, K. T. (2006). Mouse Emi2 is required to enter meiosis II by reestablishing cyclin B1 during interkinesis. *The Journal of Cell Biology, 174*(6), 791–801.

Marangos, P., FitzHarris, G., & Carroll, J. (2003). Ca^{2+} oscillations at fertilization in mammals are regulated by the formation of pronuclei. *Development, 130*(7), 1461–1472.

Markoulaki, S., Matson, S., & Ducibella, T. (2004). Fertilization stimulates long-lasting oscillations of CaMKII activity in mouse eggs. *Developmental Biology, 272*(1), 15–25.

Matsu-ura, T., Michikawa, T., Inoue, T., Miyawaki, A., Yoshida, M., & Mikoshiba K. (2006). Cytosolic inositol 1,4,5-trisphosphate dynamics during intracellular calcium oscillations in living cells. *The Journal of Cell Biology, 173*(5), 755–765.

McGuinness, O. M., Moreton, R. B., Johnson, M. H., & Berridge, M. J. (1996). A direct measurement of increased divalent cation influx in fertilised mouse oocytes. *Development, 122*(7), 2199–2206.

McPherson, P. S., & Campbell, K. P. (1993). The ryanodine receptor calcium release channel. *The Journal of Biological Chemistry, 268*(19), 13765–13768.

Mehlmann, L. M., Mikoshiba, K., & Kline, D. (1996). Redistribution and increase in cortical inositol 1,4,5-trisphosphate receptors after meiotic maturation of the mouse oocyte. *Developmental Biology, 180*(2), 489–498.

Mikoshiba, K. (1993). Inositol 1,4,5-trisphosphate receptor. *Trends in Pharmacological Sciences, 14*(3), 86–89.

Miyazaki, S., Katayama, Y., & Swann, K. (1990). Synergistic activation by serotonin and GTP analogue and inhibition by phorbol ester of cyclic Ca^{2+} rises in hamster eggs. *The Journal of Physiology, 426*, 209–227.

Niggli, V., Sigel, E., & Carafoli, E. (1982). The purified Ca^{2+} pump of human erythrocyte membranes catalyzes an electroneutral Ca^{2+}-H^{+} exchange in reconstituted liposomal systems. *The Journal of Biological Chemistry, 257*(5), 2350–2356.

Nixon, V. L., Levasseur, M., McDougall, A., & Jones, K. T. (2002). Ca^{2+} oscillations promote APC/C-dependent cyclin B1 degradation during metaphase arrest and completion of meiosis in fertilizing mouse eggs. *Current Biology, 12*(9), 746–750.

Nixon, V. L., McDougall, A., & Jones, K. T. (2000). Ca^{2+} oscillations and the cell cycle at fertilisation of mammalian and ascidian eggs. *Biology of the Cell, 92*(3–4), 187–196.

Oancea, E., & Meyer, T. (1998). Protein kinase C as a molecular machine for decoding calcium and diacylglycerol signals. *Cell, 95*(3), 307–318.

Ozil, J. P., Banrezes, B., Tóth, S., Pan, H., & Schultz, R. M. (2006). Ca^{2+} oscillatory pattern in fertilized mouse eggs affects gene expression and development to term. *Developmental Biology, 300*(2), 534–544.

Ozil, J. P., & Huneau, D. (2001). Activation of rabbit oocytes: The impact of the Ca^{2+} signal regime on development. *Development, 128*(6), 917–928.

Ozil, J. P., Markoulaki, S., Toth, S., Matson, S., Banrezes, B. et al. (2005). Egg activation events are regulated by the duration of a sustained $[Ca^{2+}]_{cyt}$ signal in the mouse. *Developmental Biology, 282*(1), 39–54.

Ozil, J. P., & Swann, K. (1995). Stimulation of repetitive calcium transients in mouse eggs. *The Journal of Physiology, 483*(Pt 2), 331–346.

Pepperell, J. R., Kommineni, K., Buradagunta, S., Smith, P. J., & Keefe, D. L. (1999). Transmembrane regulation of intracellular calcium by a plasma membrane sodium/calcium exchanger in mouse ova. *Biology of Reproduction, 60*(5), 1137–1143.

Pesty, A., Lefèvre, B., Kubiak, J., Géraud, G., Tesarik, J., & Maro, B. (1994). Mouse oocyte maturation is affected by lithium via the polyphosphoinositide metabolism and the microtubule network. *Molecular Reproduction and Development, 38*(2), 187–199.

Pikó, L., & Matsumoto, L. (1976). Number of mitochondria and some properties of mitochondrial DNA in the mouse egg. *Developmental Biology, 49*(1), 1–10.

Pozzan, T., Rizzuto, R., Volpe, P., & Meldolesi, J. (1994). Molecular and cellular physiology of intracellular calcium stores. *Physiological Reviews, 74*(3), 595–636.

Putney, Jr., J. W., Broad, L. M., Braun, F. J., Lievremont, J. P., & Bird, G. S. (2001). Mechanisms of capacitative calcium entry. *Journal of Cell Science, 114*(Pt 12), 2223–2229.

Ridgway, E. B., Gilkey, J. C., & Jaffe, L. F. (1977). Free calcium increases explosively in activating medaka eggs. *Proceedings of the National Academy of Sciences of the United States of America, 74*(2), 623–627.

Ringer, S. (1883). A further contribution regarding the influence of the different constituents of the blood on the contraction of the heart. *The Journal of Physiology, 4*(1), 29–43.

Rizzuto, R., Duchen, M. R., & Pozzan, T. (2004). Flirting in little space: The ER/mitochondria Ca^{2+} liaison. *Science Signaling, The Signal Transduction Knowledge Environment, 2004*, re1.

Saunders, C. M., Larman, M. G., Parrington, J., Cox, L. J., Royse, J. et al. (2002). PLCζ: A sperm-specific trigger of Ca^{2+} oscillations in eggs and embryo development. *Development, 129*(15), 3533–3544.

Schmidt, T., Patton, C., & Epel, D. (1982). Is there a role for the Ca^{2+} influx during fertilization of the sea urchin egg? *Developmental Biology, 90*(2), 284–290.

Schultz, R. M., & Kopf, G. S. (1995). Molecular basis of mammalian egg activation. *Current Topics in Developmental Biology, 30*, 21–62.

Shapira, H., Lupu-Meiri, M., Lipinsky, D., Oron, Y. (1996). Agonist-evoked calcium efflux from a functionally discrete compartment in Xenopus oocytes. *Cell Calcium, 19*(3), 201–210.

Shiina, Y., Kaneda, M., Matsuyama, K., Tanaka, K., Hiroi, M., & Doi, K. (1993). Role of the extracellular Ca^{2+} on the intracellular Ca^{2+} changes in fertilized and activated mouse oocytes. *Journal of Reproduction and Fertility, 97*(1), 143–150.

Shirakawa, H., Ito, M., Sato, M., Umezawa, Y., & Miyazaki, S. (2006). Measurement of intracellular IP_3 during Ca^{2+} oscillations in mouse eggs with GFP-based FRET probe. *Biochemical and Biophysical Research Communications, 345*(2), 781–788.

Silvestre, F., Boni, R., Fissore, R. A., & Tosti, E. (2011). Ca^{2+} signaling during maturation of cumulus-oocyte complex in mammals. *Molecular Reproduction and Development, 78*(10–11), 744–756.

Solovyova, N., Veselovsky, N., Toescu, E. C., & Verkhratsky, A. (2002). Ca^{2+} dynamics in the lumen of the endoplasmic reticulum in sensory neurons: Direct visualization of Ca^{2+}-induced Ca^{2+} release triggered by physiological Ca^{2+} entry. *The EMBO Journal, 21*(4), 622–630.

Steinhardt, R. A., & Epel, D. (1974). Activation of sea-urchin eggs by a calcium ionophore. *Proceedings of the National Academy of Sciences of the United States of America, 71*(5), 1915–1919.

Steinhardt, R. A., Epel, D., Carroll, E. J., & Yanagimachi, R. (1974). Is calcium ionophore a universal activator for unfertilised eggs? *Nature, 252*(5478), 41–43.

Stricker, S. A. (1999). Comparative biology of calcium signaling during fertilization and egg activation in animals. *Developmental Biology, 211*(2), 157–176.

Stricker, S. A. (2006). Structural reorganizations of the endoplasmic reticulum during egg maturation and fertilization. *Seminars in Cell and Developmental Biology, 17*(2), 303–313.

Swann, K. (1990). A cytosolic sperm factor stimulates repetitive calcium increases and mimics fertilization in hamster eggs. *Development, 110*(4), 1295–1302.

Swann, K. (1992). Different triggers for calcium oscillations in mouse eggs involve a ryanodine-sensitive calcium store. *Biochemical Journal, 287*(Pt 1), 79–84.

Swann, K., & Yu, Y. (2008). The dynamics of calcium oscillations that activate mammalian eggs. *International Journal of Developmental Biology, 52*(5–6), 585–594.

Tatone, C., Delle Monache, S., Francione, A., Gioia, L., Barboni, B., & Colonna R. (2003). Ca^{2+}-independent protein kinase C signalling in mouse eggs during the early phases of fertilization. *International Journal of Developmental Biology, 47*(5), 327–333.

Tesarik, J., Sousa, M., & Testart, J. (1994). Human oocyte activation after intracytoplasmic sperm injection. *Human Reproduction, 9*(3), 511–518.

Tóth, S., Huneau, D., Banrezes, B., & Ozil, J. P. (2006). Egg activation is the result of calcium signal summation in the mouse. *Reproduction, 131*(1), 27–34.

Turner, P. R., Jaffe, L. A., & Fein, A. (1986). Regulation of cortical vesicle exocytosis in sea urchin eggs by inositol 1,4,5-trisphosphate and GTP-binding protein. *The Journal of Cell Biology, 102*(1), 70–76.

Turner, P. R., Sheetz, M. P., & Jaffe, L. A. (1984). Fertilization increases the polyphosphoinositide content of sea urchin eggs. *Nature, 310*(5976), 414–415.

Van Blerkom, J., Davis, P. W., & Lee, J. (1995). ATP content of human oocytes and developmental potential and outcome after in-vitro fertilization and embryo transfer. *Human Reproduction, 10*(2), 415–424.

Wang, C., Lee, K., Gajdócsi, E., Bali Papp, A., & Machaty, Z. (2012). Orai1 mediates store-operated Ca^{2+} entry during fertilization in mammalian oocytes. *Developmental Biology* (in press).

Whitaker, M. (1996). Control of meiotic arrest. *Reviews of Reproduction, 1*(2), 127–135.

Whitaker, M. (2006). Calcium at fertilization and in early development. *Physiological Reviews, 86*(1), 25–88.

Whitaker, M. (2008). Calcium signalling in early embryos. *Philosophical Transactions of the Royal Society B: Biological Sciences, 363*(1495), 1401–1418.

Whitaker, M., & Irvine, R. F. (1984). Inositol 1,4,5-trisphosphate microinjection activates sea urchin eggs. *Nature, 312*(5995), 636–639.

Wu, H., Smyth, J., Luzzi, V., Fukami, K., Takenawa, T. et al. (2001). Sperm factor induces intracellular free calcium oscillations by stimulating the phosphoinositide pathway. *Biology of Reproduction, 64*(5), 1338–1349.

Yao, Y., Choi, J., & Parker, I. (1995). Quantal puffs of intracellular Ca^{2+} evoked by inositol trisphosphate in Xenopus oocytes. *The Journal of Physiology, 482*(Pt 3), 533–553.

Yeromin, A. V., Zhang, S. L., Jiang, W., Yu, Y., Safrina, O., & Cahalan, M. D. (2006). Molecular identification of the CRAC channel by altered ion selectivity in a mutant of Orai. *Nature, 443*(7108), 226–229.

Yoshida, N., Amanai, M., Fukui, T., Kajikawa, E., Brahmajosyula, M. et al. (2007). Broad, ectopic expression of the sperm protein PLCZ1 induces parthenogenesis and ovarian tumours in mice. *Development, 134*(21), 3941–3952.

Zucker, R. S., & Steinhardt, R. A. (1978). Prevention of the cortical reaction in fertilized sea urchin eggs by injection of calcium-chelating ligands. *Biochimica et Biophysica Acta, 541*(4), 459–466.

9 Oocyte Metabolism and Its Relationship to Developmental Competence

Rebecca L. Krisher and Jason R. Herrick

9.1 Introduction

Oocyte competence, or quality, describes not only the ability of the oocyte to resume and complete nuclear maturation, but also to sustain early embryonic development, form a viable blastocyst, and result in birth of a healthy offspring. Oocyte metabolism plays a critical role in determining oocyte developmental competence, not only in the provision of energy, but also by controlling meiotic progression, balancing intracellular redox potential, and providing building blocks for growth. Although oocyte metabolism has been studied for many years, we are just beginning to appreciate the intricate cellular control mechanisms that lead to appropriate metabolic function and thus developmental competence (Figure 9.1). Typically, cumulus cells facilitate glucose uptake by the oocyte, as well as metabolize glucose into pyruvate and lactate, which can then be transferred to, and metabolized by, the oocyte (Gardner et al., 1996; Gardner & Leese, 1990; Leese & Barton, 1985; Saito et al., 1994; Sutton et al., 2003a). Similar to other cells, oocytes metabolize glucose via glycolysis, the pentose phosphate pathway, and the tricarboxylic acid (TCA—also known as the citric acid, or Krebs) cycle (mice: Downs & Utecht, 1999; cats: Spindler et al., 2000; cattle: Krisher, 1999; Rieger & Loskutoff, 1994; sheep:, O'Brien et al., 1996; pigs: Krisher et al., 2007). Glycolysis results in the production of pyruvate and/or lactate from glucose, which can be further metabolized via the TCA cycle for energy production. Glycolytic intermediates also enter the PPP, which provides NADPH for antioxidant defense and precursors for nucleotide and purine synthesis (Figure 9.1). However, our view of oocyte metabolism has expanded and evolved from a simple one of carbohydrate metabolism to produce ATP to include additional pathways such as the hexosamine biosynthesis pathway (HBP) and fatty acid β-oxidation (Figure 9.1). The HBP metabolizes glucose to glucosamine, which is then used to produce extracellular matrix (hyaluronic acid) during cumulus cell expansion (Sutton-McDowall et al., 2006). In addition, the HBP participates in O-linked glycosylation to control protein function. Fatty acids can be metabolized through the TCA cycle and oxidative phosphorylation to produce ATP, potentially freeing glucose to be used for other purposes. An additional glucose-metabolizing pathway not utilized by oocytes under normal conditions is the polyol pathway, which produces sorbitol and fructose. In hyperglycemic conditions, the polyol pathway is upregulated in the COC, with negative consequences (Sutton-McDowall et al., 2010). Finally, we have come to appreciate the role of metabolism in regulation of redox and osmotic balance.

Oocyte Physiology and Development in Domestic Animals, First Edition. Edited by Rebecca L. Krisher.

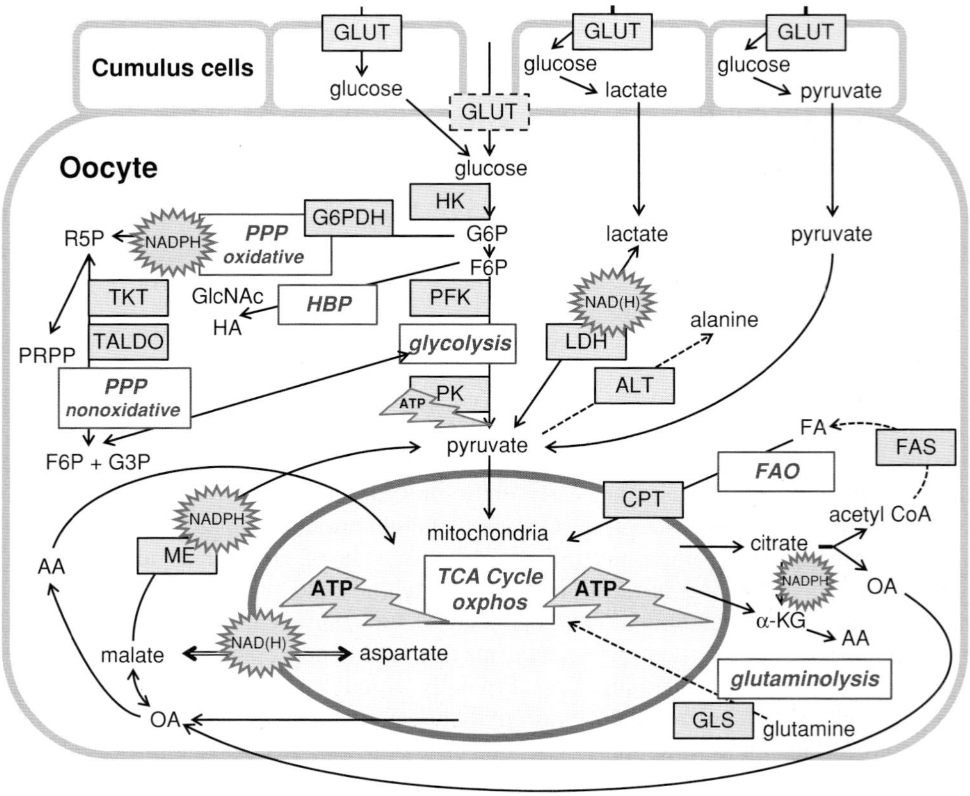

Figure 9.1 Known and potential (dashed lines) metabolic pathways operating in the mammalian oocyte. Pathways names are shown in white boxes, key regulatory enzymes are shown in blue boxes, and important end products are shown in yellow. Abbreviations: α-KG: alpha ketoglutarate; AA: amino acids; ALT: alanine transaminase; ATP: adenosine tri phosphate; CPT: carnitine palmitoyltransferase; F6P: fructose 6 phosphate; FA: fatty acids; FAO: fatty acid β-oxidation; FAS: fatty acid synthase; G3P: glyceraldehyde 3 phosphate; G6P: glucose 6 phosphate; G6PDH: glucose 6 phosphate dehydrogenase; GLS: glutaminase; GLUT: glucose transporters; HK: hexokinase; LDH: lactate dehydrogenase; ME: malic enzyme; OA: oxaloacetate; Oxphos: oxidative phosphorylation; PFK: phosphofructokinase; PK: pyruvate kinase; PRPP: phosphoribosyl pyrophosphate; R5P: ribose 5 phosphate; TALDO: transaldolase; TCA: tricarboxylic acid; TKT: transketolase. (For color detail, see color plate section.)

Oocyte maturation is often considered as two processes: nuclear and cytoplasmic maturation. However, the numerous mechanisms involved are both intimately interrelated and critical for successful embryonic development. This is true for oocyte metabolism as well; the two maturation components cannot be considered independently. Nuclear maturation is reliant on cellular metabolism, and will be discussed in detail later in this chapter. Cytoplasmic maturation, the processes supporting oocyte developmental potential after fertilization, is one of the primary limiting factors in the production of viable embryos from immature oocytes *in vitro* (Abeydeera, 2002; Crozet et al., 1995; Eppig, 1996; Krisher & Bavister, 1998). Of the many cellular changes associated with cytoplasmic maturation, metabolism has been proposed to be one of the most critical. In support of this, oocytes from prepubertal animals that have a reduced capacity for embryonic development also have altered metabolic activity relative to oocytes from adult animals (O'Brien et al., 1996; Steeves & Gardner, 1999; Steeves et al., 1999). Similarly, oocytes matured *in vitro* have altered metabolic activity and reduced developmental capacity compared to *in vivo* matured oocytes (Khurana & Niemann, 2000;

Krisher et al., 2007; Spindler et al., 2000). In addition, genes controlling metabolism are upregulated in MII compared to GV oocytes (Cui et al., 2007), and *in vitro* matured eggs have altered metabolic gene expression compared to those matured *in vivo* (Katz-Jaffe et al., 2009).

In this chapter, we will discuss oocyte metabolism from both a species and a pathway perspective. Metabolism of carbohydrates, amino acids and fatty acids will be analyzed, with special emphasis on their role in oocyte competence. Comparisons and contrasts between species will be made. The effects of metabolism on other cellular processes, particularly redox balance, will also be emphasized.

9.2 Energy Substrates, *In Vivo* and *In Vitro*

Given the importance of the environment for determining oocyte competence, it is not surprising that efforts have been made to measure and replicate the concentrations of energy sources found in the fluids of the mature follicle and the oviduct. Reported values of carbohydrates in follicular fluid from several species are summarized in Table 9.1. In general, lactate is the most abundant substrate in follicular and oviductal fluids, with concentrations of pyruvate less than or equal to that of glucose (rabbit: Leese & Barton, 1985; mouse:, Gardner & Leese, 1990; Harris et al., 2005; pig: Brad et al., 2003b; Nichol et al., 1992; human: Gardner et al., 1996; cow: Iwata et al., 2004; Iwata et al., 2006; Leroy et al., 2004; Orsi et al., 2005; goat: Herrick et al. 2006b). These findings are surprising given the reported preference of oocytes and early embryos for pyruvate (Gardner, 1998; O'Brien et al., 1996; Spindler et al., 2000; Thompson et al., 1996). This apparent discrepancy suggests substrate availability is not the only factor determining the metabolic activity of the oocyte.

Another important factor to consider when attempting to provide COC with optimal energy substrates is the presence of multiple energy sources. Consider the numerous ways that glucose and lactate interact to affect glycolysis. Cells require NAD^+ in order to metabolize both of these

Table 9.1 Reported concentrations (mM) of carbohydrates in the follicular fluid of several species.*

Species	Glucose	Lactate	Pyruvate
Bovine[1]	4.8	5.1	0.03
Bovine[2]	3.8	5.6	
Bovine[3]	1.4		
Bovine[4]	3.9		
Caprine[5]	1.4	7.1	0.002
Porcine[6]	1.8	6.1	0.01
Porcine[7]	4.8		
Mouse[8]	0.5	7.1	0.4
Human[9]	3.3	6.1	

[1](Orsi et al., 2005)
[2](Leroy et al., 2004)
[3](Iwata et al., 2006); values reported in mg/dL
[4](Iwata et al., 2004)
[5](Herrick et al., 2006)
[6](Brad et al., 2003)
[7](Chang et al., 1976); values reported in mg/dL
[8](Harris et al., 2005)
[9](Leese & Lenton, 1990)
*where specified, values are for large or preovulatory follicles

substrates. In the absence of functional mitochondrial shuttles, which have not been investigated in oocytes or cumulus cells, pools of cytosolic NAD^+ are maintained through the conversion of pyruvate into lactate (Lane & Gardner 2000a). In the presence of high concentrations of both glucose and lactate, NAD+ would be consumed as both substrates are metabolized to pyruvate, leading to decreased concentrations of NAD^+ and, potentially, depressed glycolysis (Edwards et al., 1998; Lane & Gardner, 2000a, 2000b). In support of this model, increasing concentrations of lactate and glucose inhibited the ability of mature goat oocytes to metabolize glucose via glycolysis (Herrick et al., 2006b). Similarly, lactate in the culture medium alters the amount of pyruvate metabolized by the zygote (Lane & Gardner, 2000a) and, conversely, the amount of aspartate in the medium affects the early embryo's ability to metabolize lactate (Lane & Gardner, 2005; Mitchell et al., 2009).

Porcine, feline, and bovine oocytes also metabolize lipids during maturation (Ferguson & Leese, 1999; Spindler et al., 2000; Sturmey & Leese, 2003), providing another alternative energy source to carbohydrates and potentially changing how glucose is utilized within the oocyte. In addition, bovine, feline, and ovine oocytes actively metabolize amino acids, and rabbit oocytes can mature with glutamine and BSA as the only energy sources (Bae & Foote, 1975; O'Brien et al., 1996; Rieger & Loskutoff, 1994; Spindler et al., 2000; Steeves & Gardner, 1999). In mice, the most abundant amino acids in the female reproductive tract are taurine, glycine, alanine, glutamine, and glutamate (Harris et al., 2005). In bovine, glutamine, arginine, and asparagine are the amino acids most taken up by COC in the final 6 hours of maturation *in vitro* (Hemmings et al., 2012). Alanine and glycine content of follicular fluid are predictive of bovine COC morphology and subsequent blastocyst development (Sinclair et al., 2008). Similarly, the consumption and production of glutamine and alanine, respectively, as well as utilization of a panel of amino acids including alanine, arginine, glutamine, leucine, and tryptophan, were predictive of the ability of bovine oocytes to cleave and develop to the blastocyst stage (Hemmings et al., 2012).

Although the amount of substrates present in the environment can affect oocyte and COC metabolism, the high metabolic activity of the intact COC can also affect the concentrations of energy substrates in the environment (Herrick et al., 2006b; Leese & Lenton, 1990; Sutton-McDowall et al., 2010). This is especially problematic for the translation of follicular/oviductal fluid data into *in vitro* maturation (IVM) media. Are the concentrations of various substrates measured in physiological fluids tightly controlled and maintained within narrow ranges by blood flow, or are they the result of granulosa and/or cumulus cell metabolism during oocyte maturation? During IVM, should COC be provided with concentrations of substrates that match those found in follicular fluid, or, should the final composition of the medium at the conclusion of maturation be similar to follicular fluid? Should media be changed during the culture period to maintain desired substrate concentrations? Metabolism of bovine COCs in 50 or 100 μl drops accounted for a decrease of ~3-4 mM glucose during culture, and a proportional increase in lactate concentration (Sutton-McDowall et al., 2004). Similarly, media containing 5.6 mM glucose better supported nuclear maturation of bovine COC compared to media supplemented with 2.3 mM glucose, even though follicular fluid concentrations are closer to 2 mM (Sutton-McDowall et al., 2004; Sutton-McDowall et al. 2010). If follicular fluid concentrations of glucose (2.3 mM) are provided initially, there may be little, if any, glucose remaining in the medium at the conclusion of maturation. However, if elevated concentrations of glucose (5.6 mM) are provided at the initiation of culture, the amount of glucose remaining at the end of culture would be similar to follicular fluid concentrations. This consideration may be especially important for porcine oocytes because COCs are typically cultured for 40–44 h for IVM.

9.3 Limitations of Oocyte Metabolism Assessment

Measurement of cellular metabolism in single oocytes, or even in single cumulus oocyte complexes (COC), challenges our technical ability. Metabolism of radiolabeled substrates, as well as fluorescent measurements of metabolites, has been widely used to study the basic cellular metabolic pathways, usually in isolation, and provide a reasonable level of sensitivity. More recently, metabolomics approaches are being investigated that could provide a much broader, more complex, interrelated and interesting view of oocyte metabolism on a more physiological scale. Of importance, a metabolomics approach can provide information about not only suspected pathways of importance, but also about unknown regulatory mechanisms and metabolic intermediates. In addition, it offers a non-invasive method of evaluating oocyte and/or COC metabolism as a potential method of predicting oocyte quality (Singh & Sinclair 2007).

Regardless of the method used to measure oocyte metabolism, several important factors must be considered when interpreting the data produced. First, the metabolic pathways used by the oocyte change during the maturation process (Rieger & Loskutoff, 1994; Songsasen et al., 2007; Spindler et al., 2000; Steeves & Gardner, 1999). Metabolic profiles of a GV oocyte will be different from that of an MII oocyte, and a "maturing" oocyte (actively progressing from GV to MI to MII) may have a different profile than a mature oocyte (arrested at MII). Second, metabolic measurements have not been conducted in follicular fluid, so all metabolic assessments have essentially been done under abnormal conditions. Aside from the concentrations of select substrates measured in follicular fluid, we know very little about the metabolism of the oocyte (or the COC) under *in vivo* conditions. Third, the environment in which the oocytes are matured, whether *in vivo* in follicular fluid or *in vitro* in a variety of culture media, is often different from the medium in which metabolism is actually assessed. It is not uncommon for the media used in metabolism assays to contain the radiolabeled substrate in the absence of other substrates, or in the presence of nonphysiological concentrations of other substrates that facilitate detection of changing concentrations. Although simpler media facilitate data interpretation, they also produce potentially abnormal results due to culture in nonphysiological conditions. In embryos, carbohydrate metabolism can be affected by brief exposure to media lacking amino acids and vitamins (Lane & Gardner, 1998). Similarly, decreasing the concentration of pyruvate in the assessment medium decreased the rate of glycolysis in oocytes matured under identical conditions (Krisher & Bavister, 1999). Therefore, oocyte metabolism is influenced by conditions before and/or during metabolic assessment, which result in ongoing adaptation by the oocyte to a changing external environment. However, some metabolic changes are retained regardless of the environment used for metabolic assessment. For example, embryos that had been exposed to EDTA retain an altered metabolic profile when metabolism is evaluated in the absence of this chemical (Gardner et al., 2000; Lane & Gardner, 2001). Similarly, *in vivo*– and *in vitro*–derived oocytes or embryos exhibit different metabolic profiles when evaluated under identical conditions(Gardner & Leese, 1990; Khurana & Niemann, 2000; Lane & Gardner, 1998; Spindler et al., 2000; Swain et al., 2002; Thompson et al., 1991). Glucose and lactate in the maturation medium alter glucose metabolism in porcine oocytes when metabolism is evaluated under identical conditions (Brad et al. 2003b; Krisher et al., 2007).

Evaluating metabolic activity and substrate preferences of the oocyte is also complicated by the interactions between the oocyte and the cumulus cells that surround it throughout maturation (Downs et al., 2002). The oocyte and its cumulus mass are in reality a functional metabolic unit, with dynamic communication between the two cell types. The cumulus cells are metabolically active (Gardner & Leese, 1990; Leese & Barton, 1985; Sutton et al., 2003a), and can pass a variety of metabolites to the oocytes via gap junctions (Biggers et al., 1967; Downs, 1995; Downs et al., 1986; Downs et al., 2002; Saito et al., 1994). In addition, the interaction between the oocyte and the surrounding cumulus cells is also important in maintaining developmental competence (Eppig, 1991; Heller et al., 1981). It is of interest to know the metabolic activity of both the intact COC and the oocyte itself, but one must take into account that the simple act of removing the cumulus cells may cause changes in metabolic activity in that oocyte to compensate for a change in substrate availability coming from the cumulus cells via gap junctions (Downs, 1995; Eppig, 1991; Heller et al., 1981; Saito et al., 1994). For example, the dependence of the oocyte on pyruvate is specific to the denuded oocyte, because cumulus-enclosed oocytes can mature and be fertilized in the presence of a number of other substrates if the cumulus cells are intact (Biggers et al., 1967). Similarly, oxidative metabolism of pyruvate is essential for oogenesis, maturation, and oocyte competence, but a deficit in oocyte pyruvate oxidation can be compensated for by cumulus cell metabolism at some stages (Johnson et al., 2007).

Despite the close contact between oocytes and COC, these two cell types can have different metabolic profiles when evaluated in similar culture conditions, highlighting the interrelationships between cell types and the complexities of metabolic control mechanisms within the COC (Downs et al., 2002; Downs & Utecht, 1999; Sutton et al., 2003a; Zuelke & Brackett, 1992). Enzyme activities of phosphofructokinase, lactate dehydrogenase, glucose 6-phosphate dehydrogenase, and lipase are different between cumulus cells and the oocytes, as are profiles of pyruvate uptake and oxidation, glucose uptake, lactate production, and oxygen consumption (Cetica et al., 2002; Cetica et al. 1999; Downs et al., 2002; Sutton et al., 2003a). In goats, the oocyte accounts for $\sim$25% of the protein content of the COC, but is responsible for $<$0.1% of glucose and pyruvate metabolism (Herrick et al., 2006b). In murine and bovine COCs, a mathematical model of oxygen diffusion suggests oxygen in the follicular fluid moves through the cumulus layer without being used, to reach the oocyte for use in oxidative phosphorylation (Clark et al., 2006). As a result, bovine cumulus cells actively take up and utilize glucose and exhibit little oxidative metabolism, whereas oocytes actively utilize oxidative phosphorylation and require only limited glucose metabolism (Thompson et al., 2007).

Regardless of these limitations, studies of oocyte metabolism have provided us with solid basic information to begin to understand the metabolic processes normally undertaken by oocytes that result in a high level of developmental competence. This has informed the development of media that successfully support *in vitro* oocyte maturation. Excitingly, this knowledge is constantly evolving.

9.4 Mitochondrial Function in the Oocyte

Mitochondria play a central role in oocyte metabolism as the location of the TCA cycle and oxidative phosphorylation, and thus the site of ATP production. Mitochondria in the oocyte exist in an immature form with few cristae, and are derived from a founder population in the female germ cell (Dumollard et al., 2006). Despite the relative inactivity of individual mitochondria, due to their large numbers in the oocyte they contribute significantly to cellular metabolism. In fact, it has been proposed that deficiencies in aerobic mitochondrial metabolism may be the root cause of variability in oocyte competence (Wilding et al., 2009). During oocyte growth, mitochondrial numbers increase as follicle diameter increases, but there is a large variation in mitochondrial copy number in mature oocytes (Van Blerkom, 2004; Zeng et al., 2009). Mitochondria also relocate within the oocyte during oocyte maturation, from near the germinal vesicle to the oocyte periphery, possibly to support critical metabolic processes at that location (Dumollard et al., 2006; Van Blerkom, 2004;

Van Blerkom et al., 2003). Mitochondria are typically found in close proximity to the endoplasmic reticulum, and are thought to create localized metabolic foci.

Due to the prominent role of mitochondria in cellular metabolism, the number and location of mitochondria appear to be an important component of oocyte quality. Low mitochondrial number in mature oocytes is associated with poor oocyte competence (May-Panloup et al., 2007). Similarly, if germinal vesicles from oocytes with damaged mitochondria are transplanted into normal oocytes, they can be rescued and go on to produce blastocysts (Takeuchi et al., 2005). Addition of TGFα to pig oocyte maturation medium results in oocytes with more homogeneous lipid droplets, associated mitochondria and endoplasmic reticulum, and increased blastocyst development, more similar to *in vivo* matured controls (Mito et al., 2009). In rodents, ovarian stimulation (hamsters) or *in vivo* maturation (rats) results in an increase in the oocyte's mitochondrial population and/or an increase in ATP content (Lee et al., 2006). However, there is no conclusive evidence to suggest that the effect of mitochondrial number on oocyte competence has its basis in ATP production (Brad et al., 2003a; Shoubridge & Wai, 2007).

These data suggest that mitochondrial number, activity, and distribution may be related to oocyte quality, and that these processes may be perturbed during assisted reproduction. Certainly evidence points to the ability of mitochondria to produce an adequate level of ATP in the oocyte as being important for oocyte competence. However, it may be that the spatial organization of mitochondria and localized ATP provision is more important than the net ATP content of oocytes (Van Blerkom, 2011). Ultimately, the association of mitochondria with oocyte competence is essentially unknown (Van Blerkom, 2004). Mitochondrial dysfunction may also play a major role in the decline of oocyte quality with age. If older female mice were supplemented with CoQ10 prior to ovarian stimulation and oocyte recovery, oocyte numbers and mitochondrial activity were improved to the point that they were similar to that in young mice (Bentov et al., 2010).

9.5 Cattle Oocyte Metabolism

Of the domestic species, bovine oocyte metabolism is probably the best characterized. In bovine COC, glucose metabolism via glycolysis is the main metabolic pathway in cumulus cells, while the oocyte actively metabolizes pyruvate and lactate from the cumulus cells, as well as alanine and aspartate, aerobically (Cetica et al., 2003; Sutton-McDowall et al., 2010). The cumulus cells of the COC consume many times more glucose than the oocyte, converting it primarily to lactate, while the oocyte uses three times more oxygen than do cumulus cells (Thompson, 2006). Bidirectional communication between the cumulus cells and oocyte is critical for correct metabolic regulation (Sutton-McDowall et al., 2010; Sutton et al., 2003a). However, metabolism of the cumulus cells does not depend on, nor does it indicate, metabolic activity of the enclosed oocyte.

In cattle COC, the consumption of oxygen, pyruvate, and glucose increases throughout maturation, without an increase in lactate production (Sutton et al., 2003a). Inclusion of glucose in the maturation medium improves nuclear and cytoplasmic maturation, but excessive glucose concentrations can increase reactive oxygen species and decrease intracellular glutathione content (Hashimoto et al., 2000b). A delicate balance of glucose concentration, neither too high nor too low, regulates both nuclear and cytoplasmic maturation (Sutton et al., 2003b). High glucose concentrations during maturation can result in excess ROS production, increased O-linked glycosylation, and decreased GSH, all of which negatively affect competence, whereas low glucose may limit substrate availability for nucleic acid synthesis, energy production, and oxygen radical defense (Sutton-McDowall et al., 2010).

Glucose is primarily metabolized by glycolysis in bovine oocytes, although the pentose phosphate pathway (PPP) is also active albeit at a lower rate than glycolysis (Rieger & Loskutoff 1994). Supplementation of IVM media with LH to enhance nuclear maturation increases the activity of both glycolysis and the TCA cycle in COC, while decreasing activity of the PPP (Zuelke & Brackett, 1992). Glycolytic activity has been reported to both remain constant (Rieger & Loskutoff, 1994), or to increase (Steeves & Gardner, 1999) throughout maturation. These differing reports are probably related to the different media used for metabolic evaluation (0.5 versus 5.5–8.7 mM glucose). It should be noted that the glucose concentration in the media used by Steeves and Gardner (1999: 0.5 mM) may be more physiological. Other differences in the metabolic assessment medium, including the presence of amino acids, may also have resulted in the differences observed between these two studies.

Glucose is also metabolized by the hexosamine biosynthesis pathway (HBP) in bovine oocytes, particularly in the later stages of meiotic maturation. Stimulated by FSH, the main end-product of glucose metabolism via the HBP in COC is hyaluronic acid (HA), which is a critical component of the extracellular matrix of the expanded cumulus mass (Sutton-McDowall et al., 2004; Thompson, 2006). Inclusion of glucosamine decreases glucose consumption in bovine COC by providing an alternative source of substrate for HA production (Sutton-McDowall et al., 2004). However, an upregulation of HBP or inclusion of glucosamine during maturation can be detrimental to oocyte competence, potentially by increasing O-linked glycosylation of signaling proteins (Thompson, 2006).

In oocytes, metabolism of pyruvate through the TCA cycle peaks at 12 hr of culture when most of the oocytes are entering MI (Rieger & Loskutoff, 1994; Steeves & Gardner, 1999). The rate of pyruvate metabolism was nearly 10-fold higher than all other substrates at all time points, indicating that pyruvate may be the preferred energy substrate of cattle oocytes (Rieger & Loskutoff, 1994). However, this finding may be an artifact due to the concentrations of substrate used. Using a lower pyruvate concentration (0.33 mM versus 6.9 mM), Steeves and Gardner (Steeves & Gardner, 1999) did not observe such a preference.

Oxygen (O_2) plays an important role in the regulation of COC metabolism. ATP production and the proportion of oocytes reaching MII are influenced by O_2 content of the atmosphere (20% $>$ 5% O_2). Under low O_2 conditions, increasing the concentration of glucose can support high levels of maturation via glycolytic ATP production (Hashimoto et al., 2000a). A high oxygen environment during bovine IVM resulted in better oocyte competence than low oxygen, although in low oxygen the addition of EGF and cysteamine and increased glucose concentrations can improve development post-fertilization, possibly by increasing GSH production (Oyamada & Fukui, 2004). However, Bermejo-Alvarez et al. (2010) reported a beneficial effect of maturing bovine COC in low oxygen, and suggested that the beneficial effect observed was due to an increase in glucose metabolism via anaerobic glycolysis. Mouse oocytes matured in low oxygen have oxygen-uptake profiles, mitochondrial membrane potentials, and metabolism similar to *in vivo* matured oocytes, and the resulting blastocysts had higher cell numbers, compared to oocytes matured in high oxygen (Preis et al., 2007).

In addition to carbohydrates, glutamine appears to play a significant role in bovine oocyte metabolism. Glutamine, but not glucose, lactate, or pyruvate, can promote oocyte maturation to MII when added to a basal media, although all of these substrates do promote GVBD (Bilodeau-Goeseels, 2006). The metabolism of glutamine through the TCA cycle increases steadily throughout maturation (Rieger & Loskutoff, 1994; Steeves & Gardner, 1999). Similar to glucose, LH increases oxidation of glutamine by both COC and oocytes; this increased glutamine oxidation by the oocyte depends on the presence of cumulus cells during IVM (Zuelke & Brackett, 1993).

These patterns of metabolism during bovine oocyte maturation *in vitro* are accompanied by numerous changes in metabolic enzyme activity in the COC and the oocyte. In the oocyte, the activities of phosphofructokinase, glucose-6 phosphate dehydrogenase, and lipase increase during maturation when standardized for protein content, while activities of these same enzymes decrease in the cumulus cells (Cetica et al., 2002). In contrast, the activities of phosphofructokinase, glucose-6 phosphate dehydrogenase, and lactate dehydrogenase do not change during maturation in either the oocyte or the cumulus when expressed on a per oocyte or per COC basis (Cetica et al., 1999). The activities of phosphofructokinase, glucose-6 phosphate dehydrogenase, and lactate dehydrogenase were higher in COC than in oocytes, while lipase activity was higher in oocytes than COC (Cetica et al., 2002; Cetica et al., 1999).

9.6 Pig Oocyte Metabolism

Glucose also plays a critical role in pig oocyte maturation. Work from our laboratory suggests that glucose is the primary energy substrate for pig oocytes (Krisher et al., 2007), which differs from other species. Maturation of porcine oocytes *in vitro*, similar to other domestic species, results in significant metabolic alterations compared to those matured *in vivo*, most likely indicating inadequate maturation conditions. *In vitro*-matured porcine oocytes metabolized more glucose via the glycolytic pathway than glucose via PPP, pyruvate via the TCA cycle, or glutamine via the TCA cycle. In contrast, *in vivo*–matured oocytes metabolized equivalent amounts of glucose via glycolysis and the PPP, with both of these pathways used preferentially over either pyruvate or glutamine via the TCA cycle (Krisher et al., 2007). The importance of the PPP in porcine oocytes is confirmed by other studies (Flood & Wiebold, 1988; Sato et al., 2007) and suggests a critical and unique role for this pathway during oocyte maturation in pigs.

Another finding of interest is that pyruvate oxidation was reduced in *in vitro*–matured porcine oocytes compared with those matured *in vivo* (Krisher et al., 2007). Decreased pyruvate metabolism was probably not due to insufficient amounts of pyruvate in the medium, as concentrations were greater than that found in the pig oviduct (Nichol et al., 1992). In addition, cumulus cell metabolism provides significant quantities of lactate and pyruvate during bovine and caprine oocyte maturation (Herrick et al., 2006a; Sutton-McDowall et al., 2004). Instead, the reduced level of pyruvate metabolism observed may be considered an indication of abnormal cytoplasmic maturation or deficiencies in other metabolic activities. The oocyte may take up pyruvate from the medium, but utilize it in pathways other than the TCA cycle, such as amino acid synthesis. A similar phenomenon has been reported in feline oocytes, in which the majority of the pyruvate taken up by the oocyte is not metabolized by the TCA cycle (Spindler et al., 2000). Alternatively, transcription of genes responsible for key regulatory enzymes in pyruvate metabolism may not be appropriate, resulting in compromised pyruvate metabolism. The inability of oocytes to produce these enzymes or regulate their activity may have negative effects not only on oocyte maturation, but also on subsequent fertilization and embryo development.

9.7 Mouse Oocyte Metabolism

In the mouse, the oocyte appears to prefer pyruvate as an energy source, as this substrate alone supports maturation, maintains viability, and promotes cleavage of the fertilized oocyte in the absence of cumulus cells (Biggers et al., 1967; Downs & Hudson, 2000; Downs & Mastropolo,

1994; Fagbohun & Downs, 1992; Leese & Barton, 1984). Pyruvate can also overcome the arresting action of cAMP and hypoxanthine on COC, as can lactate and oxaloacetate, but this effect is not dependent on metabolism through the TCA cycle (Downs & Mastropolo, 1994; Fagbohun & Downs, 1992). The metabolism of pyruvate is related to the stage of meiosis, with arrested oocytes (GV or MII) metabolizing less pyruvate than oocytes progressing through meiosis (Downs et al., 2002). In addition, pyruvate metabolism is differentially regulated in the cumulus cells and the oocyte. Although COC consume more pyruvate than oocytes, oocytes oxidize more pyruvate than COC (Downs et al., 2002).

In general, glucose is metabolized more by the cumulus cells than by the oocyte in the mouse. Cumulus cells take up glucose via glucose transporters (GLUTs) and then transfer glucose and glucose 6-phosphate, as well as lactate and pyruvate, into the oocyte via gap junctions (Downs & Utecht, 1999; Saito et al., 1994). Glucose can also enter the oocyte directly, although this may not be mediated by GLUTs (Wang et al., 2012). Cumulus cells, but not oocytes, display insulin-stimulated glucose uptake (Purcell et al., 2012). The presence of glucose affects pyruvate consumption, and the consumption of glucose is influenced by the presence of pyruvate in COC (Downs et al., 1997; Downs & Utecht, 1999). The addition of glucose inhibits pyruvate oxidation in COC, without affecting pyruvate metabolism by the oocyte (Downs et al., 2002). In fact, it is the oocyte that regulates cumulus cell glucose metabolism, via the oocyte secreted factors BMP15 and GDF9 (Su et al., 2008). Despite the primary role of pyruvate in mouse oocyte metabolism, metabolism of glucose via the PPP has been associated with increased competence (Downs et al., 1998).

9.8 Oocyte Metabolism in Other Species

One of the most complete single studies of oocyte metabolism was done in the domestic cat (Spindler et al., 2000). In feline oocytes, the rates of glycolysis, glucose oxidation, and lactate oxidation increase throughout maturation (GVBD to MII). In contrast, oxidation of glutamine and palmitate increase at the time of GVBD, but then remain constant through MII (Spindler et al., 2000). Pyruvate was the preferred substrate at all stages of maturation, but only a small proportion of this was oxidized. In the mature oocyte, glutamine, lactate, and pyruvate were metabolized more than glucose or palmitate (Spindler et al., 2000).

As in the mouse, cow, and cat, the mature sheep oocyte seems to prefer pyruvate (O'Brien et al., 1996). Glutamine is also important for sheep oocytes, as this substrate was metabolized at a higher rate than glucose. Although the activity of specific metabolic pathways were not measured, both glucose and pyruvate (in combination with lactate) supported maturation of rhesus monkey oocytes to MII at comparable rates (Zheng et al., 2001). However, glucose was stimulatory and pyruvate was inhibitory to cytoplasmic maturation of monkey oocytes, as measured by development to the morula or blastocyst stage (Zheng et al., 2001). In contrast to these studies, Bae and Foote (1975) found that rabbit oocytes could mature to MII in the absence of carbohydrates, if provided with glutamine. It is apparent from these results that metabolism plays a critical role in oocyte maturation, but the precise role and the mechanism of its effects are species-specific.

9.9 Oocyte Metabolism of Fatty Acids

Although the metabolism of carbohydrates by oocytes and embryos has been studied extensively, the contribution of fatty acid β-oxidation (FAO) has been relatively ignored until recently. Several species, including cow, pig, and cat, have large stores of intracellular lipids, whereas the mouse has very little (McEvoy et al., 2000). The amount of lipids present in the oocyte is reflected by the color of the ooplasm: oocytes with uniform dark cytoplasm have more lipids than oocytes with granulated or pale cytoplasm (Leroy et al., 2005a). In bovine oocytes, those with a dark ooplasm have more mitochondria and higher developmental competence than those with a pale or light ooplasm (Jeong et al., 2009). In domestic ruminant oocytes, palmitic, stearic, and oleic are the most abundant fatty acids, although pig oocytes contain greater polyunsaturated fatty acids, particularly linoleic acid (Homa et al., 1986; McEvoy et al., 2000). Palmitic is the most abundant fatty acid in bovine oocytes, with oleic acid being more abundant in good quality and stearic acid more abundant in low-quality oocytes (Kim et al., 2001). In human oocytes that failed fertilization, the majority of fatty acids were saturated, of which stearic was the most prominent, followed by palmitic (Matorras et al., 1998). In human embryos, those that developed further had higher concentrations of linoleic and oleic (unsaturated fatty acids), and lower total saturated fatty acids, than those that did not (Haggarty et al., 2006). Under conditions of heat stress in the summer, when oocyte competence is poor, bovine oocytes have an increased percentage of saturated FA, whereas in the winter season when oocyte competence is better, percentages of mono- and polyunsaturated fatty acids are increased (Zeron et al., 2001). These studies imply that lipid composition may reflect oocyte competence. The fatty acid content of follicular fluid, however, does not seem to be associated with oocyte competence (Sinclair et al., 2008).

Recent evidence suggests that lipid metabolism may play a role in energy production in the oocyte and embryo, even in species with minimal amounts of intracellular lipid (Downs et al., 2009; Dunning et al., 2010). The mature mouse oocyte expresses Cpt1b and Cpt2, enzymes essential to fatty acid metabolism (Dunning et al., 2010; Gentile et al., 2004). Perilipin-2, a lipid droplet protein regulating the storage of intracellular lipids, is induced along with lipid droplet reorganization during oocyte maturation (Cerri et al., 2009). In addition, FAO is upregulated in maturing mouse COC, and inhibition of FAO during maturation reduces, while stimulation of FAO during maturation increases, subsequent blastocyst development (Dunning et al., 2010). Mouse embryo development and cell number was decreased when FAO was blocked during culture (Hewitson et al., 1996). Indirect evidence suggests that pig and cow oocytes and embryos metabolize fatty acids as well (Sturmey et al., 2009). Inhibition of fatty acid metabolism in both porcine and bovine oocytes reduces subsequent development (Ferguson & Leese 2006; Sturmey et al., 2006). However, de-lipated pig embryos can develop successfully to blastocyst, although with fewer cell numbers suggesting reduced viability (Yoneda et al., 2004). Interestingly, although glucose uptake did not change in this experiment, lactate production was decreased, suggesting a different fate for glucose when ATP is generated from lipid metabolism. In pig embryos, although development was not reduced when FAO was inhibited, glucose metabolism was upregulated (Sturmey & Leese, 2008), again suggesting an adjustment of glucose metabolism, potentially toward oxidative phosphorylation to compensate for the loss of optimal ATP production via FAO. Cow embryos behave similarly when FAO is inhibited, although development is compromised (Ferguson & Leese 2006). Interestingly, bovine, but not mouse oocytes, can develop after maturation without exogenous energy substrates only if fatty acid metabolism is active, highlighting the importance of endogenous fatty acid oxidation in domestic species (Ferguson & Leese, 1999; Ferguson & Leese, 2006).

The addition of fatty acids (FA), or carnitine to stimulate FAO, to oocyte and embryo culture medium has primarily shown positive effects on development, although results are variable due in part to differences in type and concentration of FA used (Dunning et al., 2010; Leroy et al., 2005b; Marei et al., 2010; Somfai et al., 2011; Spindler et al., 2000; Van Hoeck et al., 2011; Wu et al., 2011). Oocytes are capable of taking up exogenous fatty acids from the culture environment

(Ferguson & Leese, 1999). Supplementation of in vitro follicle culture with L-carnitine results in oocytes with increased developmental potential (Dunning et al., 2011). Treatment of pig oocytes with L-carnitine during maturation increases oocyte GSH, reduces ROS, and results in higher blastocyst development after fertilization (You et al., 2012). Carnitine supplementation during porcine IVM increases completion of meiosis and embryonic cleavage after fertilization, as well as increasing oocyte mitochondrial density while decreasing lipid droplet density, suggesting enhanced mitochondrial function (Somfai et al., 2011). In cattle, the addition of linoleic acid, the most abundant fatty acid in bovine follicular fluid, to IVM medium reduces oocyte progression to MII as well as developmental competence (Homa & Brown, 1992; Marei et al., 2010). In humans, elevated free fatty acids in follicular fluid are associated with poor morphology COCs (Jungheim et al., 2011). Of note, culture with serum changes the lipid content of oocytes (Ferreira et al., 2010; Leroy et al., 2005a), as do changes in oxygen concentration during maturation (Ferreira et al., 2010).

The role of fatty acid metabolism may be especially critical in the pig oocyte, as it has large endogenous stores of lipids (McEvoy et al., 2000; Sturmey et al., 2009). In porcine oocytes, mitochondria and lipid droplets are located very close together, suggesting a physiologically relevant metabolic role for lipids during oocyte maturation (Sturmey et al., 2006). In addition, during maturation mitochondria translocate to the periphery, where oxygen is more accessible, and become associated with lipid droplets, supporting the hypothesis of a metabolic role for fatty acids (Sturmey et al., 2006; Sturmey et al., 2009). Indirect evidence also suggests a functional role, as triglycerides decrease during *in vitro* pig and cow oocyte maturation, with a corresponding decrease in oxygen consumption. Glucose in maturation medium may be important to prime the TCA cycle with pyruvate and oxaloacetate (Sturmey & Leese, 2003) for FAO.

The lactating dairy cow has provided an interesting model in which to examine the effects of high non-esterified fatty acids (NEFA) on oocyte competence. Due to negative energy balance (NEB), high-producing dairy cows have increased levels of NEFA in follicular fluid, predominantly oleic, palmitic, and stearic acids (Leroy et al., 2005b). This situation is similar to that experienced by women with obesity or diabetes, although the end result is arrived at quite differently metabolically. If elevated NEFA are present during maturation, the resulting blastocysts have reduced oxygen, pyruvate, and glucose consumption; elevated lactate production; and high amino acid metabolism; as well as lower cell numbers, altered gene expression, and increased apoptosis (Van Hoeck et al., 2011). When palmitic and stearic acids were added to bovine oocyte maturation medium, they reduced oocyte competence, whereas oleic acid had no effect (Leroy et al., 2005b). However, this model is complicated by low glucose concentrations in the follicular fluid of cows in NEB, which may itself negatively impact oocyte competence (Leroy et al., 2006). The end result seems to be that negative energy conditions during follicle growth are expressed several months later as ovulation of a suboptimal oocyte, due to both hypoglycemia and high NEFA in follicular fluid (Leroy et al., 2008a; Leroy et al., 2008b). The majority of studies point to the conclusion that an excess of NEFA during oocyte maturation can negatively influence oocyte developmental competence.

9.10 Oocyte Metabolism Controls Meiosis: A View across Species

The role of oocyte metabolism in the control of resumption and completion of meiosis has been best studied in the mouse. In mice, gonadotropin-induced meiosis is dependent on the presence of glucose (Downs & Mastropolo, 1994; Fagbohun & Downs, 1992). However, glucose can have both stimulatory and inhibitory roles. Resumption of meiosis is associated with elevated activity of glycolysis and the PPP, as well as increased activity of hexokinase (glycolysis and PPP), phosphofructokinase (glycolysis), glucose-6 phosphate dehydrogenase (PPP), and 6-phosphogluconate dehydrogenase (PPP) within the cytoplasm of the oocyte (Cetica et al., 2002; Downs et al., 1996; Downs & Utecht, 1999; Tsutsumi et al., 1992). It appears, though, that the flux of glucose through PPP rather than glycolysis is more critical in controlling meiosis in mice (Sutton-McDowall et al., 2010). Conversely, glucose helps to maintain meiotic arrest through activity of the PPP and subsequent purine synthesis, and can inhibit the stimulatory effects of pyruvate on GVBD in hypoxanthine-arrested COC (Downs, 1998; Downs & Mastropolo, 1994; Downs & Verhoeven, 2003; Fagbohun & Downs, 1992). The arresting effects of glucose appear to be dependent on gap junction communication between the cumulus cells and the oocyte and glycolytic production of ATP, although not related to lactate production (Downs, 1995; Downs & Mastropolo, 1994). However, glucose is also required for FSH-induced maturation of hypoxanthine- or cAMP-arrested COC. FSH increases hexokinase activity, glycolysis, pentose phosphate pathway activity, glucose consumption, and lactate production, as well as decreasing oxygen consumption by the COC (Downs et al., 1997; Downs et al., 1996; Downs & Utecht, 1999). FSH also causes an increase in glucose consumption and lactate production via PI3K mediated translocation of GLUT4 to the cumulus cell membranes in mouse COC (Roberts et al., 2004). As with the inhibitory effects of glucose, the stimulatory role of glucose in FSH-induced GVBD involves metabolism through the PPP and purine synthesis. Both FSH and PPP stimulators increase glucose consumption and metabolism of glucose through glycolysis and the PPP in the COC, without affecting glucose oxidation in the COC, and increase GVBD (Downs et al., 1998; Downs & Utecht, 1999). Ribose 5-phosphate produced by the PPP is converted to phosphoribosyl pyrophosphate (PRPP), which participates in purine production via de novo synthesis or the salvage of hypoxanthine, and resumption of meiosis (Downs, 1993; Downs, 1997; Downs et al., 1998). Ribose and PRPP both stimulate GVBD in mouse oocytes (Downs et al., 1998). Similarly, ribose, glucose, and FSH lead to an increase in PRPP synthesis (Downs et al., 1998). In addition to the beneficial effects of glucose on GVBD, glucose is also stimulatory to progression to MII in COC (Downs & Hudson, 2000). Glucose is also essential to fertilization in the mouse, specifically glucose metabolism via PPP (Urner & Sakkas, 2005).

It may be that increased activity of the PPP is one of the initial stimuli for the resumption of meiosis, whereas increased glycolytic activity occurs as a result of increased energy demands during the progression of meiosis. Stimulation of the PPP either with chemical agents (pyrroline carboxylate or phenazine ethosulfate) or FSH induces a significant increase in GVBD in murine oocytes (Downs et al., 1998). In porcine oocytes, a chemical inhibitor of PPP (diphenylene iodonium; DPI) inhibited oocyte maturation, as well as the activity of both glycolysis and PPP. It is possible that inhibition of the PPP in porcine oocytes prevented nuclear maturation and the associated increase in glycolytic activity. However, whether low glucose metabolism is a cause or a consequence of low maturation remains to be determined. We have undertaken preliminary studies that demonstrate the addition of PPP cofactors or end-products (NADP, PRPP, and R5P-produced downstream of DPI action) to a medium containing DPI results in reversal of meiotic arrest and restoration of glycolytic and PPP activity in pig oocytes (Tubman et al., 2006).

Increased metabolism of glucose through one or more metabolic pathways also occurs simultaneously with the progression of meiosis to metaphase II in oocytes from cats (Spindler et al., 2000) and cattle (Rieger & Loskutoff, 1994; Steeves & Gardner, 1999; Zuelke & Brackett, 1992). In primate oocytes, glucose is necessary for cytoplasmic maturation, although nuclear maturation can occur in the absence of carbohydrates (Zheng et al., 2001; Zheng et al., 2007). In the dog, neither pyruvate nor glutamine metabolism influences meiotic maturation, although both are metabolized by the COC (Songsasen et al., 2007). In pigs, metabolism of glucose through PPP and/or glycolysis is involved in the control of nuclear maturation (Herrick et al., 2006a). These findings highlight the importance of glucose metabolism in oocyte maturation across species, and the interactions between nuclear and cytoplasmic maturation.

Recently, fatty acid oxidation has been found to play a role in controlling meiotic maturation in mice, where inhibiting fatty acid oxidation results in meiotic arrest, whereas stimulating FAO promoted meiotic resumption (Downs et al., 2009). We have conducted experiments to evaluate the importance of FA during oocyte maturation in the cow, pig, and mouse, and found that FAO is essential to oocyte nuclear maturation in all three species (Paczkowski et al., 2011). Pig oocytes were the most sensitive to inhibition of FAO (10 uM etomoxir caused a significant reduction in mature metaphase II oocytes), whereas cow oocytes were intermediate (100 uM etomoxir), and mouse oocytes were the least sensitive (250 uM etomoxir). The sensitivity of oocyte nuclear maturation to inhibition of FAO parallels the amount of fatty acids in the cytoplasm of these three species, suggesting that the amount of intracellular lipids may indicate the relative importance of fatty acid metabolism to oocyte nuclear maturation. Additional studies on the relationship between oocyte fatty acid metabolism and nuclear and cytoplasmic maturation are underway.

9.11 Oocyte Metabolism and Redox Balance

The interplay between oocyte metabolism and redox balance is multifaceted, as metabolism both contributes to the accumulation of ROS as normal byproducts of metabolic processes and produces critical factors needed for antioxidant defense (Dumollard et al., 2009). The oocyte is flexible in the substrates that it utilizes for oxidative phosphorylation and ATP production, depending at least in part on the existing intracellular redox state (Cetica et al., 2003). In turn, changes in mitochondrial function result in changes in redox state (Dumollard et al., 2009). Redox balance has been implicated in proliferation, transformation, and apoptosis in multiple cell types, including oocytes and embryos (Dalvit et al., 2005; Harvey et al., 2002; Harvey et al., 2007; Morado et al., 2009). Expression of redox-related genes is associated with pig oocyte developmental competence (Yuan et al., 2012). Oocytes, like other cell types, neutralize ROS with a variety of antioxidant defense systems. Under conditions of increased ROS, the antioxidant defenses of the oocyte can become overwhelmed, leading to oxidative stress and reduced oocyte competence (Combelles et al., 2009). However, it is likely that physiological levels of ROS are required for cell signaling during oocyte maturation and are an important component of oocyte competence. Thus, the oocyte must control a delicate balance between oxidative stress and antioxidant defense. In pig oocytes, a maturation environment that is either too oxidized or too reduced results in decreased developmental competence (Yuan et al., 2012).

Multiple defense mechanisms exist within the oocyte, including glutathione (GSH) and superoxide dismutase. Metabolic pathways, such as PPP, produce cofactors such as NADPH that are necessary for these systems to function properly (Dumollard et al., 2007). In addition, the interconversion of lactate and pyruvate contribute significantly to the redox state of the cell by regulating NAD/NADH ratios (Dumollard et al., 2009). In fact, the oocyte utilizes pyruvate and PPP activity to regulate intracellular redox state and inhibit post-ovulatory oocyte aging (Li et al., 2011).

An important aspect of oocyte maturation is the accumulation of glutathione (GSH), one of the oocyte's primary antioxidant defenses. This tripepetide (Glutamate-Glycine-Cysteine) is involved in many aspects of oocyte physiology including cumulus cell expansion, sperm decondensation, male pronuclear formation, and embryonic development (cattle: de Matos et al., 1995; Furnus et al., 1998; Sutovsky & Schatten, 1997; sheep: de Matos et al., 2002; pigs: Yoshida, 1993; Yoshida et al., 1993); hamsters, (Zuelke et al., 2003)]. Intracellular concentrations of GSH can be maintained in two ways: reduction of GSSG by glutathione reductase or de novo synthesis of GSH. Synthesis of GSH, including the addition of cysteine to the glutamate-glycine dipepetide, are dependent on ATP, and therefore dependent on the mitochondrial metabolism of the oocyte, as well as on an adequate supply of cysteine (Dumollard et al., 2009). In the cell, GSH cycles between reduced (GSH) and oxidized (GSSG) forms through the actions of glutathione reductase and glutathione peroxidase. To minimize oxidative damage, the peroxidase transfers electrons from GSH to oxidized molecules within the cytoplasm, resulting in the production of GSSG (Guerin et al., 2001). The cellular pool of GSH is maintained by reduction of GSSG to GSH by glutathione reductase, as well as further GSH synthesis. Glutathione reductase then reduces of GSSG to GSH, which requires NADPH from the PPP. Therefore, intracellular concentrations of GSH, and the many processes affected by GSH, are tightly linked to the metabolic activity of the oocyte.

Glutathione concentrations increase during *in vitro* maturation in hamster, bovine, and porcine oocytes (Furnus et al., 2008; Yuan et al., 2012; Zuelke et al., 2003). Glucose metabolism and GSH accumulation are important for successful oocyte maturation and embryonic development (Abeydeera et al., 2000; Brad et al., 2003a; Herrick et al., 2003; Krisher et al., 2007; Sutton et al., 2003b). However, the interactions among nuclear maturation, glucose metabolism, GSH concentration, and embryonic development are not entirely clear. In bovine oocytes, the presence of glucose in the medium does not alter oocyte GSH content (Furnus et al., 2008). However, decreased GSH following inhibition of glucose metabolism in pigs demonstrates that these two critical aspects of oocyte maturation are interrelated (Herrick et al., 2006a). A key regulator in this relationship may be p53, which controls glycolytic activity, as well as regulating ROS levels and GSH synthesis via GLS2 and glutamine metabolism (Thompson et al., 2007).

9.12 The Relationship between Oocyte Metabolism and Oocyte Quality

The metabolic profiles of oocytes have been correlated with developmental competence in a number of species, further reinforcing the importance of oocyte metabolism. In both cats and cattle, the rate of glycolysis of the mature oocyte is positively correlated with its ability to develop to the blastocyst stage *in vitro* (Krisher & Bavister, 1999; Spindler et al., 2000), despite the fact that cumulus cells metabolize many times more glucose through this pathway than does the oocyte (Sutton et al., 2003b). Similarly, lactate production, indicative of glycolysis, at the beginning of maturation was positively correlated with developmental potential in bovine oocytes derived from different size follicles (Lequarre et al., 2005). In bovine oocytes, increased pyruvate (immature and mature oocytes) and lactate (mature oocytes) oxidation was also positively associated with the developmental potential. In pigs, PPP activity of *in vitro*–matured oocytes is lower than those matured *in vivo*, which have higher developmental competence (Krisher et al., 2007). Pharmacological stimulation of the pentose phosphate pathway in bovine oocytes resulted in increased blastocyst formation (Krisher, 1999), suggesting that this pathway is also important in cytoplasmic maturation. Cat oocytes that have been matured *in vitro* (low oocyte quality) have lower rates of glycolysis, glucose oxidation, palmitate oxidation, and reduced developmental potential compared to their *in vivo*–matured counterparts (Spindler et al., 2000). In sheep, metabolism of glutamine was higher in oocytes from adult ewes with better oocyte quality, as measured by blastocyst formation, compared to prepubertal ewes following maturation (O'Brien et al., 1996). Thus, most studies point to a clear association between oocyte metabolism and developmental competence, across species and oocyte quality models.

9.13 Maternal Diet and Disease Can Alter Oocyte Metabolism

In a rodent model of diabetes, the importance of metabolic regulation in the COC is highlighted. In this model, metabolic coupling and gap junctional communication in the COC are reduced, resulting in altered growth, maturation, mitochondrial function, and metabolism of oocytes from diabetic mice, and thus poor oocyte competence (Wang & Moley, 2010). Specifically, maternal diabetes alters oocyte metabolism in this mouse model by decreasing ATP levels and delaying GVBD due to altered AMPK activity (Ratchford et al., 2007). In conditions of hyperglycemia, the polyol pathway is activated, disrupting metabolism, gene expression and cellular communication (Sutton-McDowall et al., 2010). A similar condition is observed in obese mice, with alterations in insulin, glucose, and free fatty acid concentrations in the follicle, resulting in altered oocyte metabolism and reduced oocyte competence (Minge et al., 2008; Purcell & Moley, 2011). Oocytes from obese mice also have a higher mitochondrial membrane potential and an oxidized redox status (Igosheva et al., 2010). Treatment of mice with rosiglitazone, an insulin-sensitizing drug, resulted in weight loss and normalization of glucose metabolism, and ultimately improved oocyte quality, possibly through PPARγ-mediated mechanisms (Minge et al., 2008). Finally, in women with polycystic ovary syndrome, oocytes have increased glucose and pyruvate consumption during maturation, as well as abnormally high NADPH content (Harris et al., 2010). Metformin, an insulin-sensitizing drug, can lower the abnormal pyruvate consumption observed in these oocytes (Harris et al., 2010). In cumulus cells from high-fat–fed, insulin-resistant mice and PCOS women, impaired insulin-stimulated glucose uptake was observed, demonstrating that insulin resistance occurs in cumulus cells under systemic insulin-resistant conditions (Purcell et al., 2012). In addition, *in vivo* hyperglycemia and *in vitro* culture in high-glucose conditions result in elevated glucose levels within the oocyte, mediated by gap junction transport (Wang et al., 2012). These findings suggest that in obese or insulin-resistant disease states oocyte metabolism, and thus competence, may be negatively affected.

Dietary fatty acids may also affect oocyte quality. In dairy cows, a high-fat diet improved blastocyst development and quality, whereas a low-fat diet resulted in reduced blastocyst development and increased non-esterified fatty acids (NEFA) concentrations (Fouladi-Nashta et al., 2007). Cows fed a diet high in flax, resulting in a lowered n6:n3 ratio, produced embryos with increased cleavage competence compared to controls (Zachut et al., 2010). In contrast, another study did not detect any effect of feeding heifers omega-3 polyunsaturated fatty acids (PUFA) and reducing n3:n6 ratios on superovulation, embryo recovery, embryo quality, or embryo gene expression (Childs et al., 2008). In sheep, omega-3 and -6 fatty acids, when fed prior to ovarian stimulation, can alter the fatty acid composition of both granulosa cells and oocytes, and a diet high in n-6 PUFA reduced oocyte competence (Wonnacott et al., 2010). Supplementation of omega-3 fatty acids during the preconception period improves embryo morphology after IVF in women (Hammiche et al., 2011). In mice, feeding a diet high in omega-3 fatty acids resulted in altered mitochondrial distribution and increased ROS in oocytes after IVM, as well as lower blastocyst development after IVF (Wakefield et al., 2008). Cows supplemented with unsaturated fatty acids had improved embryo development compared to those fed saturated fatty acids (Cerri et al., 2009; Thangavelu et al., 2007). However, culturing embryos with PUFA enriched serum and albumin increases the proportion of embryos with poor morphology, and increases oxidative stress in the resulting embryos (Hughes et al., 2011). In dairy heifers, body condition score (BCS) affects how diet affects oocyte quality. When the heifers were fed a diet high in starch and fatty acids, fewer blastocysts were recovered from low-BCS heifers, but not moderate-BCS heifers (Adamiak et al., 2006). In ewes, peri-conception nutrition altered cellular allocation in blastocysts (Kakar et al., 2005).

9.14 Oocytes and the Warburg Effect

Recently, we proposed a novel model for oocyte and embryo metabolism (Krisher & Prather, 2012). Typically in cellular metabolism, glucose is metabolized through glycolysis to pyruvate, which then enters the TCA cycle and is oxidized to produce ATP. An alternative form of metabolism is utilized by cancer cells and is known as the Warburg Effect (WE), or aerobic glycolysis (Warburg, 1956). In this metabolic strategy, pyruvate is directed away from the TCA cycle and metabolized to lactate. Importantly, conversion of pyruvate to lactate produces NAD^+, supporting further glycolytic activity. Although this is typically an anaerobic process, cancer cells utilize this pathway even when oxygen is abundant. This is an inefficient way to generate ATP, particularly when compared to oxidative phosphorylation. However, mitochondria are still functional, although they may use alternative substrates (Cairns et al., 2011; Locasale & Cantley, 2010). In this strategy, glucose metabolism and oxidative phosphorylation are independent (Locasale & Cantley 2010). Of particular interest, WE is not unique to cancer cells, but is observed in many rapidly proliferating cell types (Lopez-Lazaro, 2008).

Proliferating cells may utilize WE to meet needs other than ATP production. The requirement for macromolecular synthesis to create new biomass, including DNA, proteins, and lipids, may be a critical need in these rapidly dividing cell types. Production of ribose-5-phosphate (R5P) for nucleic acid synthesis, fatty acids (FA) for lipid synthesis, and redox control may comprise the most important outcomes of glucose metabolism in these cells (Cairns et al., 2011). These processes use glucose as a carbon source, consume TCA cycle intermediates such that they must be replenished, and require NADPH as reductive power (Deberardinis et al., 2008). To produce R5P and NADPH, glucose is shunted into the PPP. In short, the majority of cellular carbon is required for biosynthesis rather than ATP production (Vander Heiden et al., 2009). Thus, utilization of WE gives proliferating cells a selective growth advantage. In this strategy, adequate alternative energy sources, such as FA or amino acids, must be available to support basal TCA activity and ATP production. In addition to an increased use of glucose, cancer cells also take up and metabolize high levels of glutamine, as do oocytes (Rieger & Loskutoff, 1994). Glutamine can replenish TCA cycle intermediates potentially removed for fatty acid synthesis, to produce NADPH via malic enzyme (ME), a process called glutaminolysis, and for GSH synthesis (Deberardinis et al., 2008; Vousden & Ryan, 2009).

We hypothesize that oocytes may utilize a metabolic strategy similar to WE, in preparation for rapid embryonic growth after fertilization. In this scenario, aerobic glycolysis would generate glycolytic intermediates for R5P and NADPH production, as well as glutaminolysis for NADPH generation. Pyruvate is converted to lactate to maintain NAD^+ levels to support elevated glycolysis. The TCA cycle and oxidative phosphorylation are active, relying on FAO to generate ATP as well as replenish cycle intermediates. Although glucose uptake is high, control mechanisms actually slow glycolysis to increase intermediates, which are then shuttled to other metabolic fates. Utilization of WE by oocytes, and embryos, is a tantalizing possibility, and one that will require further investigation.

9.15 Conclusions

Despite our knowledge of metabolic pathways within the oocyte, and the well-established hypothesis that metabolism is related to oocyte competence, few attempts have been made to identify and meet the specific metabolic needs of oocytes from domestic species *in vitro*, in order to produce a metabolic profile more like that of an *in vivo*–matured oocyte. Particularly *in vitro*, understanding oocyte metabolism and metabolic regulation is critical, as the health and competence of the oocyte following *in vitro* maturation sets the stage for the success or failure of all future events in preimplantation embryo development. Although we are typically very effective in supporting resumption and completion of meiosis, we are less successful in sustaining developmental competence during IVM. Further investigations into the metabolic needs and control mechanisms within the cumulus cells, as well as the relationships between oocyte and cumulus cell metabolism, will facilitate the development of media that satisfies the needs of both cell types. This may consist of a dynamic approach to oocyte maturation, including replenishment of media with adjusted nutrients during the maturation period to meet the changing needs of the COC (Thompson et al., 2007). Such novel approaches will likely result in more normal oocyte metabolism. A better understanding of and appreciation for oocyte metabolism should lead to development of maturation systems that better support oocyte developmental competence.

References

Abeydeera, L. R. (2002). In vitro production of embryos in swine. *Theriogenology, 57*(1), 256–273.

Abeydeera, L. R., Wang, W, H., Cantley, T. C., Rieke, A., Murphy, C. N. et al. (2000). Development and viability of pig oocytes matured in a protein-free medium containing epidermal growth factor. *Theriogenology, 54*, 787–797.

Adamiak, S. J., Powell, K., Rooke, J. A., Webb, R., & Sinclair, K. D. (2006). Body composition, dietary carbohydrates and fatty acids determine post-fertilisation development of bovine oocytes in vitro. *Reproduction, 131*(2), 247–258.

Bae, I. H., & Foote, R. H. (1975). Carbohydrate and amino acid requirements and ammonia production of rabbit follicular oocytes matured in vitro. *Experimental Cell Research, 91*, 113–118.

Bentov, Y., Esfandiari, N., Burstein, E., & Casper, R. F. (2010). The use of mitochondrial nutrients to improve the outcome of infertility treatment in older patients. *Fertility and Sterility, 93*(1), 272–275.

Bermejo-Alvarez, P., Lonergan, P., Rizos, D., & Gutierrez-Adan, A. (2010). Low oxygen tension during IVM improves bovine oocyte competence and enhances anaerobic glycolysis. *Reproductive BioMedicine Online, 20*(3), 341–349.

Biggers, J. D., D. G., W., & Donahue, R. P. (1967). The pattern of energy metabolism in the mouse oocyte and zygote. *The Proceedings of the National Academy of Sciences, 58*, 560–567.

Bilodeau-Goeseels, S. (2006). Effects of culture media and eneregy sources on the inhibition of nuclear maturation in bovine oocytes. *Theriogenology, 66*, 297–306.

Brad, A. M., Bormann, C. L., Swain, J. E., Durkin, R. E., Johnson, A. E. et al. (2003a). Glutathione and adenosine triphosphate content of in vivo and in vitro matured porcine oocytes. *Molecular Reproduction and Development, 64*(4), 492–498.

Brad, A. M., Herrick, J. R., Lane, M., Gardner, D. K., & Krisher, R. L. (2003b). Glucose and lactate concentrations affect the metabolism of in vitro matured porcine oocytes. *Biology of Reproduction, 68*(Suppl. 1), 356.

Cairns, R. A., Harris, I. S., & Mak, T. W. (2011). Regulation of cancer cell metabolism. *Nature Reviews Cancer, 11*(2), 85–95.

Cerri, R. L. A., Juchem, S. O., Chebel, R. C., Rutigliano, H. M., Bruno, R. G. S. et al. (2009). Effect of fat source differing in fatty acid profile on metabolic parameters, fertilization, and embryo quality in high-producing dairy cows. *Journal of Dairy Science, 92*(4), 1520–1531.

Cetica, P., Pintos, L., Dalvit, G., & Beconi, M. (2002). Activity of key enzymes involved in glucose and triglyceride catabolism during bovine oocyte maturation in vitro. *Reproduction, 124*, 675–681.

Cetica, P., Pintos, L., Dalvit, G., & Beconi, M. (2003). Involvement of enzymes of amino acid metabolism and tricarboxylic acid cycle in bovine oocyte maturation in vitro. *Reproduction, 126*, 753–763.

Cetica, P., Pintos, L. N., Dalvit, G. C., & Beconi M. T. (1999). Effect of lactate dehydrogenase activity and isoenzyme localization in bovine oocytes and utilization of oxidative substrates on on vitro maturation. *Theriogenology, 51*, 541–550.

Chang, S. C. S., Jones, J. D., Ellefson, R. D., & Ryan, R. J. (1976). The porcine ovarian follicle: I. Selected chemical analysis of folicular fluid at different developmental stages. *Biology of Reproduction, 15*, 321–328.

Childs, S., Carter, F., Lynch, C. O., Sreenan, J. M., Lonergan, P. et al. 2008. Embryo yield and quality following dietary supplementation of beef heifers with n-3 polyunsaturated fatty acids (PUFA). *Theriogenology, 70*(6), 992–1003.

Clark, A. R., Stokes, Y. M., Lane, M., & Thompson, J. G. (2006). Mathematical modelling of oxygen concentration in bovine and murine cumulus-oocyte complexes. *Reproduction, 131*(6), 999–1006.

Combelles, C. M. H., Gupta, S., & Agarwal, A. (2009). Could oxidative stress influence the in-vitro maturation of oocytes? *Reproductive BioMedicine Online, 18*(6), 864–880.

Crozet, N., Ahmed-Ali, M., & Dubos, M. P. (1995). Developmental competence of goat oocytes from follicles of different size categories following maturation, fertilization and culture in vitro. *Journal of Reproduction and Fertility, 103*(2), 293–298.

Cui, X. S., Li, X. Y., Yin, X. J., Kong, I. K., Kang, J. J., & Kim, N. H. (2007). Maternal gene transcription in mouse oocytes: genes implicated in oocyte maturation and fertilization. *Journal of Reproduction and Development, 53*(2), 405–418.

Dalvit, G. C., Cetica, P. D., Pintos, L. N., & Beconi, M. T. (2005). Reactive oxygen species in bovine embryo in vitro production. *Biocell, 29*(2), 209–212.

de Matos, D. G., Furnus, C. C., Moses, D. F., & Baldassarre, H. (1995). Effect of cysteamine on glutathione level and developmental capacity of bovine oocyte matured in vitro. *Molecular Reproduction and Development, 42*(4), 432–436.

de Matos, D. G., Gasparrini, B., Pasqualini, S. R., & Thompson, J. G. (2002). Effect of glutathione synthesis stimulation during in vitro maturation of ovine oocytes on embryo development and intracellular peroxide content. *Theriogenology, 57*(5), 1443–1451.

Deberardinis, R. J., Sayed, N., Ditsworth, D., & Thompson, C. B. (2008). Brick by brick: metabolism and tumor cell growth. *Current Opinion in Genetics and Development, 18*(1), 54–61.

Downs, S. M. (1993). Purine control of mouse oocyte maturation: evidence that nonmetabolized hypoxanthine maintains meiotic arrest. *Molecular Reproduction and Development, 35*(1), 82–94.

Downs, S. M. (1995). The influence of glucose, cumulus cells, and metabolic coupling on ATP levels and meiotic control in the isolated mouse oocyte. *Developmental Biology, 167*(2), 502–512.

Downs, S. M. (1997). Involvement of purine nucleotide synthetic pathways in gonadotropin-induced meiotic maturation in mouse cumulus cell-enclosed oocytes. *Molecular Reproduction and Development, 46*(2), 155–167.

Downs, S. M. (1998). Precursors of the purine backbone augment the inhibitory action of hypoxanthine and dibutyryl cAMP on mouse oocyte maturation. *Journal of Experimental Zoology, 282*(3), 376–384.

Downs, S. M., Coleman, D. L., & Eppig, J. J. (1986). Maintenance of murine oocyte meiotic arrest: uptake and metabolism of hypoxanthine and adenosine by cumulus cell-enclosed and denuded oocytes. *Developmental Biology, 117*(1), 174–183.

Downs, S. M., Houghton, F. D., Humpherson, P. G., & Leese, H. J. (1997). Substrate utilization and maturation of cumulus cell-enclosed mouse oocytes: evidence that pyruvate oxidation does not mediate meiotic induction. *Journal of Reproduction and Fertility, 110*(1), 1–10.

Downs, S. M. & Hudson, E. D. (2000). Energy substrates and the completion of spontaneous meiotic maturation. *Zygote, 8*, 339–351.

Downs, S. M., Humpherson, P. G., & Leese, H. J. (1998). Meiotic induction in cumulus cell-enclosed mouse oocytes: involvement of the pentose phosphate pathway. *Biology of Reproduction, 58*(4), 1084–1094.

Downs, S. M., Humpherson, P. G., & Leese, H. J. (2002). Pyruvate utilization by mouse oocytes is influenced by meiotic status and the cumulus oophorus. *Molecular Reproduction and Development, 62*, 113–123.

Downs, S. M., Humpherson, P. G., Martin, K. L., & Leese, H. J. (1996). Glucose utilization during gonadotropin-induced meiotic maturation in cumulus cell-enclosed mouse oocytes. *Molecular Reproduction and Development, 44*(1), 121–131.

Downs, S. M. & Mastropolo, A. M. (1994). The participation of energy substrates in the control of meiotic maturation in murine oocytes. *Delopmental Biology, 162*(1), 154–168.

Downs, S. M., Mosey, J. L., & Klinger, J. (2009). Fatty acid oxidation and meiotic resumption in mouse oocytes. *Molecular Reproduction and Development, 76*(9), 844–853.

Downs, S. M. & Utecht, A. M. (1999). Metabolism of radiolabeled glucose by mouse oocytes and oocyte-cumulus cell complexes. *Biology of Reproduction, 60*, 1446–1452.

Downs, S. M. & Verhoeven, A. (2003). Glutamine and the maintenance of meiotic arrest in mouse oocytes: influence of culture medium, glucose, and cumulus cells. *Molecular Reproduction and Development, 66*, 90–97.

Dumollard, R., Carroll, J., Duchen, M. R., Campbell, K., & Swann, K. (2009). Mitochondrial function and redox state in mammalian embryos. *Seminars in Cell and Developmental Biology, 20*(3), 346–353.

Dumollard, R., Duchen, M., & Sardet, C. (2006). Calcium signals and mitochondria at fertilisation. *Seminars in Cell and Developmental Biology, 17*(2), 314–323.

Dumollard, R., Ward, Z., Carroll, J., & Duchen, M. R. (2007). Regulation of redox metabolism in the mouse oocyte and embryo. *Development, 134*(3), 455–465.

Dunning, K. R., Akison, L. K., Russell, D. L., Norman, R. J., & Robker, R. L. (2011). Increased beta-oxidation and improved oocyte developmental competence in response to l-carnitine during ovarian in vitro follicle development in mice. *Biology of Reproduction, 85*(3), 548–555.

Dunning, K. R., Cashman, K., Russell, D. L., Thompson, J. G., Norman, R. J., & Robker, R. L. (2010). Beta-oxidation is essential for mouse oocyte developmental competence and early embryo development. *Biology of Reproduction, 83*(6), 909–918.

Edwards, L. J., Williams, D. A., & Gardner, D. K. (1998). Intracellular pH of the preimplantation mouse embryo: effects of extracellular pH and weak acids. *Molecular Reproduction and Developpment, 50*, 434–442.

Eppig, J. J. (1991). Intercommunication between mammalian oocytes and companion somatic cells. *Bioessays, 13*(11), 569–574.

Eppig, J. J. (1996). Coordination of nuclear and cytoplasmic oocyte maturation in eutherian mammals. *Reproduction Fertility and Development, 8*(4), 485–489.

Fagbohun, C. F. & Downs, S. M. (1992). Requirement for glucose in ligand-stimulated meiotic maturation of cumulus cell-enclosed mouse oocytes. *Journal of Reproduction and Fertility, 96*(2), 681–697.

Ferguson, E. M. & Leese, H. J. (1999). Triglyceride content of bovine ocytes and early embryos. *Journal of Reproduction and Fertility, 116*, 373–378.

Ferguson, E. M. & Leese, H. J. (2006). A potential role for triglyceride as an energy source during bovine oocyte maturation and early embryo development. *Molecular Reproduction and Development, 73*, 1195–1201.

Ferreira. C. R., Saraiva, S. A., Catharino, R. R., Garcia, J. S., Gozzo, F. C. et al. (2010). Single embryo and oocyte lipid fingerprinting by mass spectrometry. *Journal of Lipid Research, 51*(5), 1218–1227.

Flood, M. R. & Wiebold, J. L. (1988). Glucose metabolism by preimplantation pig embryos. *Journal of Reproduction and Fertility, 84*, 7–12.

Fouladi-Nashta, A. A., Gutierrez, C. G., Gong, J. G., Garnsworthy, P. C., & Webb, R. (2007). Impact of dietary fatty acids on oocyte quality and development in lactating dairy cows. *Biology of Reproduction, 77*(1), 9–17.

Furnus, C. C., de Matos, D. G., & Moses, D. F. (1998). Cumulus expansion during in vitro maturation of bovine oocytes: relationship with intracellular glutathione level and its role on subsequent embryo development. *Molecular Reproduction and Development, 51*(1), 76–83.

Furnus, C. C., de Matos, D. G., Picco, S., Garcia, P. P., Inda, A. M. et al. (2008). Metabolic requirements associated with GSH synthesis during in vitro maturation of cattle oocytes. *Animal Reproduction Science, 109*(1-4), 88–99.

Gardner, D. K. (1998). Changes in requirements and utilization of nutrients during mammalian preimplantation embryo development and their significance in embryo culture. *Theriogenology, 49*(1), 83–102.

Gardner, D. K., Lane, M., Calderon, I., & Leeton, J. (1996). Environment of the preimplantation human embryo in vivo: metabolite analysis of oviduct and uterine fluids and metabolism of cumulus cells.[see comment]. *Fertility and Sterility, 65*(2), 349–353.

Gardner, D. K., Lane, M. W., & Lane, M. (2000). EDTA stimulates cleavage stage bovine embryo development in culture but inhibits blastocyst development and differentiation. *Molecular Reproduction and Development, 57*(3), 256–261.

Gardner, D. K. & Leese, H. J. (1990). Concentrations of nutrients in mouse oviduct fluid and their effects on embryo development and metabolism in vitro. *Journal of Reproduction and Fertility, 88*(1), 361–368.

Gentile, L., Monti, M., Sebastiano, V., Merico, V., Nicolai, R. et al. (2004). Single-cell quantitative RT-PCR analysis of Cpt1b and Cpt2 gene expression in mouse antral oocytes and in preimplantation embryos. *Cytogenetic and Genome Research, 105*(2-4), 215–221.

Guerin, P., Mouatassim, S. E., & Menezo, Y. (2001). Oxidative stress and protection against reactive oxygen species in the preimplantation embryo and its surroundings. *Human Reproduction Update, 7*(2), 175–189.

Haggarty, P., Wood, M., Ferguson, E., Hoad, G., Srikantharajah, A. et al. (2006). Fatty acid metabolism in human preimplantation embryos. *Human Reproduction, 21*(3), 766–773.

Hammiche, F., Vujkovic, M., Wijburg, W., de Vries, J. H. M., Macklon, N. S. et al. (2011). Increased preconception omega-3 polyunsaturated fatty acid intake improves embryo morphology. *Fertility and Sterility, 95*(5), 1820–1823.

Harris, S. E., Gopichandran, N., Picton, H. M., Leese, H. J., & Orsi, N. M. (2005). Nutrient concentrations in murine follicular fluid and the female reproductive tract. *Theriogenology, 64*(4), 992–1006.

Harris, S. E., Maruthini, D., Tang, T., Balen, A. H., & Picton, H. M. (2010). Metabolism and karyotype analysis of oocytes from patients with polycystic ovary syndrome. *Human Reproduction, 25*(9), 2305–2315.

Harvey, A. J., Kind, K. L., & Thompson, J. G. (2002). REDOX regulation of early embryonic development. *Reproduction, 123*, 479–486.

Harvey, A. J., Kind, K. L., & Thompson, J. G. (2007). Regulation of gene expression in bovine blastocysts in response to oxygen and the iron chelator desferrioxamine. *Biology of Reproduction, 77*(1), 93–101.

Hashimoto, S., Minami, N., Takakura, R., Yamada, M., Imai, H., & Kashima, N. (2000a). Low oxygen tension during in vitro maturation is beneficial for supporting the subsequent development of bovine cumulus oocyte complexes. *Molecular Reproduction and Development, 57*, 353–360.

Hashimoto, S., Minami, N., Yamada, M., & Imai, H. (2000b). Excessive concentration of glucose during in vitro maturation impairs developmental competence of bovine oocytes after in vitro fertilization: relevance to intracellular reactive oxygen species and glutathione contents. *Molecular Reproduction and Development, 56*, 520–526.

Heller, D. T., Cahill, D. M., & Schultz, R. M. (1981). Biochemical studies of mammalian oogenesis: metabolic cooperativity between granulosa cells and growing mouse oocytes. *Developmental Biology, 84*(2), 455–464.

Hemmings, K. E., J., L. H., & Picton, H. M. (2012). Amino acid turnover by bovine oocytes provides an index of oocyte developmental competence in vitro. *Biology of Reproduction, 86*(5), 165, 161–112.

Herrick, J. R., Brad, A. M., & Krisher, R. L. (2006a). Chemical manipulation of glucose metabolism in porcine oocytes: effects on nuclear and cytoplasmic maturation in vitro. *Reproduction, 131*(2), 289–298.

Herrick, J. R., Brad, A. M., Krisher, R. L., & Pope, W. F. (2003). Intracellular adenosine triphosphate and glutathione concentrations in oocytes from first estrous, multi-estrous, and testosterone-treated gilts. *Animal Reproduction Science, 78*(1-2), 123–131.

Herrick, J. R., Lane, M., Gardner, D. K., Behboodi, E., Memili, E. et al. (2006b). Metabolism, protein content, and in vitro embryonic development of goat cumulus-oocyte complexes matured with physiological concentrations of glucose and L-lactate. *Molecular Reproduction and Development, 73*(2), 256–266.

Hewitson, L. C., Martin, K. L., & Leese, H. J. (1996). Effects of metabolic inhibitors on mouse preimplantation embryo development and the energy metabolism of isolated inner cell masses. *Molecular Reproduction and Development, 43*(3), 323–330.

Homa, S. T. & Brown, C. A. (1992). Changes in linoleic acid during follicular development and inhibition of spontaneous breakdown of germinal vesicles in cumulus-free bovine oocytes. *Journal of Reproduction and Fertility, 94*(1), 153–160.

Homa, S. T., Racowsky, C., & McGaughey, R. W. (1986). Lipid analysis of immature pig oocytes. *Journal of Reproduction and Fertility, 77*(2), 425–434.

Hughes, J., Kwong, W. Y., Li, D., Salter, A. M., Lea, R. G., & Sinclair, K. D. (2011). Effects of omega-3 and -6 polyunsaturated fatty acids on ovine follicular cell steroidogenesis, embryo development and molecular markers of fatty acid metabolism. *Reproduction, 141*(1), 105–118.

Igosheva, N., Abramov, A. Y., Poston, L., Eckert, J. J., Fleming, T. P. et al. (2010). Maternal diet-induced obesity alters mitochondrial activity and redox status in mouse oocytes and zygotes. *PLoS ONE [Electronic Resource], 5*(4), e10074.

Iwata, H., Hashimoto, S., Ohota, M., Kimura, K., Shibano, K., & Miyake, M. (2004). Effects of follicle size and electrolytes and glucose in maturation medium on nuclear maturation and developmental competence of bovine oocytes. *Reproduction, 127*(2), 159–164.

Iwata, H., Inoue, J., Kimura, K., Kuge, T., Kuwayama, T., & Monji, Y. (2006). Comparison between the characteristics of follicular fluid and the developmental competence of bovine oocytes. *Animal Reproduction Science, 91*(3–4), 215–223.

Jeong, W. J., Cho, S. J., Lee, H. S., Deb, G. K., Lee, Y. S. et al. (2009). Effect of cytoplasmic lipid content on in vitro developmental efficiency of bovine IVP embryos. *Theriogenology, 72*(4), 584–589.

Johnson, M. T., Freeman, E. A., Gardner, D. K., & Hunt, P. A. (2007). Oxidative metabolism of pyruvate is required for meiotic maturation of murine oocytes in vivo. *Biology of Reproduction, 77*(1), 2–8.

Jungheim, E. S., Macones, G. A., Odem, R. R., Patterson, B. W., Lanzendorf, S. E. et al. (2011). Associations between free fatty acids, cumulus oocyte complex morphology and ovarian function during in vitro fertilization. *Fertility and Sterility, 95*(6), 1970–1974.

Kakar, M. A., Maddocks, S., Lorimer, M. F., Kleemann, D. O., Rudiger, S. R. et al. (2005). The effect of peri-conception nutrition on embryo quality in the superovulated ewe. *Theriogenology, 64*(5), 1090–1103.

Katz-Jaffe, M. G., McCallie, B. R., Preis, K. A., Filipovits, J., & Gardner, D. K. (2009). Transcriptome analysis of in vivo and in vitro matured bovine MII oocytes. *Theriogenology, 71*(6), 939–946.

Khurana, N. K. & Niemann, H. (2000). Energy metabolism in preimplantation bovine embryps derived in vitro or in vivo. *Biolgy of Reproduction, 62*, 847–856.

Kim, J. Y., Kinoshita, M., Ohnishi, M., & Fukui, Y. (2001). Lipid and fatty acid analysis of fresh and frozen-thawed immature and in vitro matured bovine oocytes. *Reproduction, 122*(1), 131–138.

Krisher, R. (1999). Exposure of bovine oocytes to perturbants of the pentose phosphate pathway affects subsequent embryonic development. *Biology of Reproduction, 60* (Suppl. 1), 390.

Krisher, R. L. & Bavister, B. D. (1998). Responses of oocytes and embryos to the culture environment. *Theriogenology, 49*(1), 103–114.

Krisher, R. L. & Bavister, B. D. (1999). Enhanced glycolysis after maturation of bovine oocytes in vitro is associated with increased developmental competence. *Molecular Reproduction and Development, 53*(1), 19–26.

Krisher, R. L., Brad, A. M., Herrick, J. R., Sparman, M. L., & Swain, J. E. (2007). A comparative analysis of metabolism and viability in porcine oocytes during in vitro maturation. *Animal Reproduction Science, 98*(1-2), 72–96.

Krisher, R. L. & Prather, R. S. (2012). A Role for the Warburg Effect in Preimplantation Embryo Development:Metabolic Modification to Support Rapid Cell Proliferation. *Molecular Reproduction and Development, 79*, 311–320.

Lane, M. & Gardner, D. K. (1998). Amino acids and vitamins prevent culture-induced metabolic perturbations and associated loss of viability of mouse blastocysts. *Human Reproduction, 13*, 991–997.

Lane, M. & Gardner, D. K. (2000a). Lactate regulates pyruvate uptake and metabolism in the preimplantation mouse embryo. *Biology of Reproduction, 62*, 16–22.

Lane, M. & Gardner, D. K. (2000b). Regulation of ionic homeostasis by mammalian embryos. *Seminars in Reproductive Medicine, 18*(2), 195–204.

Lane, M. & Gardner, D. K. (2001). Inhibiting 3-phosphoglycerate kinase by EDTA stimulates the development of the cleavage stage mouse embryo. *Molecular Reproduction and Development, 60*, 233–240.

Lane, M. & Gardner, D. K. (2005). Mitochondrial malate-aspartate shuttle regulates mouse embryo nutrient consumption. *Journal of Biological Chemistry, 280*(18), 18361–18367.

Lee, S. T., Oh, S. J., Lee, E. J., Han, H. J., & Lim, J. M. (2006). Adenosine triphosphate synthesis, mitochondrial number and activity, and pyruvate uptake in oocytes after gonadotropin injections. *Fertility and Sterility, 86*(4 Suppl), 1164–1169.

Leese, H. J. & Barton, A. M. (1984). Pyruvate and glucose uptake by mouse ova and preimplantation embryos. *Journal of Reproduction and Fertililty, 72*(1), 9–13.

Leese, H. J. & Barton, A. M. (1985). Production of pyruvate by isolated mouse cumulus cells. *Journal of Experimental Zoology, 234*(2), 231–236.

Leese, H. J. & Lenton, E. A. (1990). Glucose and lactate in human follicular fluid: concentrations and interrelationships. *Human Reproduction, 5*(8), 915–919.

Lequarre, A-S., Vigneron, C., Ribaucour, F., Holm, P., Donnay, I. et al. (2005). Influence of antral follicle size on oocyte characteristics and embryo development in the bovine. *Theriogenology, 63*(3), 841–859.

Leroy, J. L. M. R., Genicot, G., Donnay, I., & Van Soom, A. (2005a). Evaluation of the lipid content in bovine oocytes and embryos with nile red: a practical approach. *Reproduction in Domestic Animals, 40*(1), 76–78.

Leroy, J. L. M. R., Opsomer, G., Van Soom, A., Goovaerts, I. G. F., & Bols, P. E. J. (2008a). Reduced fertility in high-yielding dairy cows: are the oocyte and embryo in danger? Part I. The importance of negative energy balance and altered corpus luteum function to the reduction of oocyte and embryo quality in high-yielding dairy cows. *Reproduction in Domestic Animals, 43*(5), 612–622.

Leroy, J. L. M. R., Vanholder, T., Delanghe, J. R., Opsomer, G., Van Soom, A. et al. (2004). Metabolite and ionic composition of follicular fluid from different-sized follicles and their relationship to serum concentrations in dairy cows. *Animal Reproduction Science, 80*(3-4), 201–211.

Leroy, J. L. M. R., Vanholder, T., Mateusen, B., Christophe, A., Opsomer, G. et al. (2005b). Non-esterified fatty acids in follicular fluid of dairy cows and their effect on developmental capacity of bovine oocytes in vitro. *Reproduction, 130*, 485–495.

Leroy, J. L. M. R., Vanholder, T., Opsomer, G., Van Soom, A., & de Kruif, A. (2006). The in vitro development of bovine oocytes after maturation in glucose and beta-hydroxybutyrate concentrations associated with negative energy balance in dairy cows. *Reproduction in Domestic Animals, 41*(2), 119–123.

Leroy, J. L. M. R., Vanholder, T., Van Knegsel, A. T. M., Garcia-Ispierto, I., & Bols, P. E. J. (2008b). Nutrient prioritization in dairy cows early postpartum: mismatch between metabolism and fertility? *Reproduction in Domestic Animals, 43*(Suppl 2), 96–103.

Li, Q., Miao, D-Q., Zhou, P., Wu, Y-G., Gao, D. et al. (2011). Glucose metabolism in mouse cumulus cells prevents oocyte aging by maintaining both energy supply and the intracellular redox potential. *Biology of Reproduction, 84*(6), 1111–1118.

Locasale, J. W. & Cantley, L. C. (2010). Altered metabolism in cancer. *BMC Biology, 8*, 88.

Lopez-Lazaro, M. (2008). The warburg effect: why and how do cancer cells activate glycolysis in the presence of oxygen? *Anti-Cancer Agents in Medicinal Chemistry, 8*(3), 305–312.

Marei, W. F., Wathes, D. C., & Fouladi-Nashta, A. A. (2010). Impact of linoleic acid on bovine oocyte maturation and embryo development. *Reproduction, 139*(6), 979–988.

Matorras, R., Ruiz, J. I., Mendoza, R., Ruiz, N., Sanjurjo, P., & Rodriguez-Escudero, F. J. (1998). Fatty acid composition of fertilization-failed human oocytes. *Human Reproduction, 13*(8), 2227–2230.

May-Panloup, P., Chretien, M-F., Malthiery, Y., & Reynier, P. (2007). Mitochondrial DNA in the oocyte and the developing embryo. *Current Topics in Developmental Biology, 77*, 51–83.

McEvoy, T. G., Coull, G. D., Broadbent, P. J., Hutchinson, J. S., & Speake, B. K. (2000). Fatty acid composition of lipids in immature cattle, pig and sheep oocytes with intact zona pellucida. *Journal of Reproduction and Fertility, 118*(1), 163–170.

Minge, C. E., Bennett, B. D., Norman, R. J., & Robker, R. L. (2008). Peroxisome proliferator-activated receptor-gamma agonist rosiglitazone reverses the adverse effects of diet-induced obesity on oocyte quality. *Endocrinology, 149*(5), 2646–2656.

Mitchell, M., Cashman, K. S., Gardner, D. K., Thompson, J. G., & Lane, M. (2009). Disruption of mitochondrial malate-aspartate shuttle activity in mouse blastocysts impairs viability and fetal growth. *Biology of Reproduction, 80*(2), 295–301.

Mito, T., Yoshioka, K., Nagano, M., Suzuki, C., Yamashita, S., & Hoshi, H. (2009). Transforming growth factor-alpha in a defined medium during in vitro maturation of porcine oocytes improves their developmental competence and intracellular ultrastructure. *Theriogenology, 72*(6), 841–850.

Morado, S. A., Cetica, P. D., Beconi, M. T., & Dalvit, G. C. (2009). Reactive oxygen species in bovine oocyte maturation in vitro. *Reproduction Fertility and Development, 21*(4), 608–614.

Nichol, R., Hunter, R. H., Gardner, D. K., Leese, H. J., & Cooke, G. M. (1992). Concentrations of energy substrates in oviductal fluid and blood plasma of pigs during the peri-ovulatory period. *Journal of Reproduction and Fertility, 96*(2), 699–707.

O'Brien, J. K., Dwarte, D., Ryan, J. P., & Maxwell, W. M. C., & Evans, G. (1996). Developmental capacity, energy metabolism and ultrastructure of mature oocytes from prepubertal and adult sheep. *Reproduction Fertililty and Development, 8*, 1029–1037.

Orsi, N. M., Gopichandran, N., Leese, H. J., Picton, H. M., & Harris, S. E. (2005). Fluctuations in bovine ovarian follicular fluid composition throughout the oestrous cycle. *Reproduction, 129*(2), 219–228.

Oyamada, T. & Fukui, Y. (2004). Oxygen tension and medium supplements for in vitro maturation of bovine oocytes cultured individually in a chemically defined medium. *Journal of Reproduction and Development, 50*(1), 107–117.

Paczkowski, M., Silva, E., Schoolcraft, W. B., & Krisher, R. L. (2011). Comparative importance of fatty acid oxidation to oocyte nuclear maturation in murine, bovine and porcine species. *Proceedings, Society for the Study of Reproduction.* Portland, OR, USA. p 598.

Preis, K. A., Seidel, G. E., Jr., & Gardner, D. K. (2007). Reduced oxygen concentration improves the developmental competence of mouse oocytes following in vitro maturation. *Molecular Reproduction and Development, 74*(7), 893–903.

Purcell, S. H., Chi, M. M., & Moley, K. H. (2012). Insulin-stimulated glucose uptake occurs in specialized cells within the cumulus oocyte complex. *Endocrinology, 153*(5), 2444–2454.

Purcell, S. H. & Moley, K. H. (2011). The impact of obesity on egg quality. *Journal of Assisted Reproduction and Genetics, 28*(6), 517–524.

Ratchford, A. M., Chang, A. S., Chi, M. M. Y., Sheridan, R., & Moley, K. H. (2007). Maternal diabetes adversely affects AMP-activated protein kinase activity and cellular metabolism in murine oocytes. *American Journal of Physiology Endocrinology and Metabolism, 293*(5), E1198–1206.

Rieger, D. & Loskutoff, N. M. (1994). Changes in the metabolism of glucose, pyruvate, glutamine and glycine during maturation of cattle oocytes in vitro. *Journal of Reproduction and Fertility, 100*(1), 257–262.

Roberts, R., Stark, J., Iatropoulou, A., Becker, D. L., Franks, S., & Hardy, K. (2004). Energy substrate metabolism of mouse cumulus oocyte complexes: Response to follicle stimulating hormone is mediated by the phosphatidylinositol 3-kinase pathway and is associated with oocyte maturation. *Biology of Reproduction, 71*, 199–209.

Saito, T., Hiroi, M., & Kato, T. (1994). Development of glucose utilization studied in single oocytes and preimplantation embryos from mice. *Biology of Reproduction, 50*, 266–270.

Sato, H., Iwata, H., Hayashi, T., Kimura, K., Kuwayama, T., & Monji, Y. (2007). The effect of glucose on the progression of the nuclear maturation of pig oocytes. *Animal Reproduction Science, 99*, 299–305.

Shoubridge, E. A. & Wai, T. (2007). Mitochondrial DNA and the mammalian oocyte. *Current Topics in Developmental Biology, 77*, 87–111.

Sinclair, K. D., Lunn, L. A., Kwong, W. Y., Wonnacott, K., Linforth, R. S. T., & Craigon J. (2008). Amino acid and fatty acid composition of follicular fluid as predictors of in-vitro embryo development. *Reproductive BioMedicine Online, 16*(6), 859–868.

Singh, R. & Sinclair, K. D. (2007). Metabolomics: approaches to assessing oocyte and embryo quality. *Theriogenology, 68*(Suppl 1), S56–62.

Somfai, T., Kaneda, M., Akagi, S., Watanabe, S., Haraguchi, S. et al. (2011). Enhancement of lipid metabolism with L-carnitine during in vitro maturation improves nuclear maturation and cleavage ability of follicular porcine oocytes. *Reproduction Fertility and Development, 23*(7), 912–920.

Songsasen, N., Spindler, R. E., & Wildt, D. E. (2007). Requirement for, and patterns of, pyruvate and glutamine metabolism in the domestic dog oocyte in vitro. *Molecular Reproduction and Development, 74*(7), 870–877.

Spindler, R. E., Pukazhenthi, B. S., & Wildt, D. E. (2000). Oocyte metabolism predicts the development of cat embryos to blastocyst in vitro. *Molecular Reproduction and Development, 56*, 163–171.

Steeves, T. E. & Gardner, D. K. (1999). Metabolism of glucose, pyruvate, and glutamine during the maturation of oocytes derived from pre-pubertal and adult cows. *Molecular Reproduction and Development, 54*(1), 92–101.

Steeves, T. E., Gardner, D. K., Zuelke, K. A., Squires, T. S., & Fry, R. C. (1999). In vitro development and nutrient uptake by embryos derived from oocytes of pre-pubertal and adult cows. *Molecular Reproduction and Development, 54*(1), 49–56.

Sturmey, R. G. & Leese, H. J. (2003). Energy metabolism in pig oocytes and early embryos. *Reproduction, 126*(2), 197–204.

Sturmey, R. G. & Leese, H. J. (2008). Role of glucose and fatty acid metabolism in porcine early embryo development. *Reproduction Fertility and Development, 20*, 149.

Sturmey, R. G., O'Toole, P. J., & Leese, H. J. (2006). Fluorescence resonance energy transfer analysis of mitochondrial:lipid association in the porcine oocyte. *Reproduction, 132*(6), 829–837.

Sturmey, R. G., Reis, A., Leese, H. J., & McEvoy, T. G. (2009). Role of fatty acids in energy provision during oocyte maturation and early embryo development. *Reproduction in Domestic Animals, 44*(Suppl 3), 50–58.

Su Y-Q., Sugiura, K., Wigglesworth, K., O'Brien, M. J., Affourtit, J. P. et al. (2008). Oocyte regulation of metabolic cooperativity between mouse cumulus cells and oocytes: BMP15 and GDF9 control cholesterol biosynthesis in cumulus cells. *Development, 135*(1), 111–121.

Sutovsky, P. & Schatten, G. (1997). Depletion of glutathione during bovine oocyte maturation reversibly blocks the decondensation of the male pronucleus and pronuclear apposition during fertilization. *Biology of Reproduction, 56*, 1503–1512.

Sutton-McDowall, M. L., Gilchrist, R. B., & Thompson, J. G. (2004). Cumulus expansion and glucose utilisation by bovine cumulus-oocyte complexes during in vitro maturation: the influence of glucosamine and follicle-stimulating hormone. *Reproduction, 128*(3), 313–319.

Sutton-McDowall, M. L, Gilchrist, R. B., & Thompson, J. G. (2010). The pivotal role of glucose metabolism in determining oocyte developmental competence. *Reproduction, 139*(4), 685–695.

Sutton-McDowall, M. L., Mitchell, M., Cetica, P., Dalvit, G., Pantaleon, M. et al. (2006). Glucosamine supplementation during in vitro maturation inhibits subsequent embryo development: possible role of the hexosamine pathway as a regulator of developmental competence. *Biology of Reproduction, 74*(5), 881–888.

Sutton, M. L., Cetica, P. D., Beconi, M. T., Kind, K. L., Gilchrist, R. B., & Thompson J. G. (2003a). Influence of oocyte-secreted factors and culture duration on the metabolic activity of bovine cumulus cell complexes. *Reproduction, 126*, 27–34.

Sutton, M. L., Gilchrist, R. B., & Thompson, J. G. (2003b). Effects of in-vivo and in-vitro environments on the metabolism of the cumulus-oocyte complex and its influence on oocyte developmental capacity. *Human Reproduction Update, 9*(1), 35–48.

Swain, J. E., Bormann, C. L., Clark, S. G., Walters, E. M., Wheeler, M. B., & Krisher R. L. (2002). Use of energy substrates by various stage preimplantation pig embryos produced in vivo and in vitro. *Reproduction, 123*(2), 253–260.

Takeuchi, T., Neri, Q. V., Katagiri, Y., Rosenwaks, Z., & Palermo, G. D. (2005). Effect of treating induced mitochondrial damage on embryonic development and epigenesis. *Biology of Reproduction, 72*(3), 584–592.

Thangavelu, G., Colazo, M. G., Ambrose, D. J., Oba, M., Okine, E. K., & Dyck, M. K. 2007. Diets enriched in unsaturated fatty acids enhance early embryonic development in lactating Holstein cows. *Theriogenology, 68*(7), 949–957.

Thompson, J. G. (2006). The impact of nutrition of the cumulus oocyte complex and embryo on subsequent development in ruminants. *Journal of Reproduction and Development, 52*(1), 169–175.

Thompson, J. G., Lane, M., & Gilchrist, R. B. (2007). Metabolism of the bovine cumulus-oocyte complex and influence on subsequent developmental competence. *Society for Reproduction and Fertility, Suppl 64*, 179–190.

Thompson, J. G., Partridge, R. J., Houghton, F. D., Cox, C. I., & Leese, H. J. (1996). Oxygen uptake and carbohydrate metabolism by in vitro derived bovine embryos. *Journal of Reproduction and Fertility, 106*(2), 299–306.

Thompson, J. G., Simpson, A. C., Pugh, P. A., Wright, R. W., Jr., & Tervit, H. R. (1991). Glucose utilization by sheep embryos derived in vivo and in vitro. *Reproduction Fertility and Development, 3*(5), 571–576.

Tsutsumi, O., Satoh, K., Taketani, Y., & Kato, T. (1992). Determination of enzyme activities of energy metabolism in the maturing rat oocyte. *Molecular Reproduction and Development, 33*, 333–337.

Tubman, L., Peter, A., & Krisher, R. (2006). Pentose phosphate pathway activity controls nuclear maturation of porcine oocytes. *Reproduction Fertility and Development, 18*(1), abst 344, 279–280.

Urner, F. & Sakkas, D. (2005). Involvement of the pentose phosphate pathway and redox regulation in fertilization in the mouse. *Molecular Reproduction and Development, 70*(4), 494–503.

Van Blerkom, J. (2004). Mitochondria in human oogenesis and preimplantation embryogenesis: engines of metabolism, ionic regulation and developmental competence. *Reproduction, 128*(3), 269–280.

Van Blerkom, J. (2011). Mitochondrial function in the human oocyte and embryo and their role in developmental competence. *Mitochondrion, 11*(5), 797–813.

Van Blerkom, J., Davis, P., & Alexander, S. (2003). Inner mitochondrial membrane potential (DeltaPsim), cytoplasmic ATP content and free Ca2 + levels in metaphase II mouse oocytes. *Human Reproduction, 18*(11), 2429–2440.

Van Hoeck, V., Sturmey, R. G., Bermejo-Alvarez, P., Rizos, D., Gutierrez-Adan, A. et al. (2011). Elevated non-esterified fatty acid concentrations during bovine oocyte maturation compromise early embryo physiology. *PLoS ONE [Electronic Resource] 6*(8), e23183.

Vander Heiden, M. G., Cantley, L. C., & Thompson, C. B. (2009). Understanding the Warburg effect: the metabolic requirements of cell proliferation. *Science, 324*(5930), 1029–1033.

Vousden, K. H. & Ryan, K. M. (2009). p53 and metabolism. *Nature Reviews Cancer, 9*(10), 691–700.

Wakefield, S. L., Lane, M., Schulz, S. J., Hebart, M. L., Thompson, J. G., & Mitchell, M. (2008). Maternal supply of omega-3 polyunsaturated fatty acids alter mechanisms involved in oocyte and early embryo development in the mouse. *American Journal of Physiology Endocrinology and Metabolism, 294*(2), E425–434.

Wang, Q., Chi, M. M., Schedl, T., & Moley, K. H. (2012). An intercellular pathway for glucose transport into mouse oocytes. *American Journal of Physiology Endocrinology and Metabolism, 302*, E1511–E1518.

Wang, Q. & Moley, K. H. (2010). Maternal diabetes and oocyte quality. *Mitochondrion, 10*(5), 403–410.

Warburg, O. (1956). On the origin of cancer cells. *Science, 123*(3191), 309–314.

Wilding, M., Coppola, G., Dale, B., & Di Matteo, L. (2009). Mitochondria and human preimplantation embryo development. *Reproduction, 137*(4), 619–624.

Wonnacott, K. E., Kwong, W. Y., Hughes, J., Salter, A. M., Lea, R. G. et al. (2010). Dietary omega-3 and -6 polyunsaturated fatty acids affect the composition and development of sheep granulosa cells, oocytes and embryos. *Reproduction, 139*(1), 57–69.

Wu, G. Q., Jia, B. Y., Li, J. J., Fu, X. W., Zhou, G. B. et al. (2011). L-carnitine enhances oocyte maturation and development of parthenogenetic embryos in pigs. *Theriogenology, 76*, 785–793.

Yoneda, A., Suzuki, K., Mori, T., Ueda, J., & Watanabe, T. (2004). Effects of delipidation and oxygen concentration on in vitro development of porcine embryos. *Journal of Reproduction and Development, 50*(3), 287–295.

Yoshida, M. (1993). Role of glutathione in the maturation and fertilization of pig oocytes in vitro. *Molecular Reproduction and Development, 35*(1), 76–81.

Yoshida, M., Ishigaki, K., Nagai, T., Chikyu, M., & Pursel, V. G. (1993). Glutathione concentration during maturation and after fertilization in pig oocytes: relevance to the ability of oocytes to form male pronucleus. *Biology of Reproduction, 49*(1), 89–94.

You, J., Lee, J., Hyun, S-H., & Lee, E. (2012). L-carnitine treatment during oocyte maturation improves in vitro development of cloned pig embryos by influencing intracellular glutathione synthesis and embryonic gene expression. *Theriogenology, 78*(2), 235–243.

Yuan, Y., Wheeler, M. B., & Krisher, R. L. (2012). Disrupted redox homeostasis and aberrant redox gene expression in porcine oocytes contribute to decreased developmental competence. *Biology of Reproduction.*

Zachut, M., Dekel, I., Lehrer, H., Arieli, A., Arav, A. et al. (2010). Effects of dietary fats differing in n-6:n-3 ratio fed to high-yielding dairy cows on fatty acid composition of ovarian compartments, follicular status, and oocyte quality. *Journal of Dairy Science, 93*(2), 529–545.

Zeng, H-T., Yeung, W. S. B., Cheung, M. P. L., Ho, P-C., Lee, C. K. F. et al. (2009). In vitro-matured rat oocytes have low mitochondrial deoxyribonucleic acid and adenosine triphosphate contents and have abnormal mitochondrial redistribution. *Fertility and Sterility, 91*(3), 900–907.

Zeron, Y., Ocheretny, A., Kedar, O., Borochov, A., Sklan, D., & Arav, A. (2001). Seasonal changes in bovine fertility: relation to developmental competence of oocytes, membrane properties and fatty acid composition of follicles. *Reproduction, 121*(3), 447–454.

Zheng, P., Bavister, B. D., & Ji, W. (2001). Energy substrate requirement for in vitro maturation of oocytes from unstimulated adult rhesus monkeys. *Molecular Reproduction and Development, 58*, 348–355.

Zheng, P., Vassena, R., & Latham, K. E. (2007). Effects of in vitro oocyte maturation and embryo culture on the expression of glucose transporters, glucose metabolism and insulin signaling genes in rhesus monkey oocytes and preimplantation embryos. *Molecular Human Reproduction, 13*(6), 361–371.

Zuelke, K. A. & Brackett, B. G. (1992). Effects of luteinizing hormone on glucose metabolism in cumulus-enclosed bovine oocytes matured in vitro. *Endocrinology, 131*(6), 2690–2696.

Zuelke, K. A. & Brackett, B. G. (1993). Increased glutamine metabolism in bovine cumulus cell-enclosed and denuded oocytes after in vitro maturation with luteinizing hormone. *Biology of Reproduction, 48*(4), 815–820.

Zuelke, K. A., Jeffay, S. C., Zucker, R. M., & Perreault, S. D. (2003). Glutathione (GSH) concentrations vary with the cell cycle in maturing hamster oocytes, zygotes, and pre-implantation stage embryos. *Molecular Reproduction and Development, 64*(1), 106–112.

10 Screening for Oocyte Competence

Marc-André Sirard and Mourad Assidi

10.1 Introduction

Oocyte competence has become a crucial concern in assisted reproductive approaches in both human and livestock species. Assessment of this competence is considered a prerequisite to improve ART efficiency and safety. However, the available literature has clearly shown that this competence is a multifactorial process that is difficult to accurately assess using only the morphological parameters currently available worldwide. The identification of additional strategies to predict oocyte competence is therefore a field priority. This chapter focuses on the concept of oocyte competence and some of the crucial factors involved in this process. The new promising tools and strategies to improve and accurately assess oocyte competence using biomarkers are also discussed.

10.2 Concept of Oocyte Competence

Oocyte competence is the ability of the oocyte to fulfil maturation, undergo successful fertilization, reach the blastocyst stage, and yield a viable and healthy progeny. It has been established that this competence, measured until now by the final output, is a complex process that involves many successive, concomitant, and interdependent steps, including oocyte maturation, its fertilization, subsequent embryo development, and ultimately the delivery of a healthy offspring. This developmental ability is acquired by the oocyte through the biosynthesis and/or storage of many key molecules needed for competence process achievement. To dissect the cumulative competence mechanism, six levels/steps can be discussed.

10.2.1 Oocyte Molecules Storage

During oocyte growth, several molecules that will contribute to its maturation, fertilization, and early embryo development are produced and accumulated. These molecules are the result of key molecular processes that occur both in the nucleus and ooplasma during oocyte growth and maturation. The storage of a suitable stockpile of these molecules is referred to as molecular maturation (Sirard et al., 2006). Still poorly defined, molecular maturation is the process whereby a suitable supply (in a time-, space-, and dose-dependent manner) of the required factors needed for the successful achievement

Oocyte Physiology and Development in Domestic Animals, First Edition. Edited by Rebecca L. Krisher.

of the aforesaid steps of its competence are stored in the oocyte. This accumulation does not imply a specific phenotype, explaining why under similar visible properties of oocytes and cumulus-oocyte complexes (COC), some oocytes are developmentally more competent than others. These observations show the limits of the already used morphological criteria and highlight the importance of investigating these molecular events to unravel the intricate process of oocyte competence and improve the accuracy of its prediction. In this context, the oocyte volume is reported to increase 300-fold in the mouse. This tremendous cell enlargement is associated with intense metabolic and gene expression activities to provide sufficient reserves (mRNA, ribosomes, proteins, etc.) for the oocyte's developmental journey (Wassarman & Albertini, 1994; Mermillod & Marchal, 1999). In mammals, a positive and direct relationship between oocyte diameter, its maturation, and subsequent developmental competence was reported (Fair et al. 1995; Otoi et al., 1997; Miyano & Manabe, 2007). Several mechanisms are behind the oocyte diameter increase (molecules stockroom). Among the major molecular mechanisms involved in the oocyte competence process, we will focus herein on gene transcription and protein synthesis within the oocyte and even its follicular environment (Barnes & First, 1991, Hyttel et al., 1997, Sirard & Trounson, 2003). This transcriptional activity increases significantly during growth of the bovine oocyte and decreases around the beginning of the antral stage when the diameter of the oocyte reaches 110 μm (Fair et al., 1995, Memili et al., 1998). Moreover, it is well established that the oocyte is able to store its maternal mRNAs in an inactive translational state for long periods without being degraded. Despite its very small fraction compared to total RNA, oocyte mRNA has the property of having a variable length of poly(A) tail and RNA binding proteins that allow their storage and use (translation) at the appropriate stage/time (Memili et al., 1998; Lequarre et al., 2004, Tremblay et al., 2005). Some molecules such as the messenger ribonucleoprotein complexes (mRNP) were shown to be required to store the mRNA in the *Xenopus* oocyte. This complex is thought to protect the transcripts from both the translation and degradation machineries (Weston & Sommerville, 2006).

Protein synthesis is also crucial to the oocyte resumption of meiosis. Following GVBD, protein synthesis is achieved through a selective process based mainly on the poly(A) tail length of the stored mRNA (Kastrop et al., 1991b; Levesque & Sirard, 1996). The untranslated region (3'UTR) also appears to influence the stability, storage, and translation of mRNA (Henrion et al., 2000). This translational activity was maximal around the GVBD stage (approximately 6 h of IVM) in the bovine oocyte in vitro, about three times more than its level at GV, before decreasing again around the MII stage (Tomek et al., 2002a). Additional proteins were also synthesized and accumulated around the GVBD such as ribosomal and mitochondrial proteins, the ZP glycoproteins, histones, and kinases (Wassarman & Albertini, 1994; Massicotte et al., 2006). In addition to translation of new proteins, other post-translational modifications, mainly phosphorylation, were reported during the G2/M transition in the bovine oocyte (Kastrop et al., 1991a; Kallous et al., 1993).

In addition to transcription and translation, the metabolic activity within the oocyte is marked by a significant increase in glycolysis including pyruvate and lactate metabolism, which seems to have a direct effect on nuclear maturation and competence development of the oocyte (Coffer et al., 1998; Cetica et al., 1999; Krisher & Bavister, 1999; Fan & Sun, 2004). Interestingly, the types (quality and quantity) of stored molecules are reflective of oocyte quality, which is crucial to its subsequent developmental competence. Under similar morphological and ultrastructural properties, some oocytes have acquired/produced and accumulated the proper molecular tools for successful embryonic genome activation and the developmental competence due to special induction from their follicular environment (Sirard et al., 2006). Understanding the molecular processes underlying this competence will be a powerful tool to better assess the intrinsic competence of each oocyte.

10.2.2 Meiotic Competence

Meiotic competence is the most visible and widely studied process in mammalian oocytes. It refers to nuclear maturation and describes the progression of the oocyte nucleus from the germinal vesicle stage (GV), and through the GVBD (meiosis resumption), the association of the homologous chromosomes with the first polar body extrusion, until the second meiotic arrest at MII. It is important to note that from GVBD (oocyte diameter reaches around 110 μm in bovine and human oocytes) until the maternal-to-embryonic transition (MET) in the early embryo development, there is an arrest of most if not all transcriptional activity (Fair et al., 1997, Tremblay et al., 2005). The second meiotic resumption is induced by the male pronucleus and leads to the fulfilment of meiosis and the second polar body extrusion (Hyttel et al., 1986). This meiotic progression is very exceptional in terms of its duration as well as its multiple arrests. Despite the easy assessment of nuclear maturation (microscope), its success unfortunately does not guarantee full oocyte competence.

10.2.3 Cytoplasmic Competence

Cytoplasmic competence includes all the ultrastructural modifications that occur in the ooplasma during oocyte pre-maturation (before the LH surge) and final maturation to allow successful fertilization, cleavage, and embryo development (Kruip et al., 1983; Hyttel et al., 1986; Hyttel et al., 1997). Among the main aspects associated with the cytoplasmic signs of competence is the increase of mitochondria number and their migration close to the endoplasmic reticulum and the oolema (Michaels et al., 1982). Additionally, the Golgi apparatus becomes associated with numerous lipid vesicles and moves to the subcortical region of the oocyte where it seems to be involved in cortical granule formation. This proper organelle relocation is concomitant with two other important visible criteria of cytoplasmic competence: suitable zona pellucida (ZP) thickness and appropriate perivitelline spacing (Wassarman & Albertini, 1994, Fair et al., 1995, Sun 2003). In this context, ZP thickness and differentiation increased and were maximal at the final maturation step prior to fertilization (Hyttel et al., 1986; Familiari et al., 2006).

Unfortunately, there is no available direct measure of cytoplasmic competence, which is why meiotic maturation is commonly used as a reference for cytoplasmic maturation. In addition to this indirect approach, cytoplasmic maturation can also be assessed retroactively through the early embryo development outcome. This is why it is difficult to distinguish between oocyte maturation and competence at both the cytoplasmic and molecular levels.

10.2.4 Fertilization Competence

Fertilization competence is the ability of the oocyte and the sperm cell to fuse leading to the extrusion of the second polar body and the formation of two pronuclei. This level of competence is ensured by the molecular and cellular machinery of the oocyte after sperm penetration. The high rates of successful fertilization in mammalian species (mainly in human and cow) support an intrinsic potential of the growing oocytes to properly drive the events of gamete fusion. However, a small rate of fertilization failure could mostly be associated with an incompletely decondensed sperm or asynchronized pronuclei (Sirard et al., 2006, Swain & Pool, 2008).

10.2.5 Embryo Developmental Competence

Early embryo developmental competence measures the ability of the fertilized oocyte to cleave, properly activate the embryo genome (at the maternal-to-embryonic transition), and develop into a blastocyst. This ability to reach the blastocyst stage is clearly influenced by the follicular environment and the degree of differentiation of the follicle (Dieleman et al., 1983, Blondin et al., 2002). This blastocyst outcome is the main criterion used to retroactively assess the previous steps of competence, mainly the quality and efficiency of the molecular machinery of the oocyte at the maturation stage (transcription factors, translation, signaling molecules) (Barnes & First, 1991, Vigneault et al., 2004).

10.2.6 Development-to-term Competence

Development to term and delivery of a healthy offspring are the ultimate levels of oocyte competence. Increasing efforts have been put forth to increase the efficiency and safety of ART both in humans and animals over the years. Despite their significant contribution in the recent milestones in reproduction, these reproductive technologies were weakly associated with some adverse consequences such as Large Offspring Syndrome (LOS) in cattle, higher rates of multiple pregnancies, and some epigenetic variations in human (Lonergan et al., 2003, Farin et al., 2006, Odom & Segars, 2010). The oocyte legacy of several genetic, molecular and cellular embedded programs of maturation influences its itinerary toward the delivery of a healthy newborn and beyond (McEvoy et al., 2003, Le Bouc et al., 2010). Additional studies using animal models are required to evaluate the impact of the gametes and ART manipulations on the embryo epigenome and therefore subsequent pregnancy, delivery, offspring health, and even later adulthood health. Undoubtedly, the new molecular approaches in reproduction and oocyte culture media optimization should offer promising tools that would demystify the molecular players involved in appropriate oocyte competence, therefore allowing more efficient (biomarkers, target pathways) and safer ART approaches in the upcoming years.

10.3 Influence of Follicular Parameters on Oocyte Competence

In many mammalian species, especially in large mono-ovulatory ones, it has been established that the follicular environment has a clear impact on the oocyte's capacity to develop beyond fertilization. The main follicular parameters that can be assessed are related to size, health/atresia status, effects of superstimulation, and level of differentiation.

10.3.1 The Effect of Follicle Size

It has been established that follicle size is an important parameter that influences oocyte competence. Although they lead to successful fertilization rates, smaller bovine and porcine follicles (less than 3 mm) have reduced oocyte competence (Lonergan et al., 1994, Blondin & Sirard, 1995; Marchal et al., 2002). Moreover, oocytes from human follicles smaller than 18 mm have diminished fertilization rates and embryo quality compared to those from follicles larger than 18 mm (Rosen et al., 2008). These oocytes appear to have been recovered earlier and therefore probably lacked some additional follicular factors to reach their full competence.

10.3.2 *The Effect of Health Status*

Although still not fully understood, follicular atresia starts with signs of follicular cell apoptosis from the granulosa cells that progressively move to cumulus and thereafter the embedded oocyte (Zeuner et al., 2003). The dominant follicle as well as the large subordinate ones contain no or very slight atresia, and their oocytes preserve good developmental competence (blastocyst rate) (Vassena et al., 2003). Atretic follicles marked by disrupted cumulus layers showed poor developmental competence. Interestingly, higher developmental competence was reported in COC with slight signs of apoptosis in the cumulus outer layers (Blondin & Sirard, 1995, Zeuner et al., 2003; Feng et al., 2007). The onset of these apoptotic events looks like a form of maturation-promoting/speeding signal to the oocyte to improve its competence. It may be interpreted as a gamete preservation process that ensures species maintenance.

10.3.3 *The Effect of Superstimulation*

Superstimulation programs are hormonal schedules used to induce oocyte competence via its follicular environment. Recent knowledge concerning reproductive physiology and hormonal regulation of the follicular waves has improved superovulation strategies in order to mimic the physiological/normal hormonal patterns. Gene expression analysis of the follicular compartments showed a switch from follicular growth to differentiation through the abundance of transcripts associated with early granulosa cell luteinization (Robert et al., 2001), expression of the LH receptor (Robert et al., 2003) and cumulus cell differentiation and expansion (Assidi et al., 2010). This hormonal induction is normally achieved at the plateau phase just prior to the onset of dominance at 8.5 mm of follicular diameter in the cow, therefore allowing the recovery of a homogenous population of follicles (Ginther et al., 2000; Merton et al., 2003). It should increase the number and quality of potentially differentiated follicles (dominant or early atretic) able to induce higher developmental competence (Blondin et al., 2002). Further explorations of follicle-oocyte dialog are required, mainly at the molecular level, to unravel the complexity of the follicular differentiation stage and maximize the efficiency of superovulation protocols and oocyte competence.

10.3.4 *The Effect of Follicular Level of Differentiation*

In the natural cycle of mono-ovulatory species, the dominance stage corresponds to a rapid growth of the dominant follicle compared to its subordinate wave-mates followed by increasing differentiation and steroidogenesis (Ginther, 2000). These follicular modifications occur as the antral follicle diameter increases and the oocyte developmental competence is achieved (Humblot et al., 2005). In the luteal phase of large animals such as the cow, the ovulation of such dominant follicles will be inhibited by the progesterone level unless corpus luteum degeneration is induced.

Under reduced levels of progesterone, the preovulatory stage is associated with an increase in both LH pulsatility and progesterone/estradiol ratios, and the reorganization of the oocyte's organelles (Hyttel et al., 1986). Oocytes retrieved at this stage have complete developmental competence (Hyttel et al., 1986; Humblot et al., 2005). After the LH surge, oocytes collected from dominant follicles are highly competent. The LH surge induces the expression of the progesterone receptor that supports the progression of the dominant follicle toward ovulation by preventing granulosa cell apoptosis. Interestingly, progesterone receptor antagonist treatment reactivates cell cycle arrest and apoptosis in the dominant follicle (Quirk et al., 2004). These observations confirm the role of PGR in the final differentiation of the dominant follicle, the improvement of its longevity (aging), and the maintenance of its resistance to apoptosis. Therefore, the embedded oocyte maintains and probably reinforces its developmental competence potential.

10.4 Morphological Changes of the COC Associated with Competence

Cumulus cells maintain a close and reciprocal relationship with the oocyte and were shown to support oocyte maturation, ovulation as well as the fertilization process (Tanghe et al., 2002; Thompson et al., 2007). Given their importance and their position in the oocyte vicinity, the morphological aspect of this somatic compartment has also been studied in large mammals in order to find selection criteria associated with a better developmental outcome. In this context, human and bovine oocytes surrounded by several cumulus cell layers ($n \geq 3$) have shown higher developmental competence (Blondin & Sirard, 1995; Ng et al., 1999). The grade 3 COC, with several cumulus cell layers compacted around the oocyte and with slight signs of expansion and/or in the outer layers, was positively correlated with higher rates of normal fertilization and pregnancy (Blondin & Sirard, 1995, Lasiene et al., 2009). Additionally and following expansion, the area of expanded cumulus was suggested as a parameter that could predict the competence of porcine oocytes *in vitro* (Qian et al., 2003).

In addition to cumulus morphology, some correlations were reported between the chromatin configuration at the GV stage and oocyte competence. In most mammals, the chromatin initially spreads throughout the nucleus (non-surrounded nucleolar [NSN] pattern) at the beginning of follicular growth ($>$ 1 mm in the cow) and starts a progressive and size-dependent condensation before reaching a floccular form close to the nucleoli and the nuclear membrane (F pattern) (Liu et al., 2006). This F pattern is associated with an important reduction in RNA transcription in the bovine oocyte and the acquisition of meiotic competence (ability to resume meiosis, GVBD) (Fair et al., 1995; Liu et al., 2006). Similar observations were reported in other mammals such as human and mouse (Miyara et al., 2003; Tan et al., 2009). It must be mentioned that increased oocyte competence was correlated with both gene transcription repression and at least the SN chromatin configuration (nucleoli surrounded by chromatin) in mammalian oocytes (Tan et al., 2009).

In the growing phase and before the 3-mm follicular diameter, bovine oocytes contain one to two vacuolated and fibrillogranular nucleoli indicating intensive rRNA synthesis. After that, these structures become homogenous and compact with a sharp decrease in transcriptional activity indicating meiotic competence (Crozet et al., 1986). When the oocyte resumes meiosis, the maternal nucleolus is dissolved and only reappears (reassembly of the oocyte nucleolar material) later following fertilization in the male and female pronucleus (Lefevre, 2008). It is considered a crucial factor in successful embryo development and competence because nucleoli-removed oocytes were unable to form a blastocyst (Ogushi et al., 2008). Taken together, the chromatin conformation and the nucleolar state at the GV stage could be helpful in oocyte developmental competence prediction.

10.5 Biochemical Changes within the COC Associated with Competence

10.5.1 Glucose Metabolism

Glucose metabolism is an important pathway in the oocyte during its growth and meiosis resumption in murine, porcine, and bovine species (Biggers et al., 1967; Eppig, 1976; Leese & Barton, 1985;

Mayes et al., 2007; Thompson et al., 2007). Growing oocytes were shown to exhibit efficient glucose-6-phosphate dehydrogenase (G6PDH) activity to metabolize the brilliant cresyl blue (BCB) staining and lead to colorless cytoplasm. However, the oocytes that had successfully finished their growth were characterized by decreased transcriptional activity and insufficient levels of G6PDH to metabolize the BCB. This metabolic property has been used as a marker to select fully grown oocytes. Therefore, the oocytes with blue-colored cytoplasm are associated with higher phosphorylation levels of Akt and MAP kinases, and subsequent improvement of developmental competence (Alm et al., 2005; Sutton-McDowall et al., 2010) with better quality blastocysts (Torner et al., 2008).

10.5.2 Dielectrophoretic Migration

In order to improve the efficiency of good-quality oocyte selection, the dielectrophoretic approach was investigated as a potential, noninvasive method of competence prediction that measures the speed of oocyte/zygote migration in an electric field (14 peak-to-peak volts; frequency of 4 MHz). The faster oocytes were developmentally more competent than the slower ones (Dessie et al., 2007). This variability in migration speed may be due to various factors such as membrane and/or cytoplasm permeability and cell molecular composition as well as cell size and shape. Interestingly, in the same study, several genes were also reported to be differentially expressed between the two classes of oocytes.

10.5.3 Zona Refringence

Zona pellucida (ZP) birefringence is another morphological criterion that has been used to reinforce the selection of high-quality oocytes. In fact, the inner immediate layers of CCs are connected with the ZP, which is a thick protective layer that surrounds the oocyte. The thickness of this membrane increases with oocyte growth and reaches 17 μm at the human MII stage. The oocyte ZP protein production decreases slightly until ovulation, where the ZP diameter remains relatively stable until the first steps of cleavage (Goyanes et al., 1990; Gook et al.; 2008). ZP birefringence measures ZP thickness and uniformity. In human IVF, this morphological criterion was positively correlated with oocyte developmental competence, successful embryo implantation, and subsequent pregnancy rates (Montag et al., 2008; Montag & van der Ven, 2008; Madaschi et al., 2009). This indicator may be considered in oocyte selection to reinforce the selection parameters already in use.

10.6 The Use of Coasting to Induce Competence in Large Mammals

In humans, coasting is referred to as the gonadotropin input cancelation during a superstimulation program mainly to avoid hyperstimulation syndrome where it is very helpful to stabilize serum E2 levels prior to hCG administration, whereas in animals, coasting refers to the arrest of gonadotropin support in presence of endogenous LH to stimulate follicular differentiation and oocyte competence (Sher et al., 1995, Blondin et al., 2002). This gonadotropin starvation exerts a selective pressure that eliminates the smaller follicles and increases the proportion of medium-to-large follicles (Fluker et al., 1999, Blondin et al., 2002). Moreover, this coasting may induce slight atresia in the cumulus outer layers, which is favourable to grade 3 COC formation and therefore improved developmental competence (Blondin & Sirard, 1995). These findings are confirmed by the high rates of cleavage (90%) and blastocyst (80%) following 48h of coasting in the cow (Blondin et al., 2002). In human IVF, gonadotropin withholding (fewer than four days) is also an effective way to prevent OHSS (ovarian hyperstimulation syndrome) and PCO (polycystic ovarian) complications, and to avoid harmful effects on oocyte developmental competence and/or uterine receptivity (Simon et al., 1998; Isaza et al., 2002). The coasting strategy is therefore a useful tool to improve oocyte competence through the contribution in the increase of the potential pool of oocytes to be selected.

10.7 The Use of Genomic/Gene Expression in Follicular Cells to Assess Oocyte Competence

Several morphological, ultrastructural, and metabolic criteria have been used to predict oocyte competence. Despite some improvements in pregnancy outcomes, these morphological criteria remain subjective, in some cases invasive and/or poorly correlated with oocyte competence. Therefore, the identification of other relevant biomarkers to accurately predict oocyte quality is of primary interest. In this context, the new potential offered by genomic tools allows the exploration of gene expression level in very small samples. This can therefore be applied to individual follicles and even individual cumulus complexes. Considering the crucial contribution of the somatic follicular compartment to oocyte quality through bidirectional and continuous communication, some genomic studies have focused on the follicular status correlation with gene expression. Evans et al. (2004) reported an increase in differentially expressed genes associated with apoptosis in subordinate versus dominant follicles. These gene markers were suggested as possible markers of the bovine dominant follicle. Similar work has suggested some key genes associated with follicular status progression and dominance acquisition (Bedard et al., 2003, Mihm et al., 2006, Wells & Patrizio, 2008). Other studies explored the impact of superovulation treatment on follicular transcriptome. For example, LH was shown to induce an early effect (6h after the LH surge) marked by the transcription of ovulation-related genes in bovine granulosa *in vivo*. In the case of the late LH effect (22h after LH surge), the granulosa transcriptome was characterized by large sets of luteinization-related genes (Gilbert et al., 2010). A similar study that explored the LH-gene expression effects in bovine cumulus cells has also reported numerous candidate genes that warrant further investigation (Assidi et al., 2010). For human antral follicles, the transcriptome patterns of floating granulosa cells after puncture (different from the cumulus cell patterns) showed the expression of many inflammation-related genes to prepare ovulation. In the case of cumulus cells, they express genes involved mainly in neuronal-like functions and proteolysis (Koks et al., 2010). Taken together, these findings confirm expectations regarding follicular cell behavior prior to ovulation and offer useful target genes to improve our understanding of molecular physiology and regulation. Other reports have also analyzed the gene expression patterns associated with ovarian stimulation (Grondahl et al., 2009, Gilbert et al., 2010) or *in vitro* culture (Jones et al., 2008, Tesfaye et al., 2009) using the *in vivo* context as reference. It is important to note that the exploration of the differentially expressed genes in the follicular cells associated with successful oocyte developmental competence was achieved in several species in order to find relevant markers that could be indicators of oocyte quality. The goal was to identify quantitative and noninvasive gene markers that accurately predict oocyte competence and reinforce the morphological criteria already used. Several studies performed in many mammalian species including cattle (Burns et al., 2003; Fayad et al., 2004; Bettegowda et al., 2008; Gilbert et al., 2010) and human (Feuerstein et al., 2007; Ferrari et al., 2010; Koks et al., 2010) have revealed key gene markers that could increase the efficiency and the safety of ART.

Nowadays, the culture of a group of COC or the transfer of more than one embryo in ART practices in human and livestock are challenging in terms of finding strong biomarkers to predict oocyte competence. To do this, individual assessment of follicles was performed by analyzing the gene expression of follicular cells and recording the pregnancy output of the oocyte. This individual approach is an elegant way to avoid possible bias especially in human fertility treatment. It will also be very useful in clinics where elective single embryo transfer (eSET) is recommended. Moreover, the pregnancy rates using the morphological-based SET remain low and confirm the lack of efficient oocyte selection markers. It is assumed that differential genomic profile of follicular cells from follicles leading to pregnancy could provide useful indicators of oocyte developmental competence (Hamel et al., 2008). The inter-patient (patient group with successful pregnancy versus the nonpregnant group) and intra-patient (follicle leading to pregnancy versus follicle leading to pregnancy failure) analysis approaches were used to find strong biomarkers of oocyte developmental competence (Hamel et al., 2010a). This study performed in our laboratory highlighted the importance of considering both the inter-patient and intra-patient approaches to clearly dissect the gene expression pattern variations in the context of follicular competence. Following the intra-patient analysis and gene candidate validation on separate samples, seven confirmed biomarkers were reported: phosphoglycerate kinase 1 (PGK1), regulator of G-protein signaling 2 (RGS2), regulator of G-protein signaling 3 (RGS3), aromatase (CYP19a), cell division cycle 42 (CDC42), UDP-glucose pyrophosphorylase-2 (UGP2), and pleckstrin homology-like domain, family A, member 1 (PHLDA1) (Hamel et al., 2010a; Hamel et al., 2010b). The genomic functional analysis of these markers clearly supports their involvement in specific cell signaling pathways and early follicular cell luteinization.

These markers are potentially influenced by the hormonal environment, as the majority of them have links to the different brain peptides regulating follicular differentiation (Figure 10.1).

10.8 The Use of Genomic/Gene Expression in Cumulus Cells to Assess Oocyte Competence

During folliculogenesis, the oocyte is surrounded by follicular cells that differentiate at the antral stage into mural granulosa cells and cumulus cells. The cumulus cells are highly specialized follicular cells that maintain close cell-cell connections with the oocyte. Their removal, the disruption of these intercellular communications, or the inhibition of their metabolic, transcriptional, and/or translational activities deeply reduces oocyte developmental ability (Tatemoto & Terada, 1995; Matzuk et al., 2002; Tanghe et al., 2002; Sutton et al., 2003; Tanghe et al., 2003; Atef et al., 2005). Moreover, the cumulus functions are influenced by some oocyte-secreted factors, mainly transforming growth factors (TGFs) (Elvin et al., 2000a, Eppig, 2001; Gilchrist et al., 2004) and EGF-like peptides (Tsafriri et al., 2005, Downs & Chen, 2008; Panigone et al., 2008). Consequently, the cumulus cells are considered a mirror that reflects oocyte quality. They were also shown to support crucial roles associated with oocyte competence, including oocyte maturation, ovulation, fertilization, and early embryo development (Tanghe et al., 2002, Hernandez-Gonzalez et al., 2006; Assou et al., 2008). Keeping in mind all the aforementioned cumulus cell contributions, confirmed genomic markers differentially expressed in the cumulus cells of competent oocytes should offer powerful tools that reinforce current oocyte selection strategies and result in improved pregnancy outcomes. Therefore, cumulus cells are the target of increasing genomic studies in mammalian species to establish possible correlations between the cumulus cell transcriptome and oocyte developmental competence.

In the mouse and through analysis of gene expression in cumulus cells at different follicular stages, genomic pattern variations between morphologically similar cumulus-oocyte complexes but

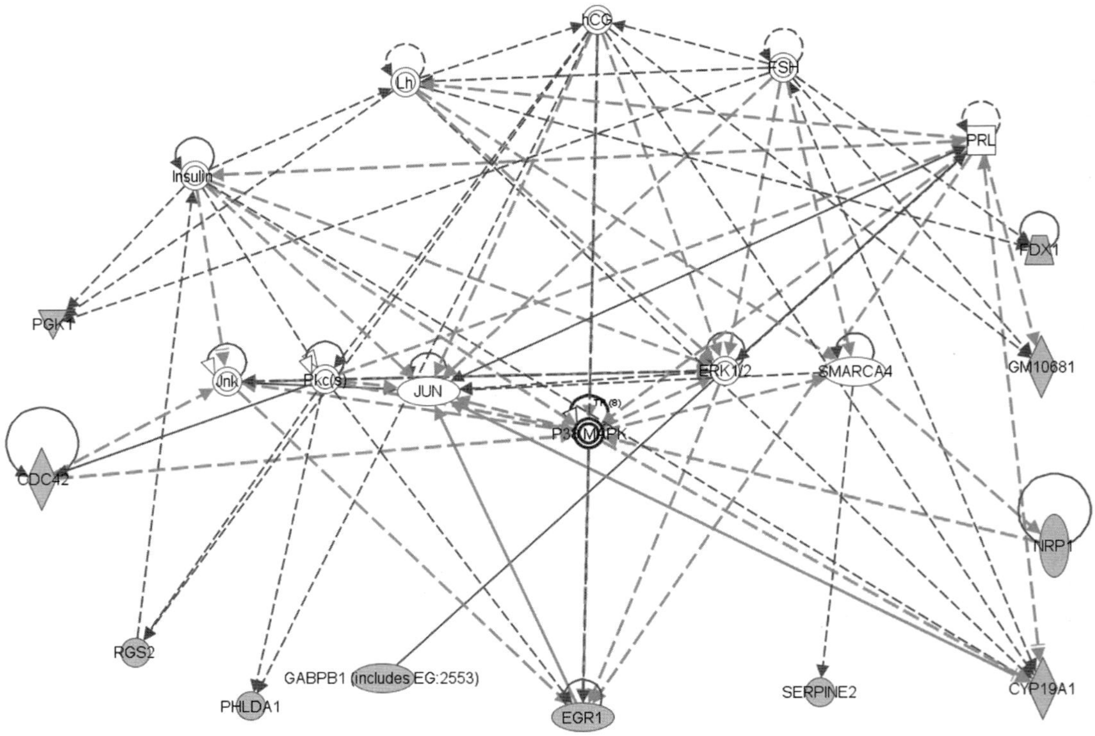

Figure 10.1 The potential signaling networks identified by the Ingenuity program in relation to the markers associated with pregnancy in humans (based on Hamel et al., 2008; Hamel et al., 2010a).

with different developmental competence were found (Adriaenssens et al., 2011). In our laboratory, several gene markers of bovine oocyte competence expressed in cumulus cells were suggested in the preovulatory stage, following a superovulation program or after *in vitro* FSH treatment (Assidi et al., 2008). These gene candidates have provided precious and meaningful data about cumulus cell behavior during follicular growth and differentiation, and offer objective and quantitative parameters of high-quality oocyte selection.

For human ART, and with the need to build efficient tools of oocyte selection, the cumulus cells have been the preferred tissue mainly due to its availability, noninvasiveness (avoiding ethical and legal restrictions) and the established contribution in the oocyte competence acquisition process. The assessment of cumulus cell gene expression should allow the identification of quantitative and noninvasive biomarkers of oocyte competence that will increase the efficiency of the SET procedure. Dr. Hamamah's group has particularly focused on this concept through the analysis of human cumulus cell gene expression profile and suggested interesting markers that require further validation (Assou et al., 2008, Assou et al., 2010). These biomarkers associated with both pregnancy success and failure have improved our understanding of the cumulus involvement in the competence process (Assou et al., 2006, Feuerstein et al., 2007, Huang & Wells, 2010; Assidi et al., 2011). Additional studies using these genomic approaches on cumulus cells from SET procedures will provide powerful predictors of oocyte competence that strengthen the morphological parameters and increase oocyte selection efficiency, and therefore improve pregnancy rates.

10.9 Signaling Pathways Involved in Competence Stimulation

Understanding the process (signaling pathways) from the follicle to the cumulus and from the cumulus to the oocyte remains the ultimate key to act on oocyte competence through different *in vitro* or *in vivo* treatments. Therefore, we consider that functional genomic analysis of the gene candidates and their matching with what is already known about follicular physiology is the way to unravel the competence enigma. Although cumulus cells improve oocyte competence, an important consideration should be addressed to oocyte feedback on its neighboring somatic cells via the TGF (transforming growth factor) family, mainly GDF9 and BMP15 (Eppig, 2001; Sugiura et al., 2007). The appropriate interplay between oocyte and follicular cells is indispensable for proper oocyte development, folliculogenesis, and progression to ovulation as well as subsequent embryo development (Eppig et al., 2002, Gilchrist et al., 2008). Interesting signaling pathways important to metabolic cumulus-oocyte cooperativity, including cholesterol synthesis (Su et al., 2008), glycolysis (Sugiura et al., 2005), and ovulation (Richards et al., 2002a; Richards, 2007), have recently been reported.

Identification of promising pathways to investigate in order to understand the competence mechanism is still a challenge. Building the functional relationships between the pathways of (1) the confirmed gene biomarkers expressed in cumulus/follicular cells, (2) the already known cumulus-oocyte metabolic and signaling pathways, and (3) the gonadotropin and growth factor pathways in a global signaling cascade will provide the key element to control developmental competence. We assume that some overlaps should exist between these signaling schemes. Such signaling overlaps will have increased chances to include the promising pathways. Valuable data reported in both human and animal models should be combined together with particular attention to inter-species physiological differences but also similarities. The identification of a central pathway of competence could allow the definition of the target pathway/gene (in culture, *in vivo* or intrafollicular) to induce/inhibit in order to safely and accurately induce oocyte competence. Consequently, the efficacy, cost, and safety of ART in both livestock industry and human infertility treatment could be improved.

10.10 Conclusion

Oocyte competence is a complex process that has yet to be fully determined. The successful achievement of each one of the levels of competence discussed in this chapter requires the ideal and well-orchestrated effect of sequences of molecules (and molecular pathways) at specific space, time, and magnitude levels. Therefore, optimal competence should be synonymous with successful nuclear and cytoplasmic maturation and functional activation of the embryonic genome followed by developmental competence to blastocyst, as well as healthy progeny. Partial or total failure of any of these steps should affect competence achievement. On the other hand, successful achievement of previous events of oocyte maturation does not guarantee completion of the following steps. Due to its complexity, there is currently no reliable way to measure oocyte competence prior to phenotype analysis following fertilization. The emergence of genomic approaches to find objective, quantitative, and noninvasive biomarkers is an interesting progress in oocyte competence assessment. Once confirmed, these reliable gene biomarkers should provide a highly discriminative tool to reinforce the morphological and biochemical factors used for prediction of oocyte developmental competence. This capacity should increase pregnancy rate, reduce multiple pregnancy incidences, and both improve the efficiency of the elective SET procedure in humans and increase the speed of genetic progress in animals.

References

Adriaenssens, T., Segers, I., Wathlet, S., & Smitz, J. (2011). The cumulus cell gene expression profile of oocytes with different nuclear maturity and potential for blastocyst formation. *Journal of Assistive Reproduction and Genetics, 28*, 31–40.

Alm, H., Torner, H., Lohrke, B., Viergutz, T., Ghoneim, I. M., & Kanitz, W. (2005). Bovine blastocyst development rate in vitro is influenced by selection of oocytes by brillant cresyl blue staining before IVM as indicator for glucose-6-phosphate dehydrogenase activity. *Theriogenology, 63*, 2194–2205.

Assidi, M., Dieleman, S. J., & Sirard, M. A. (2010). Cumulus cell gene expression following the LH surge in bovine preovulatory follicles: Potential early markers of oocyte competence. *Reproduction, 140*, 835–852.

Assidi, M., Dufort, I., Ali, A., Hamel, M., Algriany, O. et al. (2008). Identification of potential markers of oocyte competence expressed in bovine cumulus cells matured with follicle-stimulating hormone and/or phorbol myristate acetate in vitro. *Biology of Reproduction, 79*, 209–222.

Assidi, M., Montag, M., Van Der Ven, K., & Sirard, M. A. (2011). Biomarkers of human oocyte developmental competence expressed in cumulus cells before ICSI: A preliminary study. *Journal of Assistive Reproduction and Genetics* (in press).

Assou, S., Anahory, T., Pantesco, V., Le Carrour, T., Pellestor, F. et al. (2006). The human cumulus–oocyte complex gene-expression profile. *Human Reproduction, 21*, 1705–1719.

Assou, S., Haouzi, D., De Vos, J., & Hamamah, S. (2010). Human cumulus cells as biomarkers for embryo and pregnancy outcomes. *Molecular Human Reproduction*.

Assou, S., Haouzi, D., Mahmoud, K., Aouacheria, A., Guillemin, Y. et al. (2008). A non-invasive test for assessing embryo potential by gene expression profiles of human cumulus cells: a proof of concept study. *Molecular Human Reproduction, 14*, 711–719.

Atef, A., Francois, P., Christian, V., & Marc-Andre, S. (2005). The potential role of gap junction communication between cumulus cells and bovine oocytes during in vitro maturation. *Molecular Reproduction and Development, 71*, 358–367.

Barnes, F. L., & First, N. L. (1991). Embryonic transcription in in vitro cultured bovine embryos. *Molecular Reproduction and Development, 29*, 117–123.

Bedard, J., Brule, S., Price, C. A., Silversides, D. W., & Lussier, J. G. (2003). Serine protease inhibitor-E2 (SERPINE2) is differentially expressed in granulosa cells of dominant follicle in cattle. *Molecular Reproduction and Development, 64*, 152–165.

Bettegowda, A., Patel, O. V., Lee, K. B., Park, K. E., Salem, M. et al. (2008). Identification of novel bovine cumulus cell molecular markers predictive of oocyte competence: Functional and diagnostic implications. *Biology of Reproduction, 79*, 301–309.

Biggers, J. D., Whittingham, D. G., & Donahue, R. P. (1967). The pattern of energy metabolism in the mouse oocyte and zygote. *Proceedings of the National Academy of Science USA, 58*, 560–567.

Blondin, P., Bousquet, D., Twagiramungu, H., Barnes, F., & Sirard, M. A. (2002). Manipulation of follicular development to produce developmentally competent bovine oocytes. *Biology of Reproduction, 66*, 38–43.

Blondin, P., & Sirard, M. A. (1995). Oocyte and follicular morphology as determining characteristics for developmental competence in bovine oocytes. *Molecular Reproduction and Development, 41*, 54–62.

Burns, K. H., Owens, G. E., Ogbonna, S. C., Nilson, J. H., & Matzuk, M. M. (2003). Expression profiling analyses of gonadotropin responses and tumor development in the absence of inhibins. *Endocrinology, 144*, 4492–4507.

Cetica, P. D., Pintos, L. N., Dalvit, G. C., & Beconi, M. T. (1999). Effect of lactate dehydrogenase activity and isoenzyme localization in bovine oocytes and utilization of oxidative substrates on in vitro maturation. *Theriogenology, 51*, 541–550.

Coffer, P. J., Jin, J., & Woodgett, J. R. (1998). Protein kinase B (c-Akt): A multifunctional mediator of phosphatidylinositol 3-kinase activation. *Biochemistry Journal, 335*(Pt 1), 1–13.

Crozet, N., Kanka, J., Motlik, J., & Fulka, J. (1986). Nucleolar fine structure and RNA synthesis in bovine oocytes from antral follicles. *Gamete Research, 14*, 65–73.

Dessie, S. W., Rings, F., Holker, M., Gilles, M., Jennen, D. et al. (2007). Dielectrophoretic behavior of in vitro-derived bovine metaphase II oocytes and zygotes and its relation to in vitro embryonic developmental competence and mRNA expression pattern. *Reproduction, 133*, 931–946.

Dieleman, S. J., Kruip, T. A., Fontijne, P., de Jong, W. H., & van der Weyden, G. C. (1983). Changes in oestradiol, progesterone and testosterone concentrations in follicular fluid and in the micromorphology of preovulatory bovine follicles relative to the peak of luteinizing hormone. *Journal of Endocrinology, 97*, 31–42.

Downs, S. M., & Chen, J. (2008). EGF-like peptides mediate FSH-induced maturation of cumulus cell-enclosed mouse oocytes. *Molecular Reproduction and Development, 75*, 105–114.

Elvin, J. A., Yan, C., & Matzuk, M. M. (2000a). Oocyte-expressed TGF-beta superfamily members in female fertility. *Molecular Cell Endocrinology, 159*, 1–5.

Eppig, J. J. (1976). Analysis of mouse oogenesis in vitro. Oocyte isolation and the utilization of exogenous energy sources by growing oocytes. *Journal of Experimental Zoology, 198*, 375–382.

Eppig, J. J. (2001). Oocyte control of ovarian follicular development and function in mammals. *Reproduction, 122*, 829–838.

Eppig, J. J., Wigglesworth, K., & Pendola, F. L. (2002). The mammalian oocyte orchestrates the rate of ovarian follicular development. *Proceedings of the National Academy of Science USA, 99*, 2890–2894.

Evans, A. C., Ireland, J. L., Winn, M. E., Lonergan, P., Smith, G. W. et al. (2004). Identification of genes involved in apoptosis and dominant follicle development during follicular waves in cattle. *Biology of Reproduction, 70*, 1475–1484.

Fair, T., Hulshof, S. C., Hyttel, P., Greve, T., & Boland, M. (1997). Oocyte ultrastructure in bovine primordial to early tertiary follicles. *Anatomy and Embryology (Berlin), 195*, 327–336.

Fair, T., Hyttel, P., & Greve, T. (1995). Bovine oocyte diameter in relation to maturational competence and transcriptional activity. *Molecular Reproduction and Development, 42*, 437–442.

Familiari, G., Relucenti, M., Heyn, R., Micara, G., & Correr, S. (2006). Three-dimensional structure of the zona pellucida at ovulation. *Microscopic Research Techologies, 69*, 415–426.

Fan, H. Y., & Sun, Q. Y. (2004). Involvement of mitogen-activated protein kinase cascade during oocyte maturation and fertilization in mammals. *Biology of Reproduction, 70*, 535–547.

Farin, P. W., Piedrahita, J. A., & Farin, C. E. (2006). Errors in development of fetuses and placentas from in vitro-produced bovine embryos. *Theriogenology, 65*, 178–191.

Fayad, T., Levesque, V., Sirois, J., Silversides, D. W., & Lussier, J. G. (2004). Gene expression profiling of differentially expressed genes in granulosa cells of bovine dominant follicles using suppression subtractive hybridization. *Biology of Reproduction, 70*, 523–533.

Feng, W. G., Sui, H. S., Han, Z. B., Chang, Z. L., Zhou, P. et al. (2007). Effects of follicular atresia and size on the developmental competence of bovine oocytes: a study using the well-in-drop culture system. *Theriogenology, 67*, 1339–1350.

Ferrari, S., Lattuada, D., Paffoni, A., Brevini, T. A., Scarduelli, C. et al. (2010). Procedure for rapid oocyte selection based on quantitative analysis of cumulus cell gene expression. *Journal of Assistive Reproduction and Genetics.*

Feuerstein, P., Cadoret, V., Dalbies-Tran, R., Guerif, F., Bidault, R., & Royere, D. (2007). Gene expression in human cumulus cells: one approach to oocyte competence. *Human Reproduction, 22*, 3069–3077.

Fluker, M. R., Hooper, W. M., & Yuzpe, A. A. (1999). Withholding gonadotropins ("coasting") to minimize the risk of ovarian hyperstimulation during superovulation and in vitro fertilization-embryo transfer cycles. *Fertility and Sterility, 71*, 294–301.

Gilbert, I., Robert, C., Dieleman, S. J., Blondin, P., & Sirard, M. A. (2010). Transcriptional effect of the luteinizing hormone surge in bovine granulosa cells during the peri-ovulation period. *Reproduction.*

Gilchrist, R. B., Lane, M., & Thompson, J. G. (2008). Oocyte-secreted factors: regulators of cumulus cell function and oocyte quality. *Human Reproduction Update, 14*, 159–177.

Gilchrist, R. B., Ritter, L. J., & Armstrong, D. T. (2004). Oocyte-somatic cell interactions during follicle development in mammals. *Animal Reproductive Science, 82–83*, 431–446.

Ginther, O. J. (2000). Selection of the dominant follicle in cattle and horses. *Animal Reproductive Science, 60–61*, 61–79.

Ginther, O. J., Bergfelt, D. R., Kulick, L. J., & Kot, K. (2000). Selection of the dominant follicle in cattle: Role of two-way functional coupling between follicle-stimulating hormone and the follicles. *Biology of Reproduction, 62*, 920–927.

Gook, D. A., Edgar, D. H., Borg, J., & Martic, M. (2008). Detection of zona pellucida proteins during human folliculogenesis. *Human Reproduction, 23*, 394–402.

Goyanes, V. J., Ron-Corzo, A., Costas, E., & Maneiro, E. (1990). Morphometric categorization of the human oocyte and early conceptus. *Human Reproduction, 5*, 613–618.

Grondahl, M. L., Borup, R., Lee, Y. B., Myrhoj, V., Meinertz, H., & Sorensen, S. (2009). Differences in gene expression of granulosa cells from women undergoing controlled ovarian hyperstimulation with either recombinant follicle-stimulating hormone or highly purified human menopausal gonadotropin. *Fertility and Sterility, 91*, 1820–1830.

Hamel, M., Dufort, I., Robert, C., Gravel, C., Leveille, M. C. et al. (2008). Identification of differentially expressed markers in human follicular cells associated with competent oocytes. *Human Reproduction, 23*, 1118–1127.

Hamel, M., Dufort, I., Robert, C., Leveille, M. C., Leader, A., & Sirard, M. A. (2010a). Genomic assessment of follicular marker genes as pregnancy predictors for human IVF. *Molecular Human Reproduction, 16*, 87–96.

Hamel, M., Dufort, I., Robert, C., Leveille, M. C., Leader, A., & Sirard, M. A. (2010b). Identification of follicular marker genes as pregnancy predictors for human IVF: New evidence for the involvement of luteinization process. *Molecular Human Reproduction, 16*, 548–556.

Henrion, G., Renard, J. P., Chesne, P., Oudin, J. F., Maniey, D. et al. (2000). Differential regulation of the translation and the stability of two maternal transcripts in preimplantation rabbit embryos. *Molecular Reproduction and Development, 56*, 12–25.

Hernandez-Gonzalez, I., Gonzalez-Robayna, I., Shimada, M., Wayne, C. M., Ochsner, S. A. et al. (2006). Gene expression profiles of cumulus cell oocyte complexes during ovulation reveal cumulus cells express neuronal and immune-related genes: Does this expand their role in the ovulation process? *Molecular Endocrinology, 20*, 1300–1321.

Huang, Z., & Wells, D. (2010). The human oocyte and cumulus cells relationship: new insights from the cumulus cell transcriptome. *Molecular Human Reproduction.*

Humblot, P., Holm, P., Lonergan, P., Wrenzycki, C., Lequarre, A. S. et al. (2005). Effect of stage of follicular growth during superovulation on developmental competence of bovine oocytes. *Theriogenology, 63*, 1149–1166.

Hyttel, P., Callesen, H., & Greve, T. (1986). Ultrastructural features of preovulatory oocyte maturation in superovulated cattle. *Journal of Reproduction and Fertility, 76*, 645–656.

Hyttel, P., Fair, T., Callesen, H., & Greve, T. (1997). Oocyte growth, capacitation and final maturation in cattle *Theriogenology, 47*, 23–32.

Isaza, V., Garcia-Velasco, J. A., Aragones, M., Remohi, J., Simon, C., & Pellicer, A. (2002). Oocyte and embryo quality after coasting: The experience from oocyte donation. *Human Reproduction, 17*, 1777–1782.

Jones, G. M., Cram, D. S., Song, B., Magli, M. C., Gianaroli, L. et al. (2008). Gene expression profiling of human oocytes following in vivo or in vitro maturation. *Human Reproduction, 23*, 1138–1144.

Kastrop, P. M., Bevers, M. M., Destree, O. H., & Kruip, T. A. (1991a). Protein synthesis and phosphorylation patterns of bovine oocytes maturing in vivo. *Molecular Reproduction and Development, 29*, 271–275.

Kastrop, P. M., Hulshof, S. C., Bevers, M. M., Destree, O. H., & Kruip, T. A. (1991b). The effects of alpha-amanitin and cycloheximide on nuclear progression, protein synthesis, and phosphorylation during bovine oocyte maturation in vitro. *Molecular Reproduction and Development, 28*, 249–254.

Koks, S., Velthut, A., Sarapik, A., Altmae, S., Reinmaa, E. et al. (2010). The differential transcriptome and ontology profiles of floating and cumulus granulosa cells in stimulated human antral follicles. *Molecular Human Reproduction, 16*, 229–240.

Krisher, R. L., & Bavister, B. D. (1999). Enhanced glycolysis after maturation of bovine oocytes in vitro is associated with increased developmental competence. *Molecular Reproduction and Development, 53*, 19–26.

Kruip, T. A. M., Cran, D. G., van Beneden, T. H., & Dieleman, S. J. (1983) Structural changes in bovine oocytes during final maturation in vivo. *Gamete Research, 8*, 29–47.

Lasiene, K., Vitkus, A., Valanciute, A., & Lasys, V. (2009). Morphological criteria of oocyte quality. *Medicina (Kaunas), 45*, 509–515.

Le Bouc, Y., Rossignol, S., Azzi, S., Steunou, V., Netchine, I., & Gicquel, C. (2010). Epigenetics, genomic imprinting and assisted reproductive technology. *Annals of Endocrinology (Paris), 71*, 237–238.

Leese, H. J., & Barton, A. M. (1985). Production of pyruvate by isolated mouse cumulus cells. *Journal of Experimental Zoology, 234*, 231–236.

Lefevre, B. (2008). The nucleolus of the maternal gamete is essential for life. *Bioessays, 30*, 613–616.

Lequarre, A. S., Traverso, J. M., Marchandise, J., & Donnay, I. (2004). Poly(A) RNA is reduced by half during bovine oocyte maturation but increases when meiotic arrest is maintained with CDK inhibitors. *Biology of Reproduction, 71*, 425–431.

Levesque, J. T., & Sirard, M. A. (1996). Resumption of meiosis is initiated by the accumulation of cyclin B in bovine oocytes. *Biology of Reproduction, 55*, 1427–1436.

Liu, Y., Sui, H. S., Wang, H. L., Yuan, J. H., Luo, M. J. et al. (2006). Germinal vesicle chromatin configurations of bovine oocytes. *Microscopic Research Technologies, 69*, 799–807.

Lonergan, P., Monaghan, P., Rizos, D., Boland, M. P., & Gordon, I. (1994). Effect of follicle size on bovine oocyte quality and developmental competence following maturation, fertilization, and culture in vitro. *Molecular Reproduction and Development, 37*, 48–53.

Lonergan, P., Rizos, D., Kanka, J., Nemcova, L., Mbaye, A. M. et al. (2003). Temporal sensitivity of bovine embryos to culture environment after fertilization and the implications for blastocyst quality. *Reproduction, 126*, 337–346.

Madaschi, C., Aoki, T., de Almeida Ferreira Braga, D. P., de Cassia Savio Figueira, R., Semiao Francisco, L. et al. (2009). Zona pellucida birefringence score and meiotic spindle visualization in relation to embryo development and ICSI outcomes. *Reproductive Biomedicine Online, 18*, 681–686.

Marchal, R., Vigneron, C., Perreau, C., Bali-Papp, A., & Mermillod, P. (2002). Effect of follicular size on meiotic and developmental competence of porcine oocytes. *Theriogenology, 57*, 1523–1532.

Massicotte, L., Coenen, K., Mourot, M., & Sirard, M. A. (2006). Maternal housekeeping proteins translated during bovine oocyte maturation and early embryo development. *Proteomics, 6*, 3811–3820.

Matzuk, M. M., Burns, K. H., Viveiros, M. M., & Eppig, J. J. (2002). Intercellular communication in the mammalian ovary: oocytes carry the conversation. *Science, 296*, 2178–2180.

Mayes, M. A., Laforest, M. F., Guillemette, C., Gilchrist, R. B., & Richard, F. J. (2007). Adenosine 5'-monophosphate kinase-activated protein kinase (PRKA) activators delay meiotic resumption in porcine oocytes. *Biology of Reproduction, 76*, 589–597.

McEvoy, T. G., Ashworth, C. J., Rooke, J. A., & Sinclair, K. D. (2003). Consequences of manipulating gametes and embryos of ruminant species. *Reproduction Supplement, 61*, 167–182.

Memili, E., Dominko, T., & First, N. L. (1998). Onset of transcription in bovine oocytes and preimplantation embryos. *Molecular Reproduction and Development, 51*, 36–41.

Mermillod, P., & Marchal, R. (1999). La maturation de l'ovocyte de mammifères. *médecine/sciences m/s n° 2, 15*, 148–156.

Merton, J. S., de Roos, A. P., Mullaart, E., de Ruigh, L., Kaal, L. et al. (2003). Factors affecting oocyte quality and quantity in commercial application of embryo technologies in the cattle breeding industry. *Theriogenology, 59*, 651–674.

Michaels, G. S., Hauswirth, W. W., & Laipis, P. J. (1982). Mitochondrial DNA copy number in bovine oocytes and somatic cells. *Developments in Biology, 94*, 246–251.

Mihm, M., Baker, P. J., Ireland, J. L., Smith, G. W., Coussens, P. M. et al. (2006). Molecular evidence that growth of dominant follicles involves a reduction in follicle-stimulating hormone dependence and an increase in luteinizing hormone dependence in cattle. *Biology of Reproduction, 74*, 1051–1059.

Miyano, T., & Manabe, N. (2007). Oocyte growth and acquisition of meiotic competence. *Society of Reproduction and Fertility Supplement, 63*, 531–538.

Miyara, F., Migne, C., Dumont-Hassan, M., Le Meur, A., Cohen-Bacrie, P. et al. (2003). Chromatin configuration and transcriptional control in human and mouse oocytes. *Molecular Reproduction and Development, 64*, 458–470.

Montag, M., Schimming, T., Koster, M., Zhou, C., Dorn, C. et al. (2008). Oocyte zona birefringence intensity is associated with embryonic implantation potential in ICSI cycles. *Reproductive Biomedicine Online, 16*, 239–244.

Montag, M., & van der Ven, H. (2008). Symposium: Innovative techniques in human embryo viability assessment. Oocyte assessment and embryo viability prediction: Birefringence imaging. *Reproductive Biomedicine Online, 17*, 454–460.

Ng, S. T., Chang, T. H., & Wu, T. C. (1999). Prediction of the rates of fertilization, cleavage, and pregnancy success by cumulus-coronal morphology in an in vitro fertilization program. *Fertility and Sterility, 72*, 412–417.

Odom, L. N., & Segars, J. (2010) Imprinting disorders and assisted reproductive technology. *Current Opinion in Endocrinology, Diabetes, and Obesity, 17*, 517–522.

Ogushi, S., Palmieri, C., Fulka, H., Saitou, M., Miyano, T., & Fulka, Jr., J. (2008). The maternal nucleolus is essential for early embryonic development in mammals. *Science, 319*, 613–616.

Otoi, T., Yamamoto, K., Koyama, N., Tachikawa, S., & Suzuki, T. (1997). Bovine oocyte diameter in relation to developmental competence. *Theriogenology, 48*, 769–774.

Panigone, S., Hsieh, M., Fu, M., Persani, L., & Conti, M. (2008). LH signaling in preovulatory follicles involves early activation of the EGFR pathway. *Molecular Endocrinology*.

Qian, Y., Shi, W. Q., Ding, J. T., Sha, J. H., & Fan, B. Q. (2003). Predictive value of the area of expanded cumulus mass on development of porcine oocytes matured and fertilized in vitro. *Journal of Reproduction and Development, 49*, 167–174.

Quirk, S. M., Cowan, R. G., Harman, R. M., Hu, C. L., & Porter, D. A. (2004). Ovarian follicular growth and atresia: the relationship between cell proliferation and survival. *Journal of Animal Science, 82*(E-Suppl.), E40–E52.

Richards, J. S. (2007). Genetics of ovulation. *Seminars in Reproductive Medicine, 25*, 235–242.

Richards, J. S., Russell, D. L., Ochsner, S., Hsieh, M., Doyle, K. H. et al. (2002a). Novel signaling pathways that control ovarian follicular development, ovulation, and luteinization. *Recent Programs in Hormonal Research, 57*, 195–220.

Robert, C., Gagne, D., Bousquet, D., Barnes, F. L., & Sirard, M. A. (2001). Differential display and suppressive subtractive hybridization used to identify granulosa cell messenger rna associated with bovine oocyte developmental competence. *Biology of Reproduction, 64*, 1812–1820.

Robert, C., Gagne, D., Lussier, J. G., Bousquet, D., Barnes, F. L., & Sirard, M. A. (2003). Presence of LH receptor mRNA in granulosa cells as a potential marker of oocyte developmental competence and characterization of the bovine splicing isoforms. *Reproduction, 125*, 437–446.

Rosen, M. P., Shen, S., Dobson, A. T., Rinaudo, P. F., McCulloch, C. E., & Cedars, M. I. (2008). A quantitative assessment of follicle size on oocyte developmental competence. *Fertility and Sterility, 90*, 684–690.

Sher, G., Zouves, C., Feinman, M., & Maassarani, G. (1995). 'Prolonged coasting': An effective method for preventing severe ovarian hyperstimulation syndrome in patients undergoing in-vitro fertilization. *Human Reproduction, 10*, 3107–3109.

Simon, C., Garcia Velasco, J. J., Valbuena, D., Peinado, J. A., Moreno, C. et al. (1998). Increasing uterine receptivity by decreasing estradiol levels during the preimplantation period in high responders with the use of a follicle-stimulating hormone step-down regimen. *Fertility and Sterility, 70*, 234–239.

Sirard, M. A., Florman, H. M., Leibfried-Rutledge, M. L., Barnes, F. L., Sims, M. L., & First, N. L. (1989). Timing of nuclear progression and protein synthesis necessary for meiotic maturation of bovine oocytes. *Biology of Reproduction, 40*, 1257–1263.

Sirard, M. A., Richard, F., Blondin, P., & Robert, C. (2006). Contribution of the oocyte to embryo quality. *Theriogenology, 65*, 126–136.

Sirard, M. A., & Trounson, A. (2003). Follicular factors affecting oocyte maturation and developmental competence. In *Biology and Pathology of the Oocyte: Its Role in Fertility and Reproductive Medecine* (pp. 305–315).

Su, Y. Q., Sugiura, K., Wigglesworth, K., O'Brien, M. J., Affourtit, J. P. et al. (2008). Oocyte regulation of metabolic cooperativity between mouse cumulus cells and oocytes: BMP15 and GDF9 control cholesterol biosynthesis in cumulus cells. *Development, 135*, 111–121.

Sugiura, K., Pendola, F. L., & Eppig, J. J. (2005). Oocyte control of metabolic cooperativity between oocytes and companion granulosa cells: Energy metabolism. *Developments in Biology, 279*, 20–30.

Sugiura, K., Su, Y. Q., Diaz, F. J., Pangas, S. A., Sharma, S. et al. (2007). Oocyte-derived BMP15 and FGFs cooperate to promote glycolysis in cumulus cells. *Development, 134*, 2593–2603.

Sun, Q. Y. (2003). Cellular and molecular mechanisms leading to cortical reaction and polyspermy block in mammalian eggs. *Microscopic Research Technologies, 61*, 342–348.

Sutton, M. L., Gilchrist, R. B., & Thompson, J. G. (2003). Effects of in-vivo and in-vitro environments on the metabolism of the cumulus-oocyte complex and its influence on oocyte developmental capacity. *Human Reproduction Update, 9*, 35–48.

Sutton-McDowall, M., Gilchrist, R., & Thompson, J. (2010). The pivotal role of glucose metabolism in determining oocyte developmental competence. *Reproduction.*

Swain, J. E., & Pool, T. B. (2008) ART failure: Oocyte contributions to unsuccessful fertilization. *Human Reproduction Update, 14*, 431–446.

Tan, J. H., Wang, H. L., Sun, X. S., Liu, Y., Sui, H. S., & Zhang, J. (2009). Chromatin configurations in the germinal vesicle of mammalian oocytes. *Molecular Human Reproduction, 15*, 1–9.

Tanghe, S., Van Soom, A., Mehrzad, J., Maes, D., Duchateau, L., & de Kruif, A. (2003) Cumulus contributions during bovine fertilization in vitro. *Theriogenology, 60*, 135–149.

Tanghe, S., Van Soom, A., Nauwynck, H., Coryn, M., & de Kruif, A. (2002). Minireview: Functions of the cumulus oophorus during oocyte maturation, ovulation, and fertilization. *Molecular Reproduction and Development, 61*, 414–424.

Tatemoto, H., & Terada, T. (1995). Time-dependent effects of cycloheximide and alpha-amanitin on meiotic resumption and progression in bovine follicular oocytes. *Theriogenology, 43*, 1107–1113.

Tesfaye, D., Ghanem, N., Carter, F., Fair, T., Sirard, M. A. et al. (2009). Gene expression profile of cumulus cells derived from cumulus-oocyte complexes matured either in vivo or in vitro. *Reproduction and Fertility Developments, 21*, 451–461.

Thompson, J. G., Lane, M., & Gilchrist, R. B. (2007). Metabolism of the bovine cumulus-oocyte complex and influence on subsequent developmental competence. *Society of Reproduction and Fertility Supplement, 64*, 179–190.

Tomek, W., Torner, H., & Kanitz, W. (2002a). Comparative analysis of protein synthesis, transcription and cytoplasmic polyadenylation of mRNA during maturation of bovine oocytes in vitro. *Reproduction of Domestic Animals, 37*, 86–91.

Torner, H., Ghanem, N., Ambros, C., Holker, M., Tomek, W. et al. (2008). Molecular and subcellular characterisation of oocytes screened for their developmental competence based on glucose-6-phosphate dehydrogenase activity. *Reproduction, 135*, 197–212.

Tremblay, K., Vigneault, C., McGraw, S., & Sirard, M. A. (2005). Expression of cyclin B1 messenger RNA isoforms and initiation of cytoplasmic polyadenylation in the bovine oocyte. *Biology of Reproduction, 72*, 1037–1044.

Tsafriri, A., Cao, X., Ashkenazi, H., Motola, S., Popliker, M., & Pomerantz, S. H. (2005). Resumption of oocyte meiosis in mammals: on models, meiosis activating sterols, steroids and EGF-like factors. *Molecular and Cell Endocrinology, 234*, 37–45.

Vassena, R., Mapletoft, R. J., Allodi, S., Singh, J., & Adams, G. P. (2003). Morphology and developmental competence of bovine oocytes relative to follicular status. *Theriogenology, 60*, 923–932.

Vigneault, C., McGraw, S., Massicotte, L., & Sirard, M. A. (2004). Transcription factor expression patterns in bovine in vitro-derived embryos prior to maternal-zygotic transition. *Biology of Reproduction, 70*, 1701–1709.

Wassarman, P. M., & Albertini, D. F. (1994). The mammalian ovum. In *The Physiology of Reproduction* (pp. 79–122).

Wells, D., & Patrizio, P. (2008). Gene expression profiling of human oocytes at different maturational stages and after in vitro maturation. *American Journal of Obstetrics and Gynecology, 198*(455), e451–e459.

Weston, A., & Sommerville, J. (2006). Xp54 and related (DDX6-like) RNA helicases: Roles in messenger RNP assembly, translation regulation and RNA degradation. *Nucleic Acids Research, 34*, 3082–3094.

Yung, Y., Maman, E., Konopnicki, S., Cohen, B., Brengauz, M. et al. (2010). ADAMTS-1: A new human ovulatory gene and a cumulus marker for fertilization capacity. *Molecular and Cell Endocrinology, 328*, 104–108.

Zeuner, A., Muller, K., Reguszynski, K., & Jewgenow, K. (2003). Apoptosis within bovine follicular cells and its effect on oocyte development during in vitro maturation. *Theriogenology, 59*, 1421–1433.

11 *In Vitro* Maturation Environment Affects Developmental Outcome

Pat Lonergan

11.1 Introduction

In vitro embryo production (IVP) has been around in domestic mammals, particularly cattle, for more than two decades. It is probably fair to say that one of the main uses of this technology has been in research: as a tool to better understand what happens during normal *in vivo* embryo development; as a stepping stone to other technologies, such as nuclear transfer and transgenesis; and as a tool for examining what goes wrong in development following such procedures, and indeed as a model from which to extrapolate findings to other species. However, since the advent of ovum-pick up (OPU) technology in the early 1990s it has also been used to generate offspring in breeding programs from valuable females, in particular in countries such as Brazil, where OPU/in vitro fertilization (IVF) currently outstrips superovulation as a means of generating embryos for commercial transfer (Stroud, 2011).

There are distinct challenges associated with IVP technology. These challenges include, or are derived from, the fact that the origin of the oocytes used (identity or genetic merit of the donor, stage of estrous cycle, stage of follicular wave, etc.) is often unknown, and therefore oocyte quality is very variable; there are distinct differences in the ability of sperm collected from different sires to fertilize oocytes *in vitro*, and because changes in post-fertilization culture conditions can dramatically alter the quality of the embryo. It is generally accepted that embryos produced *in vitro* are inferior in quality to those derived *in vivo*. There is a significant amount of evidence in the literature to support this statement based on morphological data, cryotolerance, transcript expression profiles, and, of course, pregnancy rates after transfer (Stroud, 2011). However, this must be qualified by the caveat that there are many types of culture systems *in vitro* and some are clearly better than others. The practical significance of the difference in quality is reflected in the data from commercial embryo transfer statistics compiled each year by the International Embryo Transfer Society (www.iets.org); in 2010, the latest year for which data are available, approximately 590,561 *in vivo*–derived embryos were transferred worldwide, of which more than half (327,525; 55.5%) were transferred fresh and the remainder (263,036; 44.5%) transferred frozen. In contrast, 339,685 IVP embryos were transferred in the same year, of which the vast majority (315,715; 93%) was transferred fresh (Lonergan et al., 2006). In addition, following transfer, problems with IVP-derived fetuses and calves have been reported, including abortions, increased birth weights, dystocia, and higher rates of neonatal mortality (Farin et al., 2006; Kruip & den Daas, 1997).

Oocyte Physiology and Development in Domestic Animals, First Edition. Edited by Rebecca L. Krisher.

Rather than focus on the entire process of *in vitro* embryo production, a subject which could fill, and has filled, entire books (e.g., Gordon, 2003), the aim of this chapter is to highlight some of the issues associated with *in vitro* maturation (IVM) and some approaches that have been taken to deal with these issues. The emphasis is placed on cattle due to widespread use of IVP embryos for both research and, more recently, for the production of offspring in breeding programs.

11.2 Oocyte Maturation *in Vivo*

Oocytes in most mammals enter into the early stages of meiosis during fetal life and become arrested at the diplotene stage of prophase I (the so-called germinal vesicle stage) until they are committed to ovulation or atresia. The bovine oocyte reaches its full size when the follicle enclosing it reaches a diameter of approximately 3 mm (Fair et al., 1995). Prior to resumption of meiosis, the oocyte is 2n but 4C (i.e., it has four times the haploid complement of DNA). Resumption of meiosis and progression through maturation results in arrest at the metaphase II stage with extrusion of the first polar body, and a DNA complement of 1n2C. Penetration of the sperm (1n1C) leads to extrusion of the second polar body, establishment of a 1n1C state in the oocyte leading to a diploid embryo (2n2C) after the first mitotic division following fertilization. All subsequent divisions are mitotic resulting in two identical daughter cells (Figure 11.1).

The estrous cycle of cattle is composed of two or three waves of follicular growth involving emergence of a new wave of follicles in association with a transient rise in follicle-stimulating hormone (FSH), growth of a follicular cohort and selection of a dominant follicle (a process whereby a follicle prevents growth of other follicles, or grows in a hormonal milieu unfit for growth of other follicles). Each follicular wave culminates in development of a single nonovulatory or ovulatory dominant follicle (Ireland et al., 2000; Roche, 1996). If the dominant follicle develops at a time when progesterone (P4) concentrations are low, it is exposed to an appropriate luteinizing hormone (LH) pulsatility pattern and will go on to ovulate. The trigger for resumption of meiosis in the oocyte within the dominant follicle is the preovulatory surge of LH, which triggers breakdown of the germinal vesicle and progression to metaphase II. Approximately 24 hours after the LH surge

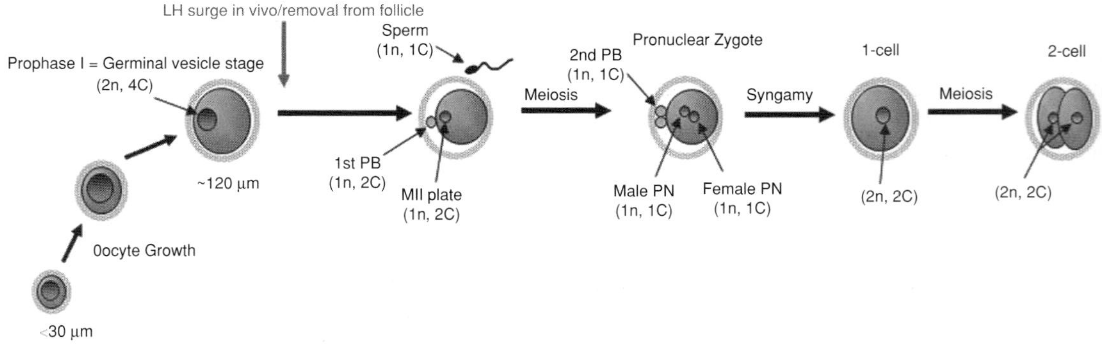

Figure 11.1 Oocyte growth and development. During development in the fetal ovary, oocytes in most mammals become arrested at the diplotene stage of prophase I (the germinal vesicle stage). Resumption of meiosis, triggered by the preovulatory surge of luteinizing hormone, and progression through maturation results in further arrest at the metaphase II stage with extrusion of the first polar body, and establishment of a DNA complement of 1n2C. Penetration of the sperm (1n1C) leads to extrusion of the second polar body and establishment of a 1n1C state in the oocyte leading to a diploid embryo (2n2C) after the first mitotic division following fertilization. All subsequent divisions are mitotic resulting in two identical daughter cells

ovulation takes place and now-matured oocyte is released for fertilization into the ampulla of the oviduct.

During the period from the preovulatory LH surge to ovulation, the follicular fluid changes from an environment dominated by oestradiol to one that is dominated by progesterone (Dieleman, 1983) as the granulosa cells luteinize in preparation for the formation of the corpus luteum after ovulation. Whether these changes in the follicle have a direct influence on the oocyte is unclear, but given that this is coincident with resumption of meiosis and maturation of the oocyte, a role in determining oocyte quality is likely. Interestingly, injection of bovine preovulatory follicles with trilostane, an inhibitor of 3 beta-hydroxysteroid dehydrogenase, an enzyme that catalyses the synthesis of P4 from pregnenolone (to block the preovulatory rise in intrafollicular P4) resulted in a significant decline in P4 production in the follicle but did not affect ovulation rate or corpus luteum function as determined by serum P4 on Days 1–9 of the subsequent luteal phase (Li et al., 2007). An effect on oocyte quality was not determined.

The central role of P4 in driving endometrial gene expression and conceptus elongation has been well demonstrated (Lonergan, 2011). Evidence of a role during oocyte maturation is less clear. The presence of both genomic and non-genomic P4 receptors in bovine oocytes and associated cumulus cells have been recently characterized as well as the effect of inhibiting P4 production by the cumulus cells *in vitro* on oocyte developmental competence (Aparicio et al., 2011). The dynamic changes observed in the protein expression of P4 receptors following *in vitro* maturation or in response to supplementation with LH, FSH, or P4 suggest a role for P4 during bovine oocyte maturation. Inhibition of cumulus cell P4 synthesis during maturation using trilostane resulted in a significant reduction in blastocyst formation rate, an effect which was abrogated by supplementation with exogenous P4 or a P4 agonist (Aparicio et al., 2011). Reduced embryo development following inhibition of P4 synthesis by cumulus cells or blocking of nuclear P4 receptor activity during *in vitro* maturation supports a role for progesterone in determining oocyte quality.

The follicular environment is responsible for both maintaining meiotic arrest of the oocyte at prophase I (GV stage) and for resumption of meiosis. High levels of intra-oocyte cAMP keep the oocyte in meiotic arrest by suppressing maturation promoting factor activity via stimulation of cAMP-dependent protein kinase A. The preovulatory gonadotrophin surge causes a drop in follicular and oocyte cGMP levels leading to upregulated oocyte phosphodiesterase activity, which causes a fall in intraoocyte cAMP and meiotic resumption (Norris et al., 2009; Vaccari et al., 2009). In addition, several changes occur at the level of the cytoplasm involving redistribution of intracellular organelles (Hyttel et al., 1989).

11.3 *In Vitro* Embryo Production

Techniques and procedures based on the use of IVF technology offer significant benefits in relation to genetic improvement and reproductive management of domestic ruminants as well as facilitating our understanding of the regulation of embryo development. The *in vitro* production of ruminant embryos is a three-step process involving oocyte maturation (IVM), oocyte fertilization (IVF), and embryo culture (IVC) (Figure 11.2). Oocyte maturation involves the collection of oocytes either from the ovaries of slaughtered heifers and cows, or from the ovaries of live animals using transvaginal ovum-pickup procedures (see below). Good quality oocytes, normally selected based on their morphological appearance, are matured for 24 hours, typically in Tissue Culture Medium 199 or a similar medium supplemented with gonadotrophins and/or growth factors in an atmosphere of 5% CO_2 in air. The first morphological sign that maturation has occurred is the expansion of the

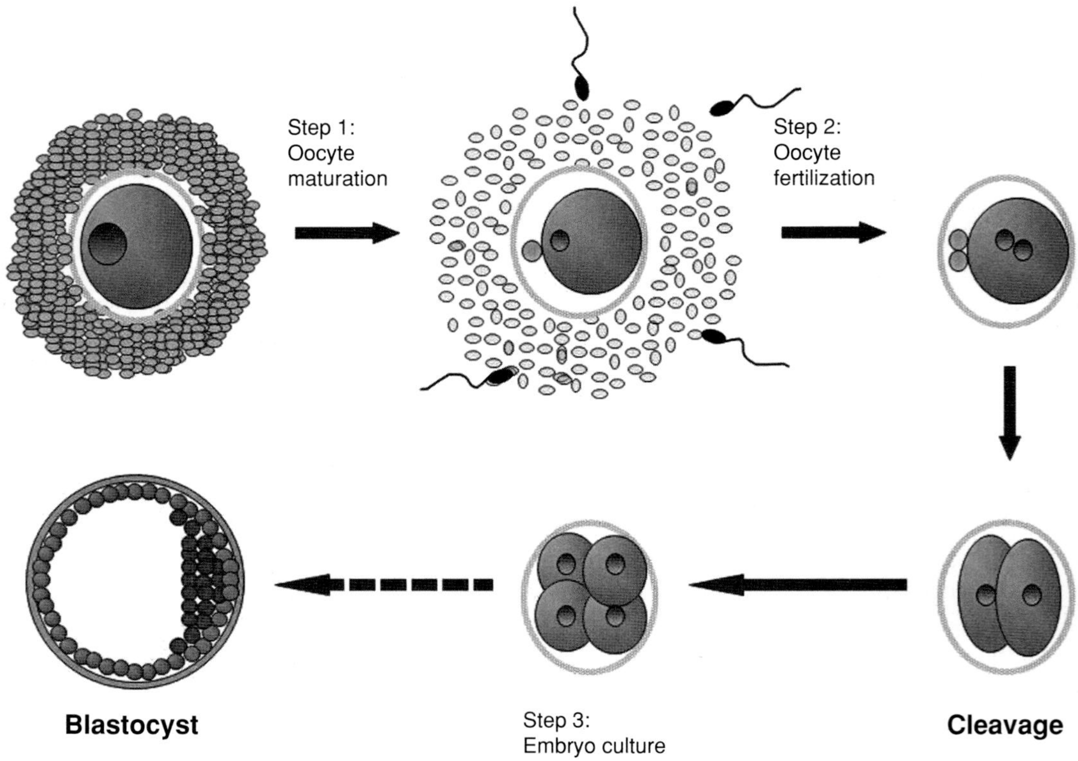

Figure 11.2 Schematic representation of the process of *in vitro* production of ruminant embryos, involving oocyte maturation, oocyte fertilization, and embryo culture.

cumulus cells away from the oocyte. Removal of the cumulus cells and more detailed examination of the oocyte itself will reveal the first polar body extruded into the perivitelline space, indicating attainment of the metaphase II stage, the stage at which the oocyte is ovulated *in vivo* and the stage at which fertilization must take place. For IVF, oocytes are inseminated *in vitro* with fresh or, more typically, frozen-thawed sperm. Gametes are co-incubated for up to 24 hours after which the presumptive zygotes are washed and cultured in one of a variety of culture media to the blastocyst stage.

In vitro, most studies terminate at the blastocyst or hatched blastocyst stage of development. This is a reflection of two things: (1) the blastocyst stage is, in cattle at least, the stage most routinely used for transfer or cryopreservation; and (2) *in vitro* systems do not support normal development of ruminant embryos through the post-hatching elongation stages. Artificial elongation has been induced *in vitro* using agarose gel tunnels (Brandao, 2004). Although such results are interesting, the technology still requires considerable refinement; for example, the embryonic disk characterizing the pre-streak stage 1 is never established (Alexopoulos et al., 2005; Vejlsted et al., 2006). Furthermore, available evidence indicates that the process of elongation is exclusively maternally driven. Up to the blastocyst stage the embryo is somewhat autonomous (i.e., it does not need contact with the maternal reproductive tract) as evidenced by the fact that blastocysts can be successfully developed *in vitro* in large numbers using IVF technology and then transferred to synchronized recipients. In contrast, development of the post-hatching and preimplantation conceptus is dependent on substances in the uterine lumen fluid, termed histotroph, that are derived from the endometrium, particularly the uterine glands, and stimulate growth and development. This is evidenced by the fact that post-hatching elongation does not occur *in vitro* (Brandao, 2004; Alexopoulos et al., 2005) and that the absence of uterine glands *in vivo* results in a failure of blastocysts to elongate (Gray et al., 2002; Spencer et al., 2006). Transfer of blastocysts to recipients and assessment of pregnancy and calving rate is the ultimate test of developmental ability. This is not feasible in most cases for practical and/or financial reasons; however, a multiple embryo transfer model can be used to assess the ability of the embryo to hatch and undergo elongation *in vivo* (Clemente et al., 2009; Forde et al., 2010). Given that most embryonic loss in cattle occurs before Day 16 (Diskin et al., 2008), such transfers allow an assessment of embryo survival through this critical window.

In terms of efficiency of IVP, in cattle, approximately 90% of immature oocytes, generally recovered from follicles at unknown stages of the estrous cycle, undergo nuclear maturation *in vitro* from prophase I to metaphase II; about 80% undergo fertilization following insemination and cleave at least once, to the two-cell stage. However, only 30–40% of such oocytes reach the blastocyst stage, at which they can be transferred to a recipient or frozen for future use. Thus, the major falloff in development occurs during the last part of the process (*in vitro* culture), between the 2-cell and blastocyst stages, suggesting that post-fertilization embryo culture is the most critical period of the process in terms of determining the blastocyst yield. This, however, is not the case. There is considerable evidence demonstrating that events further back along the developmental axis, in particular the quality of the oocyte, are crucial in determining the proportion of immature oocytes that form blastocysts, and that the post-fertilization culture environment, within certain limits, does not have a major influence on the capacity of the immature oocyte to form a blastocyst (Rizos et al., 2002). It would appear that once the oocyte has been removed from the follicle its ability to develop to the blastocyst stage is more or less sealed. Despite attempts at temporarily inhibiting resumption of meiosis to allow cytoplasmic maturation to proceed *in vitro*, thereby improving development, or modifications of maturation media, blastocyst yields *in vitro* using oocytes recovered from slaughtered heifers and cows rarely exceed 40% on a consistent basis. Therefore, it is likely that attempts at increasing oocyte competence applied before removal from the follicle (e.g., by manipulating follicle development) may be more beneficial than treatments applied after removal from the follicle.

11.4 Improving Oocyte Competence *before* Removal from the Follicle

Oocyte competence, defined as the ability of an oocyte to mature, be fertilized, and develop to the blastocyst stage, is progressively acquired during the period of oocyte growth accompanying follicular development. Pituitary gonadotropins and bidirectional local communication between the oocyte and adjacent somatic (cumulus) cells are critical for both nuclear and cytoplasmic maturation (acquisition of ability to complete meiosis, ensure monospermic fertilization, and undergo preimplantation development.

The follicular origin of the oocyte has a significant influence on its developmental potential, and it appears that once the oocyte is removed from its follicle its developmental potential is capped. Intrinsic oocyte developmental competence, as assessed by development to the blastocyst stage, has been positively associated with the size of the antral follicle from which it is recovered (Lonergan et al., 1994), the stage of the follicular wave (Hendriksen et al., 2004; Machatkova et al., 2004), and the site of maturation (in vivo vs. in vitro) (Greve et al., 1987; Leibfried-Rutledge et al., 1987;

Bordignon et al., 1997; van de Leemput et al., 1999; Dieleman et al., 2002; Rizos et al., 2002; Machatkova et al., 2004).

It is possible to manipulate follicular growth and in that way affect developmental competence of the immature oocyte after removal. A "coasting" period between hormonal stimulation and ovary collection (Blondin, Guilbault et al., 1997) as well as the time interval between ovary collection and oocyte aspiration (Blondin, Coenen et al., 1997) have been reported to significantly affect subsequent oocyte development. In an ensuing study, the best results were obtained when animals received 6 injections of FSH with a 48-hour coasting period; administering LH 6 hours before ovum pickup resulted in 80% of oocytes reaching the blastocyst stage *in vitro* (Blondin et al., 2002). More recently, the same group reported that the optimal period between FSH surge and follicle aspiration was 54 $\pm$ 7 hours and that a well-defined competence window is crucial to obtain optimal oocyte quality in ovarian-stimulated milking cows (Nivet *et al.*, 2012). These results are certainly encouraging. However, to gain widespread uptake, it would be comforting to see these data repeated by other groups given that in a retrospective study examining data collected over a 5-year period, varying the superovulatory treatment in terms of FSH type in association with a shorter or longer coasting period did not affect ovarian response or embryo developmental rates after IVF (Durocher et al., 2006).

11.5 Improving Oocyte Competence *after* Removal from the Follicle

Two broad approaches have been taken toward improving oocyte developmental competence after recovery from the follicle. The first has been the addition of putative growth- promoting substances to culture media (gonadotrophins, steroids, growth factors). Although modest improvements in development can be achieved in this way, the yield of blastocysts rarely consistently goes above 50%. Although addition of a variety of gonadotrophins, steroids, and growth factors has been reported to increase blastocyst development, the increases tend to be modest and rarely if ever approach that obtained with, for example, *in vivo* matured oocytes.

The second approach has been to try to mimic intrafollicular conditions for the oocyte where meiotic arrest is maintained. One explanation for the poor development of immature oocytes recovered from the ovaries of slaughtered heifers is that they are obtained primarily from small to medium-sized follicles that form a very heterogeneous pool and that, if selected to become dominant, are still at least several days away from possible ovulation. In contrast, the follicle that ovulates a mature oocyte at metaphase II grows to a size of 15–20 mm. Thus, typical oocytes submitted to *in vitro* maturation, although capable of high rates of nuclear maturation, have had insufficient time to undergo normal cytoplasmic maturation. In addition, the *in vivo*– matured oocyte matures in the presence of follicular fluid and a "positive" stimulus initiated by the LH surge; in contrast, the *in vitro*–matured oocyte matures following its removal from the "negative" (from a regulatory standpoint) follicular fluid environment without the positive stimulus initiated by the LH surge. It is therefore not surprising that outcomes are somewhat different.

It is well-known that dramatic changes occur in the oocyte nucleus during final growth and maturation as the follicle increases from 1 to 15–20 mm (Assey et al., 1994). Oocytes aspirated from the dominant follicle just before the LH surge display alterations in their nuclear and cytoplasmic morphology, which has been suggested as a prerequisite to the attainment of full developmental competence. This would indicate that not only final oocyte maturation (i.e., the process occurring from the LH surge to ovulation) is significant, but also that the period preceding the LH surge may be important for the establishment of developmental competence. Despite endless modifications of the IVM medium, there may be a biological limit on blastocyst yields achievable *in vitro*. It may well be that a prematuration treatment is necessary in order to allow oocytes from smaller follicles (i.e., those whose growth has been interrupted) to "catch up" with those from larger follicles or those totally matured *in vivo*. As part of the IVP procedure, IVM is initiated immediately following the removal of the immature oocyte from small antral follicles, and such oocytes may neither have the time nor the correct environment to complete the necessary changes required for subsequent successful development.

Although there is some evidence that modifying conditions of *in vitro* maturation can affect the subsequent blastocyst yield, improvements are generally modest and even in the best cases, approximately half of the oocytes fail to reach that stage. Some authors have attempted to improve oocyte developmental competence by artificially maintaining oocytes in meiotic arrest *in vitro* for varying periods (Sirard, 2001). Thus far, although it is possible to reversibly inhibit meiotic resumption without having any adverse effect on blastocyst formation rate, there is little evidence for a positive effect on oocyte competence. This may change, however, as recent publications using pharmacological compounds that allow synchronization of nuclear and cytoplasmic maturation processes within the oocyte and that prolong the oocyte maturation period in order to promote a longer interaction between the immature oocyte with adequately conditioned cumulus cells has led to promising results (Albuz et al., 2010; Gilchrist, 2011). As mentioned above, for widespread acceptance, it is important that these results are replicable by different groups worldwide.

11.6 Effect of Oocyte Environment on Embryo Gene Expression

In the cow, the fully grown oocyte is thought to be transcriptionally inactive, based on nucleolar morphology (Fair et al., 1996). Transcripts made and stored during the oocyte growth phase drive oocyte maturation, fertilization, and early embryo development, up to the switching on of the embryonic genome at the 8- to 16-cell stage (i.e., during the fourth cell cycle). There is convincing evidence to demonstrate that the environment to which the oocyte is exposed during maturation can influence the pattern of transcripts in the matured oocyte. Watson et al. (2000) examined the influence of *in vitro* maturation medium on the relative abundance of 5 transcripts in mature oocytes (*Na+/K+ ATPase* α-1 isoform, *bFGF*, *Cu/ZnSOD*, *cyclin A*, *cyclin B*). The greatest differences in relative abundance, and the lowest developmental frequencies, were observed following maturation in synthetic oviduct fluid in the absence of amino acids. Similarly, oocyte maturation *in vivo* gives rise to oocytes of superior developmental competence compared to those matured *in vitro* and this has been associated with differences in relative transcript abundance for a number of genes (Katz-Jaffe et al., 2009).

Until recently, relatively little attention had been focused on the influence of oocyte maturation conditions on gene expression in the resultant embryo. There is evidence to support the notion that the pattern of gene expression in the blastocyst is dictated for the most part by the post-fertilization conditions of culture (*in vivo* or *in vitro*). For example, several groups have reported that culture of *in vitro*–produced zygotes *in vivo* in the sheep oviduct results in embryos with a pattern of mRNA expression similar to that of true *in vivo*–derived embryos (Rizos et al., 2002). Furthermore, Knijn et al. (2002) examined transcript abundance in blastocysts derived from oocytes matured either *in vitro* or *in vivo* and found no differences for the panel of transcripts examined. This would suggest that blastocysts produced in a common post-fertilization culture environment have a similar transcript profile irrespective of the origin of the oocyte.

More recently, Fischer Russell et al. (2006) examined the impact of oocyte culture media with different protein supplements on the yield and quality of blastocysts obtained *in vitro*. Expression of a variety of transcripts involved in modulation of insulin growth factor bioavailability (IGFBP6), histone deacetylation (*Hdac*1 and *Hdac*2), trophoblast development (*Mash*2), maternal recognition of pregnancy (interferon-tau), environmental stress (*Hsp*), and trophoblast function (*Oct*-4) was significantly affected by the basal maturation medium (TCM 199 versus SOF) and the presence and type of protein supplementation (BSA or serum). In a similar study, Warzych et al. (2007) examined the effect of different protein (fatty acid–free BSA, serum) and macromolecule (PVP40) supplementation to oocyte maturation medium on the relative abundance of apoptosis and cell survival related genes in oocytes and hatched blastocysts. Only the expression of IGF-2 was altered in matured oocytes whereas transcript abundance of *Hsp*70, *IGF*IR, *IGF*2, and *IGF*2R was affected in blastocysts.

Thus, although the post-fertilization conditions of culture have a major influence on the pattern of mRNA abundance in the blastocyst, the conditions to which the oocyte is exposed during progression from prophase I to metaphase II, either *in vivo* in the follicle or *in vitro* in the maturation culture dish, can influence transcript abundance, not only in the matured oocyte but also further along the developmental axis at the blastocyst stage.

11.7 Use of IVM in Practice in Cattle

Transvaginal OPU was first developed in cattle in the early 1990s (Pieterse et al., 1991). Since then many studies have been carried out aimed at refining both technical (e.g., needle size, vacuum pressure, operator skill) and biological (e.g., hormonal stimulation, frequency of collection) factors. The advent of OPU opened up significant potential for the application of IVP technology, of which IVM is the first step, to animal breeding by allowing access to the ovaries of living high genetic merit females. Collecting the ovaries of slaughtered beef animals, although offering a fantastic source of raw material (oocytes and, after IVF, embryos) for basic research studies, has relatively little to offer in terms of genetic improvement. When coupled with semen from a high genetic merit sire, and even sex-sorted semen, the possibility of producing high quality embryos, genetically speaking, in large numbers became a reality. Nowhere is this more evident currently at a practical level than in Brazil where approximately 300,000 embryos are transferred annually, the vast majority of which are produced by OPU-IVP (Lonergan et al., 2006).

The ability to successfully culture oocytes and embryos individually facilitates the study of the relationship between follicle parameters and oocyte developmental competence, in order to identify markers of competent oocytes, as well as the ability to use small numbers of oocytes from an individual donor such as when OPU is carried out. However, culturing oocytes and embryos individually often results in lower developmental rates. Recent developments describe the culture of oocytes and embryos in groups in an individually identifiable manner using a variety of techniques (Matoba et al., 2010) (Figure 11.3). These systems of culture offer the double benefit of group culture while enabling the isolation and therefore identification of individual oocytes and embryos during the entire period of in vitro development.

In addition, IVM can be used to address physiological questions of significant practical importance. One good example is the study of the impact of metabolic conditions *in vivo* on the oocyte, such as the impact of elevated non-esterified fatty acids (NEFA) associated with negative energy balance in the postpartum dairy cow (Pieterse et al., 1991). In this study NEFA were added at

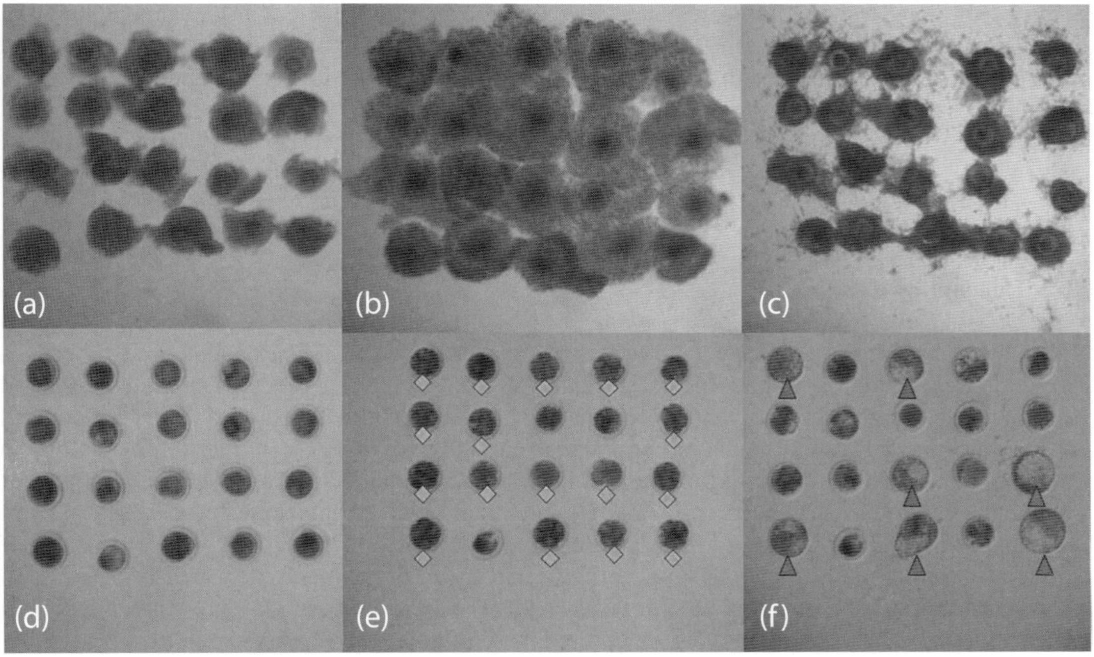

Figure 11.3 Representative images of the same batch of immature bovine oocytes cultured to Day 7 on Cell-Tak (Matoba et al., 2010). Twenty oocytes/embryos are placed on the bottom of the dish on Cell-Tak in a 5 × 4 grid formation in a 100 μl droplet of medium under mineral oil. For post-fertilization culture, embryos are placed at a maximum of 160 μm apart (d–f). (a) immature cumulus oocyte complexes (COCs), (b) *in vitro*–matured COCs, (c) presumptive zygotes 18–20 h after IVF, (d) denuded Day 1 zygotes, (e) Day 2 cleaved embryos (indicated by a diamond), and (f) Day 7 embryos. (For color detail, see color plate section.)

physiological concentrations to oocytes during IVM and were shown to negatively impact embryo development (Pieterse et al., 1991).

11.8 Long-Term Consequences of *in Vitro* Maturation

A growing body of evidence has raised concerns about the long-term health effects of assisted reproduction technologies (ART). These studies have largely been conducted in laboratory (mouse, rat) or domestic farmed species (sheep, cow), but the relevance to human health following ART is clear. The development of mammalian preimplantation embryos under suboptimal conditions can lead to the birth of offspring that display either immediate or long-term health concerns (Fleming et al., 2012). It is difficult to separate the effects of IVM (and perhaps IVF) from those of IVC (the longest part of the process of producing an embryo *in vitro* in domestic species). Specifically, it is unclear whether conditions of oocyte maturation can affect the long-term health of offspring. The finding that female mice fed a reduced protein diet specifically during the oocyte maturation period results in both physiological and behavioral aberrations in the long-term health of offspring (Watkins et al., 2008) raises concerns about *in vitro* oocyte maturation. Analyses of human offspring produced by IVM of oocytes detected no adverse health consequences (Cha et al., 2005; Mikkelsen, 2005; Soderstrom-Anttila et al., 2006). However, these studies did not report health status beyond early childhood. No long-term studies have been conducted in domestic species. Eppig et al. (2009)

examined the long-term health status and lifespans of offspring produced by *in vitro* maturation of mouse oocytes compared with that of oocytes induced to mature *in vivo* using gonadotrophin treatment. The study suggested that IVM has minimal effects on the long-term health of offspring. For example, there was no difference among the groups in lifespan or in most of the physiological and behavioral analyses. However, the pulse rate and cardiac output were slightly, but significantly, reduced in the *in vitro*–matured group compared with the *in vivo*–matured group. Surprisingly, these decreases were largely abrogated when IVM occurred in the absence of amino acids.

11.9 Concluding Comments

IVM is the starting point for many embryo technologies in domestic species and is used, albeit to a lesser extent currently, in human ART. The relatively poor blastocyst development rates observed following IVP in cattle (typically, 30–40%) are to a large extent associated with the use of oocytes that have not been exposed to the environmental conditions that occur as the dominant follicle grows toward ovulation. Thus far, successful attempts at dramatically improving developmental competence have been associated with strategies aimed at altering the oocyte before removal from the follicle. Manipulation of maturation conditions *in vitro*, including the short-term prevention of meiotic resumption to allow cytoplasmic maturation to occur, or the manipulation of IVM media in an attempt to mimic the dynamic environment of the ovulatory follicle, may prove beneficial and replicable in the future, but more work is required in these areas.

References

Albuz, F. K., Sasseville, M., Lane, M., Armstrong, D. T., Thompson, J. G., & Gilchrist, R. B. (2009). Simulated physiological oocyte maturation (SPOM): A novel in vitro maturation system that substantially improves embryo yield and pregnancy outcomes. *Human Reproduction, 25*, 2999–3011.

Alexopoulos, N. I., Vajta, G., Maddox-Hyttel, P., French, A. J., & Trounson, A. O. (2005). Stereomicroscopic and histological examination of bovine embryos following extended in vitro culture. *Reproduction and Fertility Development, 17*, 799–808.

Aparicio, I. M., Garcia-Herreros, M., O'Shea, L. C., Hensey, C., Lonergan, P., & Fair, T. (2011). Expression, regulation, and function of progesterone receptors in bovine cumulus oocyte complexes during in vitro maturation. *Biology of Reproduction*.

Assey, R. J., Hyttel, P., Greve, T., & Purwantara, B. (1994). Oocyte morphology in dominant and subordinate follicles. *Molecular Reproduction and Development, 37*, 335–344.

Blondin, P., Bousquet, D., Twagiramungu, H., Barnes, F., & Sirard, M. A. (2002). Manipulation of follicular development to produce developmentally competent bovine oocytes. *Biology of Reproduction, 66*, 38–43.

Blondin, P., Coenen, K., Guilbault, L. A., & Sirard, M. A. (1997). In vitro production of bovine embryos: Developmental competence is acquired before maturation. *Theriogenology, 47*, 1061–1075.

Blondin, P., Guilbault, L. A., & Sirard, M. A. (1997). The time interval between FSH-P administration and slaughter can influence the developmental competence of beef heifer oocytes. *Theriogenology, 48*, 803–813.

Bordignon, V., Morin, N., Durocher, J., Bousquet, D., & Smith, L. C. (1997). GnRH improves the recovery rate and the in vitro developmental competence of oocytes obtained by transvaginal follicular aspiration from superstimulated heifers. *Theriogenology, 48*, 291–298.

Brandao, D. O., Maddox-Hyttel, P., Lovendahl, P., Rumpf, R., Stringfellow, D., & Callesen, H. (2004). Post hatching development: A novel system for extended in vitro culture of bovine embryos. *Biology of Reproduction, 71*, 2048–2055.

Cha, K. Y., Chung, H. M., Lee, D. R., Kwon, H., Chung, M. K. et al. (2005). Obstetric outcome of patients with polycystic ovary syndrome treated by in vitro maturation and in vitro fertilization/embryo transfer. *Fertility and Sterility, 83*, 1461–1465.

Clemente, M., de La Fuente, J., Fair, T., Al Naib, A., Gutierrez-Adan, A. et al. (2009). Progesterone and conceptus elongation in cattle: A direct effect on the embryo or an indirect effect via the endometrium? *Reproduction, 138*, 507–517.

Dieleman, S. J., Bevers, M. M., Poortman, J., & van Tol, H. T. (1983). Steroid and pituitary hormone concentrations in the fluid of preovulatory bovine follicles relative to the peak of LH in the peripheral blood. *Journal of Reproduction and Fertility, 69*, 641–649.

Dieleman, S. J., Hendriksen, P. J., Viuff, D., Thomsen, P. D., Hyttel, P. et al. (2002). Effects of in vivo prematuration and in vivo final maturation on developmental capacity and quality of pre-implantation embryos. *Theriogenology, 57*, 5–20.

Diskin, M. G., & Morris, D. G. (2008). Embryonic and early foetal losses in cattle and other ruminants. *Reproduction of Domestic Animals, 43*(Suppl. 2), 260–267.

Durocher, J., Morin, N., & Blondin, P. (2006). Effect of hormonal stimulation on bovine follicular response and oocyte developmental competence in a commercial operation. *Theriogenology, 65*, 102–115.

Eppig, J. J., O'Brien, M. J., Wigglesworth, K., Nicholson, A., Zhang, W., & King, B. A. (2009). Effect of in vitro maturation of mouse oocytes on the health and lifespan of adult offspring. *Human Reproduction, 24*(4), 922–928.

Fair, T., Hyttel, P., & Greve, T. (1995). Bovine oocyte diameter in relation to maturational competence and transcriptional activity. *Molecular Reproduction and Development, 42*, 437–442.

Fair, T., Hyttel, P., Greve, T., & Boland, M. (1996). Nucleus structure and transcriptional activity in relation to oocyte diameter in cattle. *Molecular Reproduction and Development, 43*, 503–512.

Farin, P. W., Piedrahita, J. A., & Farin, C. E. (2006). Errors in development of fetuses and placentas from in vitro-produced bovine embryos. *Theriogenology, 65*, 178–191.

Fleming, T. P., Velazquez, M. A., Eckert, J. J., Lucas, E. S., & Watkins, A. J. (2012). Nutrition of females during the periconceptional period and effects on foetal programming and health of offspring. *Animal Reproduction Science, 130*(3–4), 193–197.

Forde, N., Beltman, M. E., Duffy, G. B., Duffy, P., Mehta, J. P. et al. (2010). Changes in the endometrial transcriptome during the bovine estrous cycle: Effect of low circulating progesterone and consequences for conceptus elongation. *Biology of Reproduction, 84*, 266–278.

Gilchrist, R. B. (2011). Recent insights into oocyte-follicle cell interactions provide opportunities for the development of new approaches to in vitro maturation. *Reproduction and Fertility Developments, 23*, 23–31.

Gordon, I. (2003). *Laboratory production of cattle embryos*. Oxford: CABI Publishing.

Gray, C. A., Burghardt, R. C., Johnson, G. A., Bazer, F. W., & Spencer, T. E. (2002). Evidence that absence of endometrial gland secretions in uterine gland knockout ewes compromises conceptus survival and elongation. *Reproduction, 124*, 289–300.

Greve, T., Xu, K. P., Callesen, H., & Hyttel, P. (1987). In vivo development of in vitro fertilized bovine oocytes matured in vivo versus in vitro. *Journal of In Vitro Fertilization and Embryo Transfer, 4*, 281–285.

Hendriksen, P. J., Steenweg, W. N., Harkema, J. C., Merton, J. S., Bevers, M. M. et al. (2004). Effect of different stages of the follicular wave on in vitro developmental competence of bovine oocytes. *Theriogenology, 61*, 909–920.

Hyttel, P., Greve, T., & Callesen, H. (1989). Ultrastructural aspects of oocyte maturation and fertilization in cattle. *Journal of Reproduction and Fertility Supplement, 38*, 35–47.

Ireland, J. J., Mihm, M., Austin, E., Diskin, M. G., & Roche, J. F. (2000). Historical perspective of turnover of dominant follicles during the bovine estrous cycle: Key concepts, studies, advancements, and terms. *Journal of Dairy Science, 83*, 1648–1658.

Katz-Jaffe, M. G., McCallie, B. R., Preis, K. A., Filipovits, J., & Gardner, D. K. (2009). Transcriptome analysis of in vivo and in vitro matured bovine MII oocytes. *Theriogenology, 71*, 939–946.

Knijn, H. M., Wrenzycki, C., Hendriksen, P. J., Vos, P. L., Herrmann, D. et al. (2002). Effects of oocyte maturation regimen on the relative abundance of gene transcripts in bovine blastocysts derived in vitro or in vivo. *Reproduction, 124*, 365–375.

Kruip, T. A. M., & den Daas, J. H. G. (1997). In vitro produced and cloned embryos: Effects on pregnancy, parturition and offspring. *Theriogenology, 47*, 43–52.

Leibfried-Rutledge, M. L., Critser, E. S., Eyestone, W. H., Northey, D. L., & First, N. L. (1987). Development potential of bovine oocytes matured in vitro or in vivo. *Biology of Reproduction, 36*, 376–383.

Leroy, J. L., Vanholder, T., Mateusen, B., Christophe, A., Opsomer, G. et al. (2005). Non-esterified fatty acids in follicular fluid of dairy cows and their effect on developmental capacity of bovine oocytes in vitro. *Reproduction, 130*, 485–495.

Li, Q., Jimenez-Krassel, F., Bettegowda, A., Ireland, J. J., & Smith, G. W. (2007). Evidence that the preovulatory rise in intrafollicular progesterone may not be required for ovulation in cattle. *Journal of Endocrinology, 192*, 473–483.

Lonergan, P. (2011). Influence of progesterone on oocyte quality and embryo development in cows. *Theriogenology, 76*, 1594–1601.

Lonergan, P., Fair, T., Corcoran, D., & Evans, A. C. (2006). Effect of culture environment on gene expression and developmental characteristics in IVF-derived embryos. *Theriogenology, 65*, 137–152.

Lonergan, P., Monaghan, P., Rizos, D., Boland, M. P., & Gordon, I. (1994). Effect of follicle size on bovine oocyte quality and developmental competence following maturation, fertilization, and culture in vitro. *Molecular Reproduction and Development, 37*, 48–53.

Machatkova, M., Krausova, K., Jokesova, E., & Tomanek, M. (2004). Developmental competence of bovine oocytes: Effects of follicle size and the phase of follicular wave on in vitro embryo production. *Theriogenology, 61*, 329–335.

Matoba, S., Fair, T., & Lonergan, P. (2010). Maturation, fertilisation and culture of bovine oocytes and embryos in an individually identifiable manner: A tool for studying oocyte developmental competence. *Reproduction and Fertility Development, 22*, 839–851.

Mikkelsen, A. L. (2005). Strategies in human in-vitro maturation and their clinical outcome. *Reproductive Biomedicine Online, 10*, 593–599.

Nivet, A. L., Bunel, A., Labrecque, R., Belanger, J., Vigneault, C. et al. (2012). FSH withdrawal improves developmental competence of oocytes in the bovine model. *Reproduction, 143*(2), 165–171.

Norris, R. P., Ratzan, W. J., Freudzon, M., Mehlmann, L. M., Krall, J. et al. (2009). Cyclic GMP from the surrounding somatic cells regulates cyclic AMP and meiosis in the mouse oocyte. *Development, 136*, 1869–1878.

Pieterse, M. C., Vos, P. L., Kruip, T. A., Wurth, Y. A., van Beneden, T. H. et al. (1991). Transvaginal ultrasound guided follicular aspiration of bovine oocytes. *Theriogenology, 35*(4), 857–862.

Rizos, D., Lonergan, P., Boland, M. P., Arroyo-Garcia, R., Pintado, B. et al. (2002). Analysis of differential messenger RNA expression between bovine blastocysts produced in different culture systems: implications for blastocyst quality. *Biology of Reproduction, 66*, 589–595.

Rizos, D., Ward, F., Duffy, P., Boland, M. P., & Lonergan, P. (2002). Consequences of bovine oocyte maturation, fertilization or early embryo development in vitro versus in vivo: Implications for blastocyst yield and blastocyst quality. *Molecular Reproduction and Development, 61*, 234–248.

Roche, J. F. (1996). Control and regulation of folliculogenesis – a symposium in perspective. *Review of Reproduction, 1*, 19–27.

Russell, D. F., Baqir, S., Bordignon, J., & Betts, D. H. (2006). The impact of oocyte maturation media on early bovine embryonic development. *Molecular Reproduction and Development, 73*, 1255–1270.

Sirard, M. A. (2001). Resumption of meiosis: Mechanism involved in meiotic progression and its relation with developmental competence. *Theriogenology, 55*, 1241–1254.

Soderstrom-Anttila, V., Salokorpi, T., Pihlaja, M., Serenius-Sirve, S., & Suikkari, A. M. (2006). Obstetric and perinatal outcome and preliminary results of development of children born after in vitro maturation of oocytes. *Human Reproduction, 21*, 1508–1513.

Spencer, T. E., & Gray, C. A. (2006). Sheep uterine gland knockout (UGKO) model. *Methods in Molecular Medicine, 121*, 85–94.

Stroud, B. (2011). IETS 2011 Statistics and Data Retrieval Committee Report. The year 2010 worldwide statistics of embryo transfer in domestic farm animals. *Embryo Transfer Newsletter, 29*(4), 14–23.

Vaccari, S., Weeks II, J. L., Hsieh, M., Menniti, F. S., & Conti, M. (2009). Cyclic GMP signaling is involved in the luteinizing hormone-dependent meiotic maturation of mouse oocytes. *Biology of Reproduction, 81*, 595-604.

van de Leemput, E. E., Vos, P. L., Zeinstra, E. C., Bevers, M. M., van der Weijden, G. C., & Dieleman, S. J. (1999). Improved in vitro embryo development using in vivo matured oocytes from heifers superovulated with a controlled preovulatory LH surge. *Theriogenology, 52*, 335–349.

Vejlsted, M., Du, Y., Vajta, G., & Maddox-Hyttel, P. (2006). Post-hatching development of the porcine and bovine embryo – defining criteria for expected development in vivo and in vitro. *Theriogenology, 65*, 153–165.

Warzych, E., Wrenzycki, C., Peippo, J., & Lechniak, D. (2007). Maturation medium supplements affect transcript level of apoptosis and cell survival related genes in bovine blastocysts produced in vitro. *Molecular Reproduction and Development, 74*, 280–289.

Watkins, A. J., Wilkins, A., Cunningham, C., Perry, V. H., Seet, M. J. et al. (2008). Low protein diet fed exclusively during mouse oocyte maturation leads to behavioural and cardiovascular abnormalities in offspring. *Journal of Physiology, 586*(8), 2231–2244.

Watson, A. J., De Sousa, P., Caveney, A., Barcroft, L. C., Natale, D. et al. (2000). Impact of bovine oocyte maturation media on oocyte transcript levels, blastocyst development, cell number, and apoptosis. *Biology of Reproduction, 62*, 355–364.

Abbreviations

α-KG	alpha ketoglutarate
γTuRC	γ-tubulin ring complexes
AA	amino acids
Ac-pre-miRNA	product of Ago2-mediated pre-miRNA cleavage
ADAM17	disinteglin metaloprotease 17
AGO	argonaute
AGT	aminogluthethimide
AHR	aryl hydrocarbon receptor
ALK	activin receptor-like kinase
ALT	alanine transaminase
AMH	anti-mullerian hormone
APC/C	anaphase-promoting complex/cyclosome
AREG	amphiregulin
ART	assisted reproductive technologies
ATM	ataxia-telangiectasia mutated homolog
ATP	adenosine tri phosphate
ATRX	X-linked alpha-thalassemia/mental retardation syndrome
BCB	brilliant cresyl blue
BCL2	B-cell leukemia/lymphoma 2 family
BCS	body condition score
BDNF	brain-derived neurotrophic factor
BHLH	basic-helix-loop-helix
BMP15	bone morphogenetic protein 15
BMPR2	BMP receptor type 2
BPA	bisphenol A
BSA	bovine serum albumin
BTC	betacellulin
BWS	Beckwith-Wiedemann-Syndrome
C	clumped chromatin configuration
Ca^{2+}	calcium
CaMKII	calmodulin-dependent protein kinase II
cAMP	cyclic AMP

Oocyte Physiology and Development in Domestic Animals, First Edition. Edited by Rebecca L. Krisher.

CC	cumulus cells
CCAN	constitutive centromere associated network
CDC42	cell division cycle 42
CDK	cyclin-dependent kinase
CEEF	cumulus expansion enabling factor
CENP-A	centromere associated protein-A
CICR	Ca^{2+}-induced Ca^{2+} release
cKO	conditional knockout
COC	cumulus oocyte complexes
CPA	cyclopiazonic acid
CPEB	cytoplasmic polyadenylation element binding protein
CPT 1b,2	carnitine palmitoyltransferase 1b, 2
CRE	cAMP response element
CREB	CRE binding protein
CX43	connexin-43
CYP19	aromatase
CYP26b1	cytochrome P450, family 26, subfamily B
DAX1	dosage sensitive sex reversal, adrenal hypoplasia critical region on chromosome X, gene 1
DAZL(A)	deleted in azoospermia-like
DMC1	disrupted meiotic cDNA 1
DNMT	DNA methyltransferases
DNMTL	DNA methyltransferase-like
DPC	days post coitum
DPI	diphenylene iodonium
DPPA3	developmental pluripotency associated 3
dsRBD	double-stranded RNA binding domain
E	embryonic day
EGF	epidermal growth factor
EGFR	EGF receptor
ER	endoplasmic reticulum
ERBB	EGF receptor family
EREG	epiregulin
ERK 1/2	extracellular signal regulated kinases 1/2
ERα	estrogen receptor α
eSET	elective single embryo transfer
EST	expressed sequence tags
F pattern	floccular form of condensed chromatin close to the nucleoli and nuclear membrane
F6P	fructose 6 phosphate
FA	fatty acids
FAO	fatty acid β-oxidation
FAS	fatty acid synthase
FGF	fibroblast growth factor
FIGLA	factor in the germ line alpha
FLOPED	factor located in oocytes permitting embryonic development
FOXI2	forkhead box I2

FSH	follicle stimulating hormone
FST	follistatin
G3P	glyceraldehyde 3 phosphate
G6P	glucose 6 phosphate
G6PDH	glucose-6-phosphate dehydrogenase
GCP	gamma complex proteins
GDF9	growth and differentiation factor 9
GDNF	glial cell line derived neurotrophic factor
GLS	glutaminase
GLUT	glucose transporters
GPR	G-protein coupled receptor
GSC	germline stem cells
GSH	glutathione, reduced
GSSG	glutathione, oxidized
GV	germinal vesicle
GVBD	germinal vesicle breakdown
H2AT119ph	histone H2A phosphorylation
H3K14ac	histone H3 acetylation
H3K4me	mono-methylated lysine residue 4 of histone H3
H3K4me2	di-methylated lysine residue 4 of histone H3
H3K4me3	tri-methylated lysine residue 4 of histone H3
H3S10ph	histone H3 phosphorylation
H4K12ac	acetylation of histone H4 at lysine 12
H4K5ac	acetylation of histone H4 at lysine 5
H4K5ac	histone H4 acetylation
HA	hyaluronic acid
HBP	hexosamine biosynthesis pathway
HDAC	histone deacetylases
HELLS	helicase lymphoid-specific
HK	hexokinase
HP1	heterochromatin protein 1
IBMX	3-isobutyl-1-methylxanthine
ICF	immunodeficiency-centromeric instability syndrome
ICM	inner cell mass cells
ICR	imprinting control regions
ICSI	intracytoplasmic sperm injection
IGF1	insulin like growth factor 1
IP_3	inositol 1,4,5-trisphosphate
IVC	in vitro culture
IVF	in vitro fertilization
IVM	in vitro maturation
IVP	in vitro embryo production
KB	kilobases
KDM1B	lysine demethylase 1B
KMT1C	H3 methyltransferase G9A
KPNA7	importin alpha 8
LC/LCC	loosely condensed chromatin configuration

LDH	lactate dehydrogenase
LH	luteinizing hormone
LHCGR	LH receptor
LHM8	LIM homeobox protein 8
LOS	large offspring syndrome
LSD1	lysine demethylase 1
LSH	lymphoid-specific helicase
m^7G	7-methyl-G cap
MAPK	mitogen-activated protein kinase
MCU	mitochondrial Ca^{2+} uniporter
ME	malic enzyme
MET	maternal-to-embryonic transition
MI	metaphase I
MII	metaphase II
miRNA	micro RNA molecules
miRNP	microRNA ribonucleoprotein
MLCK	myosin light chain kinase
MLL2	histone-lysine N-methyltransferase
MOF	multiple oocyte follicles
MPF	maturation or M-phase promoting factor
mRNA	messenger RNA
mRNP	messenger ribonucleoprotein complexes
MSH4,5	mutS homolog 4 and 5
MSY2	Y box protein 2
MTOC	microtubule organizing centers
mTORC1	target of rapamycin complex 1
N	netlike chromatin configuration
NAADP	nicotinic acid adenine dinucleotide phosphate
NAD(P)	Nicotinamide adenine dinucleotide (phosphate)
ncRNA	non-coding RNA molecules
NEB	negative energy balance
NEDD1	neuronal precursor cell expressed developmentally down-regulated protein 1
NEFA	non esterified fatty acids
NGF	nerve growth factor
NL	net-like chromatin configuration
NOBOX	newborn ovary homeobox-encoding gene
NPM	nucleoplasmin
NPPC	natriuretic peptide precursor type C
NRG1	neuregulin
NSN	non-surrounded nucleolus
NT	nucleotides
NT4	neurotrophin 4
NTRK2	neurotrophic tyrosine kinase receptor type 2
O2	oxygen
OA	oxaloacetate
OBOX	oocyte specific homeobox
OCT4	octamer binding transcription factor 4

OPU	ovum pick-up
OSF	oocyte-secreted factors
OXPHOS	oxidative phosphorylation
P4	progesterone
PARP-1	poly(ADP ribose) polymerase-1
PCM	pericentriolar material/matrix
PDE	prostaglandin E
PDE3	phosphodiesterase type III
PDPK1	3-phosphoisnositide dependent protein kinase 1
PFK	phosphofructokinase
PG	prostaglandin
PGC	primordial germ cells
PGE2	prostaglandin E2
PGK1	phosphoglycerate kinase 1
PGR	progesterone receptor
PHLDA1	pleckstrin homology-like domain, family A, member 1
PI3K	phosphatidylinositol 3 kinase
PIP_2	phosphatidylinositol 4,5-bisphosphate
PK	pyruvate kinase
PKA	protein kinase A
PKC	protein kinase C
PLC	phospholipase C
PMCA	plasma membrane Ca^{2+} ATPases
POL II	RNA polymerase II
POU	(Pit, Oct, Unc) domain
PPP	pentose phosphate pathway
pre-miRNA	precursor miRNA
pri-miRNA	primary miRNA
PRPP	phosphoribosyl pyrophosphate
PTEN	phosphatase and tensin homolog deleted on chromosome 10
PTGER 2,4	G protein–coupled receptor subtypes EP2 or EP4
PTGS2	prostaglandin synthase 2
PUFA	polyunsaturated fatty acids
PVP	polyvinyl pyrrolidone
R5P	ribose-5-phosphate
RGS 2,3	regulator of G-protein signalling 2, 3
RISC	RNA-induced silencing complex
RNAi	RNA interference
RNP	ribonucleoprotein
RRM	RNA recognition motif
RSP6	ribosomal protein s6
RSPO1	r-spondin homolog 1
S1P	sphingosine 1 phosphate
SAC	spindle assembly checkpoint
SAGE	serial analysis of gene expression
SC	singly condensed chromatin configuration
SCF	stem cell factor

SCMC	subcortical maternal complex
SCYP	synaptonemal complex protein
SERCA	sarcoplasmic/endoplasmic reticulum Ca^{2+} ATPases
shRNA	short hairpin RNA
SMAD	Sma- and Mad-related
SN	surrounded nucleolus
SNBP	sperm nuclear binding proteins
SOF	synthetic oviductal fluid
SOHLH 1,2	spermatogenesis and oogenesis-specific basic helix-loop-helix 1 and 2
SPCA	secretory pathway Ca^{2+} ATPases
SSH	suppressive subtractive hybridization
STRA8	stimulated by retinoic acid, gene 8
SWI/SNF	switch/sucrose non-fermenting family
TALDO	transaldolase
TC/TCC	tightly condensed chromatin configuration
TCA	tricarboxylic acid cycle
TCM	tissue culture medium
TGFβ	transforming growth factor β family
TKT	transketolase
TRBP	transactivating response (TAR) RNA-binding protein
TSC1	tuberosclerosis complex
TUBG1	γ-tubulin gene
UGP2	UDP-glucose pyrophosphorylase-2
UTR	untranslated region
VRK1	vaccinia related kinase
WE	Warburg effect
WT	wild type
YBX2	Y box protein 2
ZAR1	zygote arrest 1
ZP 2,3	zona pellucida protein 2,3
ZP	zona pellucida

Index

Oocyte Physiology and Development in Domestic Animals, First Edition. Edited by Rebecca L. Krisher.